CYCLES OF SOIL

Carbon, Nitrogen, Phosphorus, Sulfur, Micronutrients

Second Edition

F. J. STEVENSON AND M. A. COLE
Department of Natural Resources and Environmental Sciences
University of Illinois at Urbana-Champaign

JOHN WILEY & SONS, INC.
New York • Chichester • Weinheim • Brisbane • Toronto • Singapore

ISBN 0-471-32071-4

Printed in the United States of America

10 9 8 7 6 5 4 3 2 1

CYCLES OF SOIL

Dedicated by FJS to my wife,
with appreciation for her enduring patience

Dedicated by MAC to my friends in México,
for their encouragement; their enthusiasm,
determination, and courage have been an inspiration

CONTENTS

Summary, 363
References, 364

PREFACE

Living organisms and the transformations they perform have a profound effect on the ability of soils to provide food and fiber for an expanding world population. Soil organisms also have diverse influences on the quality of air and water. Of paramount importance is the cycling of carbon (C), nitrogen (N), phosphorus (P), sulfur (S), and the micronutrient cations (B, Cu, Fe, Mn, Mo, and Zn). An understanding of the various cycles and their interactions is essential for the intelligent use of soil as a medium for plant growth, for the rational use of natural and synthetic fertilizers, for disposal of wastes in soil, and for the prevention of soil-derived pollution of air and water. Since the biochemical cycles constitute the lifeline of planet Earth, information relative to their functioning in terrestrial soils has a direct application to other ecosystems.

A common feature of the biochemical cycles is that microorganisms are key agents in the transformations; some microorganisms have capabilities that are not seen in other lifeforms. As plant residues undergo decay in soil, N, P, S, and the micronutrient cations may appear in plant-available forms; portions are assimilated into microbial cells (the soil biomass). With time, N, P, and S are stabilized by conversion into recalcitrant organic forms and by interactions with inorganic soil components. Nutrient losses occur through erosion, leaching, and, for N and S, by volatilization. Although separate processes are operative for specific cycles (e.g., denitrification in the case of N; chelation for micronutrients), most cycles have some aspects in common, notably mineralization and immobilization of N, P, and S. Interest in the various cycles is at an all-time high, as evidenced by the large number of books that have been published in recent years. Unfortunately, most volumes represent collections of symposium papers and are of interest only to the research specialist.

A unique feature of this book is that it is devoted exclusively to the biochemical cycles in soils. Many facets of the C, N, P, S cycles, as well as micronutrient behavior, are covered, including fluxes among soil, water and air, biochemical pathways and chemical transformations, plant availability, gains, losses, recycling, and environmental pollution. Considerable latitude was taken in developing the various chapters, with the result that both panoramic and specific views have been presented for each major cycle. Because of the voluminous literature that has accumulated, exhaustive coverage of the literature was not possible, and both selection of references and depth of documentation have been arbitrary. One function of the book was to present a critical account of our general knowledge of each cycle. In doing so, the authors may have inadvertently misrepresented some of the research discussed. The authors apologize for any omissions of important work.

This second edition reflects substantial changes from the first edition. Some topics (particularly C and N dynamics in soil) have been the subjects of intense research activity since the first edition was published. Many chapters have been updated to incorporate this new knowledge, as well as descriptions of relatively new methodologies for determining pools of organic matter. New material has also been added on the determination of organic matter quality by ^{13}C–NMR spectroscopy and analytical pyrolysis. Many figures and tables and much text have been revised to improve clarity and consistency.

In a sense, the C cycle acts as a driving force for the other cycles. Accordingly, this topic is covered first (Chapters 1 to 4), followed in order by the N cycle (Chapters 5 to 8), the P cycle (Chapter 9), the S cycle (Chapter 10), and micronutrient behavior (Chapter 11). The book is intended for the soil scientist, but it will also interest researchers and students in microbiology, forestry, agronomy, horticulture, organic geochemistry, environmental science and engineering, and a host of other disciplines concerned with global cycling of C, N, P, S and micronutrients. As a companion volume to the second edition of *Humus Chemistry: Genesis, Composition, Reactions* (Wiley, 1994), this book is well suited as a graduate or advanced undergraduate reference text for courses in Soil Microbiology and Soil Biochemistry.

The authors wish to express appreciation to graduate students and staff members at the University of Illinois for encouragement in the book's preparation.

F. J. STEVENSON
M. A. COLE

Urbana, Illinois
January 1999

1

THE CARBON CYCLE

The degradation of dead plant and animal materials in soil is a fundamental biological process because carbon (C) is recirculated to the atmosphere as carbon dioxide (CO_2), nitrogen (N) is made available as ammonium (NH_4^+) and nitrate (NO_3^-), and other associated elements (phosphorus, sulfur, and various micronutrients) appear in inorganic forms required by higher plants. In the process, part of the nutrients is assimilated by microorganisms and incorporated into microbial biomass. The conversion of C, N, P, and S to inorganic (mineral) forms is called mineralization; assimilation by synthesis of microbial cells and other organisms is called immobilization.

The photosynthetic process is of primary importance in providing raw material for microbial growth and humus synthesis. By using solar energy plus nutrients derived from the soil, higher plants produce lignin, cellulose, protein, and other organic substances that make up their structures. During decay by microorganisms, much of the C is released to the atmosphere as CO_2, but a significant portion remains behind as soil organic matter and microbial components. Part of the native humus is mineralized concurrently. Results of studies with ^{14}C-labeled plant residues added to temperate zone soils have shown that approximately one-third of the applied C remains behind in the soil after the first year, with the remainder being recycled into the atmosphere. Mean residence times (MRT) of the residual C range from weeks to years for the biomass components to centuries for the humic material.

In this volume, the C, N, P, and S cycles will be examined separately. It should be noted, however, that the four cycles do not operate independently but are linked by biological processes common to each, notably mineralization and immobilization. A schematic illustration of the flow of C and nutrients through the biosphere is shown in Fig. 1.1.

The decay process is emphasized in this chapter, and results obtained using ^{14}C-labeled substrates will be stressed. A brief account is also given of the

1

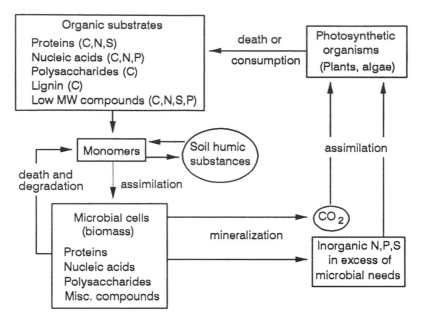

Fig. 1.1   Compartments of the global C cycle and interchanges between them.

extraction, fractionation, age, and general chemical properties of soil organic matter. The disposal of organic wastes in soil is discussed separately in Chapter 4.

Supplemental information regarding the C cycle can be found in several excellent reviews.[1-8] Global aspects of the distribution of organic C in soils are discussed in volumes edited by Bolin and Cook[6] and Andreux.[9]

GLOBAL ASPECTS

The global C cycle is a multicompartment system with terrestrial, aquatic (marine and freshwaters), and atmospheric compartments. The original source of the C was the fundamental rocks, from which CO_2 was evolved by outgassing during periods of intense volcanic activity. A secondary source was the primitive atmosphere, which probably contained appreciable amounts of methane (CH_4) gas.

Organic reservoirs and direction of C fluxes for compartments of the global C cycle are illustrated in Fig. 1.2. The distribution of C in some of the main compartments is given in Table 1.1. A unique feature of the overall C cycle is that much of the C that has passed through the terrestrial component is now preserved as organic matter in sediments.

The amount of C contained in the organic matter of terrestrial soils (30 to 50×10^{14} kg) is three to four times the C content of the atmosphere (7 ×

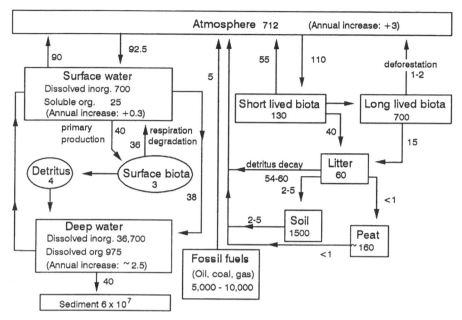

Fig. 1.2 Flow of C (and nutrients) through the biosphere. Sizes of reservoirs (10^{12} kg) and fluxes (10^{12} kg/year) are for the part of the C cycle that is in a state of comparatively rapid turnover (<1000 years). Other estimates for C in the main compartments are given in Table 1.1. Adapted from Bolin and Cook.[6]

TABLE 1.1   Distribution of C in Various Reservoirs of the Earth[a]

Reservoir	Amount of C, $\times 10^{14}$ kg
At earth's surface	
Atmosphere (as CO_2)	7
Biomass	4.8
Fresh water	2.5
Marine above thermoclime	5–8
Soil organic matter	30–50
At depths to 16 km	
Marine organic detritus	30
Coal and petroleum	100
Deep-sea solute C	345
Sediments	200,000

[a]From Bolin[3]

10^{14} kg) and five to six times the land biomass (4.8×10^{14} kg). In contrast, the amount contained in sediments (2×10^{19} kg to a depth of 16 km) is many orders of magnitude higher. The annual input of C to the soil is about 1.10 $\times 10^{14}$ kg/year, or about 15% of the atmospheric CO_2. However, an equivalent amount of C is returned to the atmosphere by decay. A small quantity of C finds its way to the sea, and a portion of it is deposited in sediments. The CO_2 content of the atmosphere has been increasing at a steady rate over the past century because of the burning of fossil fuel and deforestation, as discussed in Chapter 4.

Data for the percentage distribution of C in soil associations of the world are given in Fig. 1.3. Histosols (organic soils), because of their very high C contents, are major contributors to the total soil C (i.e., about one-third), even though they occupy a relatively small proportion of the world's surface area (compare Fig. 1.3*a* and 1.3*b*). Another factor worthy of note is that the amount of C stored in tropical soils of the world, represented by the Cambisols and Acrisols, is somewhat less than for soils of temperate zones, which can be attributed to the low organic C contents of tropical soils (see Chapter 2).

Considerable attention has been given recently to the impact of past and present soil organic C levels on global warming (due to the greenhouse effect). Among the questions that need to be answered are:

1. To what extent have changes in soil organic C through cropping—levels generally decline when soils are first placed under cultivation—affected atmospheric CO_2 levels and associated global warming?
2. Are soils currently serving as a source or sink for atmospheric CO_2?
3. Can soils be managed so as to serve as a sink for atmospheric CO_2 and thereby mitigate the harmful effect of CO_2 on global warming?
4. How will soil C levels respond to predicted increases in global warming?

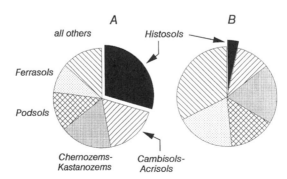

Fig. 1.3   Global distribution of soil organic C. (*a*) In soil associations of the world (total global C is 30 to 50×10^{14} kg). (*b*) World soils as a percentage of total surface area (122.2×10^6 km²). Adapted from Bohn[5] and Andreux.[9]

With regard to item 2, analyses of the global C cycle based on increases in the CO_2 content of the atmosphere indicate that there may be a net increase in the C of land biota, and if so, this should be reflected in an increase (admittedly small) in the net storage of organic C in terrestrial soils. Various aspects of global warming as affected by soil organic C are discussed by Paustian et al.[10] and in volumes edited by Bolin and Cook[6] and Scharpenseel et al.[11]

The literature pertaining to C storage in soils is extensive, and the reader is referred to recent books and reviews for additional information.[12–14] Factors affecting the organic C content of soils, including changes brought about through cropping, are discussed at length in Chapter 2.

ORGANIC CONSTITUENTS OF SOIL

Soil organic matter, or humus, consists of two major types of compounds: nonhumic substances and humic substances. The former is represented by the well-characterized classes of organic compounds, including carbohydrates, fats, waxes, and proteins. The humic substances, which represent the most active fraction of humus, consist of a series of highly acidic, yellow- to black-colored, high-molecular-weight polyelectrolytes referred to by such names as humic acid, fulvic acid, and so on.

The percentage distribution of soil organic matter in the different components is approximately as follows:

Type of material	Usual range (% by weight)
Nonhumic substances	
Lipids	1–6%
Carbohydrates	5–25%
Proteins/peptides/amino acids	9–16%
Other	trace
Humic substances	to 80%

The distinction between the two main classes is arbitrary in that some of the nonhumic material (notably proteinaceous constituents) may be covalently bound to humic substances, as shown below. *In situ,* soil organic matter (humus) can be described as a complicated, intertwined network of humic and nonhumic materials adsorbed onto mineral components and containing complexed (chelated) metal ions.

Nonhumic Substances

The soil, being a graveyard for dead microorganisms, contains essentially all of the biopolymers and biochemicals synthesized by bacteria, actinomycetes, and fungi, although many of the biochemicals will occur only in trace quantities. For example, soils contain a rather large number of amino acids that are not normally found in proteins (see Chapter 6).

Lipids Designation of compounds as "lipids" is a convenient grouping based on solubility, rather than their being a specific type of compound. Their common property is that they are soluble in moderately hydrophobic solvents, such as benzene, acetone, chloroform, and hexane; some are soluble in more polar solvents like methanol and ethanol. They are a diverse group of materials ranging from relatively simple organic acids (some of which are also water soluble) to more complex fats, waxes, and resins.

From 1.2 to 6.3% of the organic matter in most soils occurs as lipids, with somewhat higher values being reported for acid forest soils and peats.[15] A detailed review of lipids in soil is presented by Dinel et al.[16]

Hydrocarbon-based compounds are of considerable interest because many of them (e.g., aldehydes, ketones, phenolic acids, coumarins, glycosides, short-chain aliphatic acids) are phytotoxic, while others (e.g., B vitamins) stimulate plant growth. Waxes and similar materials may be responsible for the water-repellent nature of certain sands and dried peats.

Carbohydrates From 5 to 25% of the total soil organic matter occurs in the form of carbohydrates.[15] Plant debris contributes carbohydrates in the form of simple sugars, cellulose, and hemicelluloses, but these are rapidly degraded by bacteria, actinomycetes, and fungi, which in turn synthesize cell wall and extracellular polysaccharides ("gums"). The carbohydrate material in soil occurs as:

1. Low concentrations of free sugars in the soil solution
2. Complex carbohydrates that can be extracted and separated from other organic constituents
3. Polymers of various sizes and shapes that are so strongly attached to clay and/or humic colloids that they cannot be easily isolated, purified, or identified[15]

Polysaccharides in soil are significant because they bind soil particles into water-stable aggregates, with the result that a soil with a high polysaccharide content is more permeable to water and air than one with a low content.[17,18] Carbohydrates also form complexes with metal ions, which modifies the bioavailability of the metals. Other soil properties affected by polysaccharides include cation exchange capacity (attributed to –COOH groups of uronic acids) and biological activity (e.g., as an energy source for microorganisms). In certain submerged soils or soils subjected to high waste loadings, the production of microbial gums and mucilages may lead to an undesirable reduction in permeability because of blocking of soil pores.[19]

The stability of soil polysaccharides is due to a combination of factors, including structural complexity, which reduces their biodegradability, adsorption onto clay minerals or metal oxide surfaces, formation of insoluble salts or chelate complexes with polyvalent cations, and adsorption or covalent binding to humic substances.

Proteinaceous Constituents   As will be noted in Chapter 6, from 30 to 45% of the organic N in soils is found as amino acids after acid hydrolysis, a result that indicates that much of the N in soil is protein-N. On this basis, and for an assumed C/N ratio of 12 for soil organic matter, from 9 to 16% of the organic matter would exist as proteinaceous compounds.

Proteinaceous material is present in soil as:

1. Low concentrations of free amino acids in the soil solution
2. Amino acids, peptides, and proteins bound to clay minerals and humic colloids (discussed in Chapter 6)
3. Mucopeptides and teichoic acids originating from bacterial cell walls.

Humic Substances

Humic substances, which represent the most active fraction of humus, consist of a series of highly acidic, yellow- to black-colored polyelectrolytes referred to by such names as humic acid and fulvic acid. As noted below, these substances are formed by secondary synthesis reactions and have properties distinctly different from those of the biopolymers of living organisms, including the lignin of higher plants.

Humic acids are among the most widely distributed organic materials in the earth. They are found not only in soils but in marine and lake sediments, peat bogs, composts, natural waters, sewage, carbonaceous shales, brown coals, and miscellaneous other deposits. The total amount of C in the earth as humic acids,[20] estimated at 55×10^{14} kg, is an order of magnitude higher than shown in Table 1.1 for the land biomass (4.8×10^{14} kg).

Extraction and Fractionation   Humic substances are normally recovered from the soil by extraction with alkali (usually 0.1–0.5 N NaOH), although in recent years milder reagents, such as neutral sodium pyrophosphate,[15] have been used. The following fractions, based on solubility characteristics, are obtained: humic acid, soluble in alkali, insoluble in acid; fulvic acid, soluble in alkali, soluble in acid; hymatomelanic acid, alcohol-soluble part of humic acid; and humin, insoluble in alkali.

Humic acids are sometimes separated into two groups by partial precipitation with electrolyte (salt solution) under alkaline conditions. Those in the first group, the brown humic acids (Braunhuminsäure), are not coagulated by an electrolyte and are characteristic of humic acids in peat and Spodosols.

Those in the second group, the gray humic acids (Grauhuminsäure), are easily coagulated and are characteristic of humic acids in Mollisols. In the older literature, considerable attention was given to the so-called "apocrenic" and "crenic" acids, which were light-yellow fulvic acid-type substances. A typical fractionation scheme is given in Fig. 1.4.

The fulvic acid *fraction* has a straw-yellow color at low pH values that turns to wine-red at high pH's, passing through an orange color at a pH near 3.0. There is little doubt that compounds of a nonhumic nature are present. The term *fulvic acid* should be reserved as a class name for the pigmented components of the acid-soluble fraction. Attempts have been made to separate "generic" fulvic acids from the acidified extract by sorption–desorption procedures,[15] such as from an XAD-8 resin column (see Fig. 1.4).

Humic Substances as a System of Polymers   A useful concept that has evolved over the years, popularized several decades ago by the Russian scientist Kononova,[21] is that the various humic fractions represent a system of polymers that vary in a systematic way in elemental content, acidity, degree of polymerization, and molecular weight. The proposed interrelationships are shown in Fig. 1.5. No sharp difference exists between the two main fractions (humic and fulvic acids) or their subgroups. The humin fraction (material not

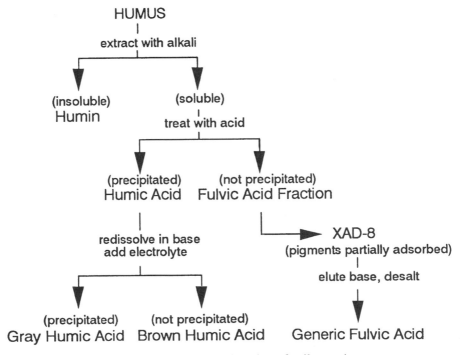

Fig. 1.4 Scheme for the fractionation of soil organic matter.

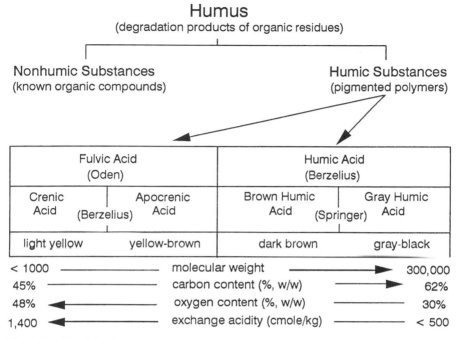

Fig. 1.5   Classification and chemical properties of humic substances. Adapted from Stevenson.[15]

extracted with alkali) is not represented, but this component may consist of one or more of the following:

1. Humic acids so intimately bound to mineral matter that the two cannot be separated
2. Highly condensed humic matter that has a high C content ($> 60\%$) and is thereby insoluble in alkali
3. Fungal melanins, which have properties similar to humic acids and that are partially insoluble in alkali.

All soils would be expected to contain a broad spectrum of humic substances, as depicted in Fig. 1.5. However, distribution patterns will vary from soil to soil and with depth in the soil profile. The humus of forest soils (Alfisols, Spodosols, and Ultisols) contains high amounts of fulvic acids; that of peat and grassland soils (Mollisols) contains high amounts of humic acid. As noted earlier, the humic acids of forest soils are mostly of the brown humic acid type; those of grassland soils are of the gray humic acid type. The approximate distribution of the three main humus fractions in the soils of four great soil groups is shown in Fig. 1.6.

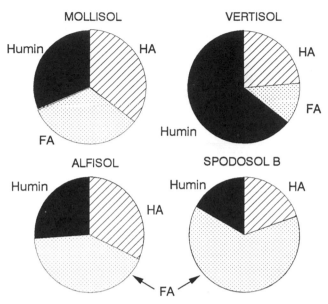

Fig. 1.6 Distribution of humus forms in the soils of four great soil groups. Values for FA are for the fulvic acid "fraction." From Stevenson,[15] reproduced by permission of John Wiley & Sons, Inc.

Biochemistry of Humus Formation

Several pathways exist for the formation of humic substances during the decay of plant and animal remains in soil, as shown in Fig. 1.7. A brief résumé follows:

1. For many years it was thought that humic substances were derived from lignin (pathway 4 of Fig. 1.7). According to this theory, lignin is incompletely utilized by microorganisms and the residuum becomes part of the soil humus.
2. In pathway 3, lignin still plays an important role, but in different ways. In this case phenolic aldehydes and acids released during microbial attack undergo enzymatic conversion to quinones, which polymerize in the presence of amino compounds to form humic-like macromolecules.
3. Pathway 2 is similar to pathway 3, except that the polyphenols are synthesized by microorganisms from nonlignin C sources (e.g., cellulose).
4. The notion that humus is formed from sugars through a nonenzymatic "browning" reaction (as occurs during dehydration of certain food products) dates back to the early days of humus chemistry.

All pathways may be involved in humus formation, but not to the same extent in all soils or in the same order of importance. For example, a lignin

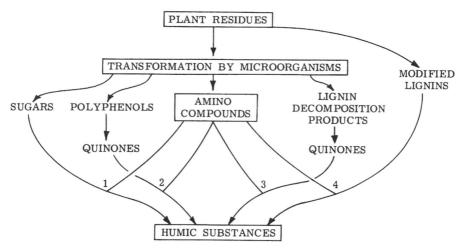

Fig. 1.7 Pathways for the formation of humic substances. Amino compounds syn thesized by microorganisms are seen to react with modified lignins (pathway 4), qui nones (pathways 2 and 3), and reducing sugars (pathway 1) to form dark-colored polymers. From Stevenson,[15] reproduced by permission of John Wiley & Sons, Inc.

pathway would be of importance in drained soils and wet sediments (swamps, etc.), whereas synthesis from polyphenols in leachates from forest litter may play a prominent role in forest soils.

The importance of lignin as a source of humic substances (pathway 4) arises from its recalcitrance to attack by microorganisms, notably bacteria.

The mechanism of lignin biodegradation is discussed by Leisola and Gar cia[22] and in the review of Haider.[23] A reexamination of the lignin theory of humus formation is given in Shevchenko and Bailey.[24]

Basic principles of microbial lignin transformations are as follows:

1. Complete degradation of lignin is suppressed under anaerobic condi tions, the main changes being alteration of peripheral groups without uncoupling of polymeric units. Changes brought about in lignin through microbial attack by bacteria (the predominant form of microorganisms in wet sediments) include losses of methoxyl groups ($-OCH_3$), with generation of phenolic OH groups, and oxidation of side chains to form $-COOH$ groups.

2. Degradation of lignin occurs rapidly in well-aerated soils, a process in which microscopic fungi of the Imperfecti group play a key role. Rates of lignin degradation are also dependent on pH.

3. Microbes do not use lignin as a sole source of C but require additional organic compounds for growth (e.g., cellulose). Specific enzymes are produced that catalyze the degradation of lignin by a free-radical scis sion mechanism.

4. The complete degradation of lignin occurs through the synergistic action of several groups of organisms. The main mechanism of microbial attack is the splitting of bonds at random in the aliphatic and/or aromatic portions of the molecule.

The demethylation and ring cleavage of lignin during attack by ligninolytic fungi are shown in Fig. 1.8.

A particularly popular theory at this time is that in many terrestrial soils humic and fulvic acids are formed primarily through pathways 2 and 3, and that the processes include:

1. Degradation of all plant components into simpler monomers
2. Metabolism of the monomers accompanied by an increase in the soil biomass
3. Repeated recycling of biomass C and N with new cell synthesis
4. Concurrent polymerization of reactive monomers (i.e., of lignin origin and newly synthesized) into high-molecular-weight polymers[15]

Haider and Martin[25] conducted incubation experiments on specifically ^{14}C-labeled phenolic carboxylic acids. Parts of the molecule that were labeled

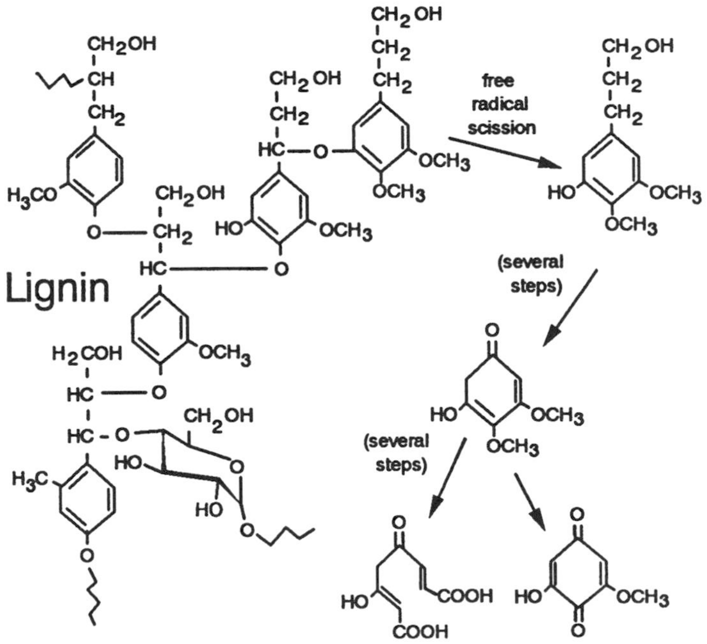

Fig. 1.8   Chain scission of lignin and production of phenols and other products during the microbial degradation of lignin.

included the –COOH group, the three-C side chain, and the aromatic ring. A summary of their findings is shown in Fig. 1.9. With the exception of –COOH attached to the aromatic ring, C of the phenolic derivatives (e.g., caffeic acid and p-hydroxybenzoic acid) was not lost (as CO_2) to the same extent as from simple organic compounds such as sugars and organic acids. When linked into polymers, the ^{14}C–COOH and ring ^{14}C of caffeic acid were highly resistant to degradation, particularly the ring [14]C. Pengra and coworkers[26] found that synthetic melanin produced from dihydroxyphenylalanine (DOPA) was also highly resistant to degradation in soil.

The relative importance of lignins and microbial synthesis as sources of polyphenols for humus synthesis is unknown and may depend upon available substrates, microbial species that are active, and environmental conditions. Because lignin is ubiquitous in most plants and is relatively resistant to microbial degradation, it is sometimes considered to be the major, if not sole, source of phenolic units. However, some microscopic soil fungi that degrade lignin produce humic acid-like substances in which the phenolic units originate both from lignin and by *de novo* biosynthesis of phenols. For example, Martin and Haider[25,27–29] found that microscopic fungi of the Imperfecti group,

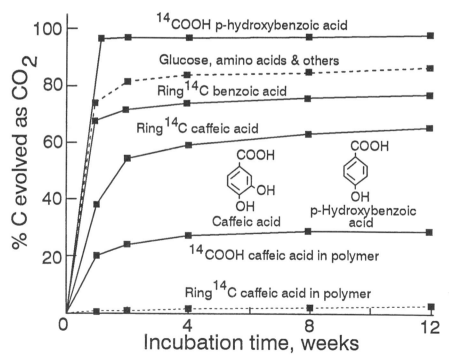

Fig. 1.9 Decomposition of several simple organic compounds, of specifically labeled ^{14}C-benzoic and caffeic acids, and of caffeic acid linked into phenolic polymers. The soil was Greenfield sandy loam. Adapted from Haider and Martin.[25]

such as *Aspergillus sydowi, Epicoccum nigrum, Hendersonula toruloidea, Stachybotrys atra,* and *S. chartarum,* degraded cellulose as well as lignin and other substrates and, in the process, synthesized appreciable amounts of humic acid-like polymers from the nonlignin precursors.

A scheme for the synthesis and formation of polyphenols by *H. toruloidea* is given in Fig. 1.10. The first products formed from the nonaromatic precursors were orselinic acid and 6-methylsalicyclic acid; in time, other polyphenols were formed. Following synthesis, the polyphenols are secreted into the external solution, where they are enzymatically oxidized to quinones, which subsequently combine with other metabolites (e.g., amino acids) to form humic polymers. Reactions postulated to occur between quinones and amino acids are outlined in Chapter 6.

The quantities of humic acid synthesized by fungi can be appreciable. Martin et al.[30] found that about one-third of the compounds synthesized by *H. toruloidea* (including biomass constituents) consisted of humic acid. Furthermore, a humic acid-type polymer could be extracted from the mycelium with 0.5 N NaOH. The microbial synthesis of polyphenols is emphasized in

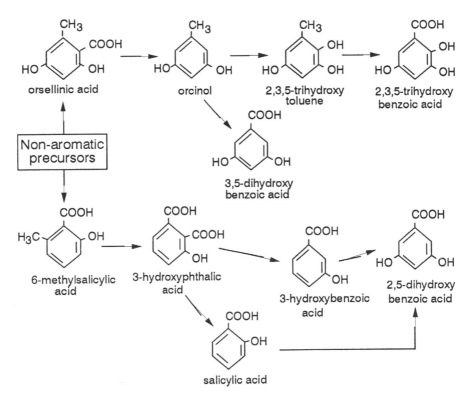

Fig. 1.10 Synthesis and possible transformations of polyphenols by *H. toruloidea.* Adapted from Martin et al.[30]

the early work of Kononova,[21] who gives a detailed account of research in which histological microscopic techniques and chemical methods were used to study plant residue degradation.

Chemical Properties and Structures of Humic and Fulvic Acids

As noted above, humic substances consist of a heterogenous mixture of compounds, with each fraction (humic acid, fulvic acid, etc.) being made up of molecules of different sizes. In contrast to humic acids, the low-molecular-weight fulvic acids have higher oxygen but lower C contents, and they contain considerably more acidic functional groups, particularly —COOH.[15]

The range of oxygen-containing functional groups in humic and fulvic acids is given in Table 1.2. For any specific group, a considerable range of values can be found, even for preparations obtained from the soils of any given great soil group. The total acidities of the fulvic acids (640–1420 cmole/kg) are considerably higher than those of the humic acids (560 890 cmole/kg). Both —COOH and acidic OH groups (presumed to be phenolic OH) contribute to the acidic nature of these substances, with —COOH being the most important. The concentration of acidic functional groups in fulvic acids would appear to be substantially higher than for any other naturally occurring organic polymer.

A "type" molecule for the "core" of the humic acid molecule probably consists of micelles of a polymeric nature, the basic structure of which is an aromatic ring of the di- or trihydroxyphenol type bridged by —O—, —CH$_2$—, —NH—, —N=, —S—, and other groups, and containing both free OH groups and the double linkages of quinones. Some of the common chromophoric groups that may be responsible for the dark color of humic substances are:

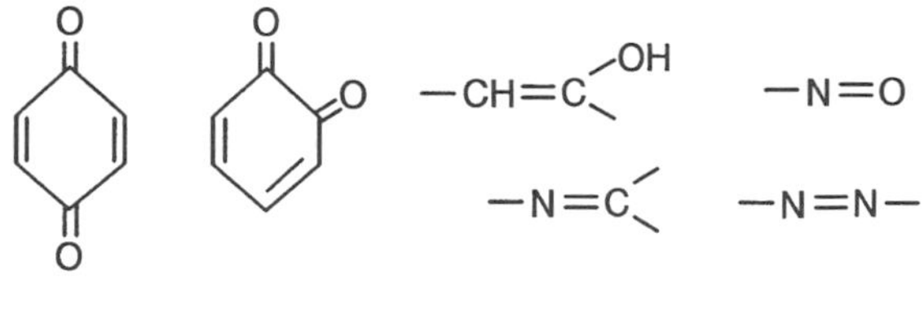

Various "type" structures have been proposed for humic and fulvic acids (see Ref. 15), but none of them can be considered as entirely satisfactory. The representative "structural unit" for a humate molecule shown in Fig. 1.11 conforms to many of the properties of humic and fulvic acids, notably a mixture of aliphatic and aromatic properties and a high content of oxygen-containing functional groups.

The structure and formation of humic materials was the subject of a symposium sponsored by the International Humic Substances Society. The reader is referred to the published volume resulting therefrom.[31]

TABLE 1.2   Range of Oxygen-Containing Functional Groups in Humic and Fulvic Acids (in cmole/kg)[a]

Functional Group	Humic Acids	Fulvic Acids
Total acidity	560–890	640–1420
COOH	150–570	520–1120
Phenolic OH	210–570	30–570
Alcoholic OH	20–490	260–950
Total carbonyl (C=O)	10–560	120–420
Methoxyl (OCH₃)	30–80	30–120

[a]See Stevenson[15] for additional details.

USE OF ¹⁴C IN SOIL ORGANIC MATTER STUDIES AND BIODEGRADATION RESEARCH

Carbon has three isotopes: ^{12}C (98.9% of all global C), ^{13}C (1.1% of all global C), and ^{14}C (trace). Both ^{12}C and ^{13}C are stable; ^{14}C is radioactive. By use of isotopes (e.g., ^{13}C or ^{14}C), it has been possible to follow the fate of specific C sources (e.g., crop residues) against the background of large amounts of soil organic matter.

Emphasis is given herein to ^{14}C; results using ^{13}C will be discussed in a later section. Isotopes of N, P, and S have also been used in soil organic matter studies, and these results are discussed in subsequent chapters.

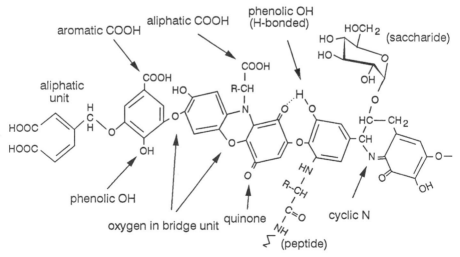

Fig. 1.11   Representative structure unit of a humate molecule showing the presence of free and bound phenolic OH groups, quinones, oxygen and N as bridge units, and —COOH groups variously placed on the molecule. Adapted from Stevenson.[15]

The traditional method of studying degradation of plant residues and other organic substrates in soil has been to measure C loss (as CO_2) from amended and unamended soil, the difference being attributed to C derived from the added substrate. This approach has serious limitations, including analytical errors in measuring long-term release of CO_2, particularly when the amount of C in the added substrate is small relative to that in the soil organic matter. Also, the assumption is made that addition of the substrate does not alter the degradation rate for the native soil organic matter, which may not be true.

Use of ^{14}C-labeled substrates has made it possible to follow degradation of added organic materials with considerable accuracy, even in the presence of relatively large amounts of soil organic C. For example, degradation of as little as 1–2 mg/kg of a pesticide can be studied in soils containing 1–50 g/kg of organic matter. It is also possible to trace the added C as it becomes incorporated into the various soil organic fractions.[32]

The initial studies using ^{14}C appear to be those of Bingeman et al.[33] and Hallam and Bartholomew,[34] who examined the effect of organic matter additions on decay of native humus. Since then, a multitude of papers have been published on degradation of ^{14}C-labeled natural and synthetic organic compounds in soil. For additional details, the reader is referred to the reviews by Jenkinson,[2] Paul and van Veen,[1] and others.[4,35–37]

Techniques and Approaches in [14]C Studies

Carbon-14 is a weak β-emitter with a half-life of about 5730 years. As with other radioactive compounds, quantitative determination is based on ionization or excitation of some substance by emitted radiation. In early work, Geiger-Müller tubes were used; more recently, liquid scintillation counting has been employed. Scintillation counters measure light emission resulting from the interaction of β particles with a compound that fluoresces when bombarded by the particles. Autoradiography relies on the interaction between a β particle and a photographic emulsion; in effect, the β particle reacts in much the same way as visible light, and radioactive areas appear as black, exposed sections on the x-ray film. Detailed procedures on radiocarbon methods for studies on soil organic C are available.[38]

For the study of C transformations under laboratory conditions, soil amended with a given quantity of ^{14}C-labeled plant tissue or a known organic compound is incubated under controlled temperature and moisture conditions. At intervals, the evolved CO_2 is trapped and total C as CO_2 and the ^{14}C content of the CO_2 are determined. Since the specific activity (proportion of ^{14}C to ^{12}C atoms) of the substrate is known, the evolved CO_2 can be partitioned between the added substrate and the indigenous soil organic matter. For field studies, where quantitative recovery of CO_2 is not feasible, the specific activity of the soil C is determined by combustion of the soil samples and analysis of the liberated CO_2 for [14]C. Changes in specific activity at any one time reflect the relative proportion of C loss from the substrate as compared to the soil organic matter.

$$\text{Relative C loss from substrate} = \frac{\text{C loss from added substrate}}{\text{C loss from soil organic matter}}$$

At the termination of an experiment, or after any given time interval, measurements can be made for residual ^{14}C in the microbial biomass. Also, the soil organic matter can be fractionated, with each component (i.e., carbohydrates, humic acid, fulvic acid) being analyzed for ^{14}C.

Preparation of ^{14}C-labeled Plant Material   Use of plant tissue *uniformly* labeled with ^{14}C is imperative in studying plant residue degradation. Plant composition varies with stage of growth, and a short labeling period can result in poor labeling of specific components. For example, lignin is formed relatively late in the growing season, and therefore using seedlings as ^{14}C-labeled plant tissue would result in poor labeling of this important constituent. For uniform labeling to be ensured, plants must be grown to maturity, or nearly so, a procedure that precludes use of soil as a growth medium because the production of ^{12}C—CO_2 through decay of soil organic matter would lead to dilution of ^{14}C—CO_2 added to the growth chamber. Several types of growth chambers have been used with apparent success to obtain uniformly labeled ^{14}C plant material. More specific labeling of plant components can be achieved by feeding of ^{14}C-labeled precursors of polymers such as lignin and cellulose into the plant's vascular system.[39]

Laboratory Incubations   Laboratory incubations are a convenient and inexpensive way to study organic transformations in soil. In a typical experiment, substrate-amended soil is placed in a sealed container with aeration and the evolved CO_2 is either trapped in alkali or determined directly by infrared spectroscopy using a CO_2 analyzer or by gas chromatography. The ^{14}C content is usually determined by liquid scintillation spectrometry after trapping of the evolved CO_2.

Field Experiments   For these studies, the residual ^{14}C content in soil is determined, with the assumption that the unrecovered ^{14}C has been lost as CO_2. The experiments are usually carried out in small cylinders of 10- to 30-cm diameter, driven into the soil to a 15-cm depth or more to prevent lateral movement of soil and added plant material and to prevent contamination to or from adjacent areas. The application rate should be similar to normal rates so as to avoid modifying the chemical or biological properties of the soil. The soil within the cylinders is removed, mixed thoroughly with labeled material, and returned to the cylinder. A 6-mm mesh screen is placed at the bottom of the cylinder prior to addition of soil. The screen delineates the boundary between treated and untreated soil while allowing movement of roots and small soil organisms. At intervals, soil subsamples are removed from the cylinder and analyzed for residual ^{14}C content, as well as specific soil fractions such as biomass or humic materials, as described elsewhere in this chapter.

Degradation of Simple Substrates and Synthesis of Labeled Microbial Components

As noted above, degradation of an organic substrate leads to incorporation of a portion of the C into microbial components and soil organic matter. A major objective of studies using ^{14}C-labeled biochemicals as a C source is to promote rapid synthesis of a wide range of microbial components and to follow their fate over time, using a technique generally referred to as a "pulse-chase" experiment. This is done by adding a C source such as glucose or acetate that will be completely transformed to CO_2 and microbial components in a few days. An example using a laboratory bacterial culture is shown in Fig. 1.12.[40] In this case, about 98% of the added ^{14}C was recovered in identifiable compounds, namely as CO_2, acetate, and biomass components.

Results of a similar study conducted in soil (but for a longer time) are shown in Fig. 1.13.[41] In contrast to the laboratory culture study, a substantially higher fraction of the residual ^{14}C could not be recovered in known compounds by acid hydrolysis (i.e., as amino acids and amino sugars) but existed in unknown forms, presumably as humic substances. Similar results have been

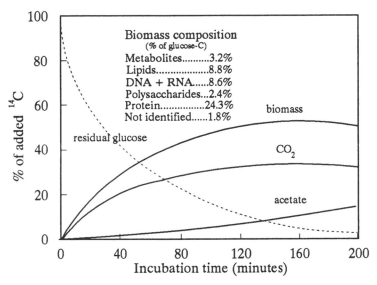

Fig. 1.12 Distribution of ^{14}C from ^{14}C-labeled glucose during growth of *Escherichia coli* in laboratory culture. From M. A. Cole, unpublished data.

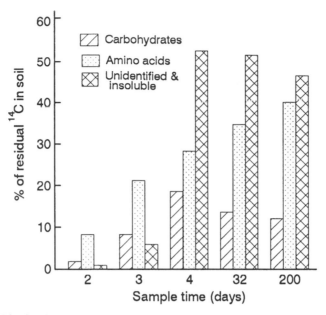

Fig. 1.13 Distribution of residual C in several soil fractions labeled with ^{14}C from ^{14}C-acetate upon long-time incubation. Values for days 2 and 3 do not total 100% because substantial amounts of ^{14}C remain as ^{14}C-acetate. All acetate had been degraded by day 4, after which time, changes were due to accumulation of unknown compounds or humic substances, with selective preservation of amino acids relative to carbohydrates (see text). Adapted from Sørensen and Paul.[41]

obtained using acid hydrolysis to characterize organic forms of N upon incubation of the soil with ^{15}N-labeled fertilizer and an energy source (see Chapter 6). Analyses conducted after the initial labeling period provide a means for establishing the MRT of residual ^{14}C from the added substrate.

The results depicted in Fig. 1.13 show several common pathways for compositional changes in newly added C over time.

1. Rapid initial formation followed by selective net loss of some metabolites (e.g., carbohydrates).
2. Net gain of some biochemical components over time (e.g., amino acids).
3. A pronounced increase in the percentage of the residual C present in unknown forms, most of which are humic substances. Selective preservation of amino acids relative to carbohydrates, as noted above, can be attributed to stabilization of amino acids by the humic material (see Chapter 6).

Various attempts have been made to fractionate the C remaining in soil after degradation of ^{14}C-labeled substrates. Ivarson and Stevenson[42] found that residual ^{14}C was widely distributed in the different humus fractions, from which they concluded that structural units of humus can be synthesized in a rather short time from simple organic compounds. A somewhat similar result was obtained by Wagner (see IAEA[36]) using ^{14}C-glucose as substrate. The high stability of the C when transformed into microbial components and metabolites was also demonstrated by Shields et al.[43]

A major difficulty in interpreting results of organic matter fractionation studies is that chemical changes can occur in the highly acidic or alkaline solutions used to extract humic materials, with conversion of metabolic products into stable humus forms. Sauerbeck and Führ (see IAEA[36]) raise a cautionary note that chemical reactions can occur between soil humus and soluble plant constituents during alkali extraction and create artifacts, an aspect of ^{14}C research that requires further study. The kinetics of initial decay and transformations of ^{14}C after incorporation into microbial cells and components of soil organic matter have substantially different time scales, depending on soil type. Retention values as recorded for ^{14}C-labeled glucose and acetate are summarized in Fig. 1.14.[44]

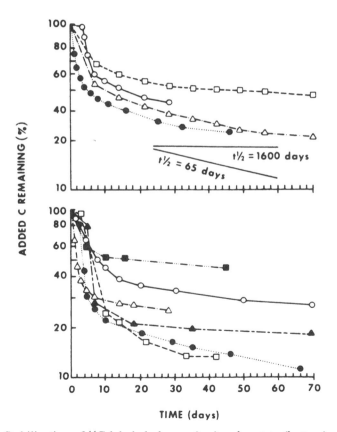

Fig. 1.14 Stabilization of ^{14}C-labeled glucose (top) and acetate (bottom) when added to soil in the laboratory. Each line represents the results of a separate experiment as recorded in the literature. From Paul and McLaren,[44] courtesy of Marcel Dekker, Inc.

A model for the data presented in Fig. 1.13 is shown in Fig. 1.15. According to this model, accumulation of biomass and other bioproducts occurs in parallel with substrate decay, followed by a period of degradation of microbial tissue with further loss of CO_2 and partial conversion of C to unidentified forms (i.e., humic substances).

The continued slow loss of ^{14}C seen in Figs. 1.12 and 1.14 can be attributed to decay of ^{14}C-labeled soil organic matter and biomass components. In most cases, a change from rapid loss of ^{14}C (as ^{14}C–CO_2) to slow release takes place when the residual ^{14}C accounts for about 30% of that initially added, and is attributed to partitioning of substrate-C between CO_2 production, biomass formation, and metabolite synthesis. Varying microbial groups process C in substantially different ways, as shown in Table 1.3.[45–47]

Decay of ^{14}C-Labeled Plant Residues

Due to regulations governing use of radioactive materials, most studies on the degradation of ^{14}C-labeled plant residues have been done in confined laboratory or greenhouse environments. In general, these studies have shown that the residues are attacked rapidly at first, but after a few months the rate slows substantially although considerable amounts of plant-derived C remain in the soil. The decreased rate over time is believed to be due not entirely to differences in the rate at which the various plant components decay (e.g., lignin is degraded quite slowly), but also because part of the C of the more readily degraded constituents (e.g., solubles and cellulose) have been resynthesized into microbial components whose extractability is similar to that of the original plant constituents, but which are more resistant to decay than the initial plant material. The general pattern for retention of ^{14}C from labeled plant residues is somewhat similar to that noted in Fig. 1.14 for biochemical compounds.

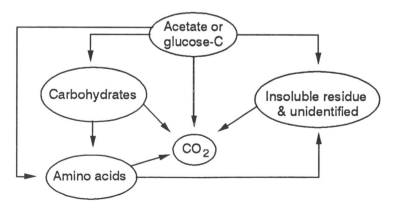

Fig. 1.15 Flow diagram for the partitioning of organic C after labeling the soil with ^{14}C-labeled acetate or glucose.

TABLE 1.3   Partitioning of Substrate C into Microbial Biomass, CO_2, and C Metabolites by Various Microbial Groups

Microbial group	% of Total C Converted to		
	Biomass	CO_2	Metabolites
Aerobic bacteria[45]	60	35–40	<5
Fermentative bacteria[46]	5–10	0–40	60–90
Aerobic fungi[47]	35	45–55	10–20
Yeasts[46]	10	30	60

Field Trials with ^{14}C-Labeled Plant Residues

Several investigations have been performed in which ^{14}C-labeled plant residues have been applied to field soils.[48-51] Results obtained for C retention after the first year (or first growing season) are summarized in Table 1.4. From 20 to 74% retention of applied C has been observed, depending on soil conditions and plant part involved. The highest values (63 to 74%) were obtained for ^{14}C-labeled roots of blue grama grass and were attributed to a high lignin content.

Results of field studies with ^{14}C-labeled plant residues show that the C becomes increasingly resistant to decay with time. In the experiments conducted by Jenkinson,[48] approximately 33% of the applied C remained behind in the soil after the first year, of which about one-third was believed to occur in microbial cells. After four years, about 20% of the labeled C still remained in the soil and only about 20% of this C was in microbial cells. The original residues decomposed rapidly with an estimated half-life of 14 to 30 days;

TABLE 1.4   Carbon Retained from ^{14}C-Labeled Plant Material Applied to Field Soils

Location	Type	Carbon Retained, %	Reference
England	Ryegrass tops and roots	Approximately 33% first year irrespective of soil type or plant material	Jenkinson[48,49]
West Germany	Wheat straw and and chaff	31% after first year for fallow and cropped soil	IAEA[36]
Canada	Wheat straw	35–45% after first growing season	Shields and Paul[50]
Colorado, U.S.	Blue grama grass a. Herbage b. Roots	43–46% after 412 days 63–74% after 412 days	Nyhan[51]

after the first year, the residual C had a half-life of about 4 years. Under the conditions specified, the native humus had a half-life of about 25 years. These relationships are shown in Fig. 1.16. It should be noted that the MRT of soil organic matter is variable but is usually somewhat longer than 25 years (discussed below).

The data given in Table 1.4 and Fig. 1.16 were obtained with fertile, temperate-zone soils. Considerably different values would be expected for soils of colder and warmer regions and for soils with a high sand content. Factors that affect the rate of degradation and retention of C include positioning of the plant material beneath or on the soil surface, soil moisture regime, temperature, and length of the growing season.

Jenkinson and Rayner[52] developed a model for C transformations in soils based on the classical long-term rotation plots at the Rothamsted Experiment Station. Using half-lives for various fractions of soil organic matter, they concluded that an annual input of 907 kg/ha of plant material for 10,000 years would lead to 10,250 kg C in physically stabilized forms and 11,065 kg C in chemically stabilized forms. Total C input over the period was 9.7 million kg C, of which only 22,000 kg was retained. These results demonstrate that, even in soil where total C content is at steady state (see Chapter 2), considerable turnover of C occurs over the long term.

From measurements for residual ^{14}C at various time intervals, MRTs have been estimated (e.g., Clark and Paul[53]), from which simulation models have been developed for predicting C and nutrient cycling (see Chapter 3). Although the nomenclature varies among investigators, the common perception is that the MRT for the C of crop residues is governed by a combination of inherent differences in biodegradability of the different plant components and differences in accessibility due to sequestering within soil aggregates and/or adsorption to soil minerals or long-lasting humic substances.

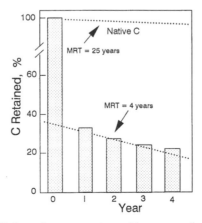

Fig. 1.16 Long-time C loss from organic residues as estimated from studies with ^{14}C-labeled ryegrass. Adapted from Jenkinson.[48]

Ladd[54] and Ladd et al.[55] carried out field studies with doubly labeled ^{14}C and ^{15}N legume material. This work is discussed in Chapter 7.

Incorporation of ^{14}C into Soil Organic Matter Components

Sørensen[56] fractionated the soil organic matter after allowing ^{14}C-labeled barley straw to decay in soil. In agreement with the observations described above, about one-third of the C remained in the soil after initial decay was complete (100 days incubation at 20°C). Residual C was found in the following soil organic matter fractions:

Component	Recovery of residual ^{14}C
Fulvic acid	11.8–18.9%
Humic acid	19.5–26.4%
Humin	49.4–54.7%
Unaccounted for	5.0–12.9%

These results, together with studies using straw-derived cellulose and hemicellulose, suggested that compounds other than lignin contributed to formation of humic substances. Sauerbeck (see IAEA[36]) found that newly formed humic acids were considerably more stable than fulvic acids and that substantial interconversion among the various humus fractions occurred over time.

The Priming Action

Newly incorporated plant residues in soil can either stimulate or retard decomposition of native humus. This change in degradation rate is referred to as "priming" and is usually positive. The effect is shown in Fig. 1.17, where the amount of native C lost through priming is taken as the difference between the amount of soil CO_2 evolved in the presence and absence of the substrate.

In evaluation of the magnitude of the priming effect, several errors can arise when ^{14}C-labeled plant materials are used to follow the decay processes. They include:

1. Some of the ^{14}C–CO_2 may undergo exchange with ^{12}C-carbonate of the soil. Also, plant C may not be uniformly labeled with ^{14}C, and calculation errors can be made in analyzing the data.
2. Differences in the microbiological environment between the untreated and amended soil may result from the organic matter addition, including changes in pH, O_2 content, moisture, and others.

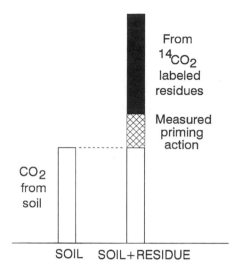

Fig. 1.17   Influence of applied plant residues labeled with ^{14}C on loss of native soil C as CO_2.

3. The ^{14}C-labeled substrate may not increase microbial biomass uniformly among members of the soil population, resulting in preferential labeling of the fastest-growing species.

One of the first investigators to suggest a priming action by addition of organic materials to soil was Löhnis.[57] From results of "green manuring" experiments in the field and greenhouse, he concluded that intensified bacterial activity accompanying incorporation of immature plant residues in soil increased mineralization of the native humus N. This view has been challenged from time to time (see IAEA[36]) and has yet to be resolved.

Unusually large priming actions were recorded by Broadbent et al.,[58,59] who concluded that decomposition of humus was increased as much as 10-fold by additions of tagged sudan grass tissues to soil. Broadbent and Norman[59] likened the process to a "forced draught on the smoldering bacterial fires of the soil." In studies carried out by Hallam and Bartholomew,[60] less total C remained in soil incubated with plant material than in soil incubated alone, from which they concluded that large, infrequent additions of plant residues were best from the standpoint of organic matter maintenance.

Assuming that the extra loss of soil C is caused by a genuine priming action, the probable explanation is that a buildup of a large and very active population of microorganisms occurs when energy-rich material is added to the soil and that these microorganisms produce enzymes that degrade the native humus as well as plant residue components. An analogous case may be stimulation of lignin degradation due to utilization of cellulose by fungi, with synthesis of lignin-degrading enzymes. The extent of organic matter loss will depend upon a variety of factors, including the size and activity of the microflora. Easily and rapidly degradable plant residues, such as young, suc-culent plant tissues, would be particularly effective in accelerating C loss.

CARBON-14 DATING OF SOILS

Results obtained with ^{14}C-labeled plant residues (discussed in the previous section) have shown that the organic C of soils consists of a pool of highly stable organic C with a mean age that may be as great as several thousand years and a relatively young (active) pool with a turnover time not exceeding 25 years. Thus, *absolute* ages of soil organic C cannot be determined using ^{14}C dating techniques. As noted above, the term *mean residence time* has been used to express the results of ^{14}C measurements for the average age of humus.

As discussed below, enrichment of the atmosphere with "bomb" radio-C has been used as the basis for determining rates of movement and turnover of organic C in soils.

Principles of ^{14}C Dating

Plant-derived organic materials formed from atmospheric CO_2 contain a small amount of ^{14}C as a result of assimilation of ^{14}C–CO_2 during photosynthesis. When the plants die and some of the C is retained as humus, radioactive decay of the ^{14}C occurs. This gradual loss of radioactivity provides a method for estimating the age of any residual C persisting in the soil (or sediment). The ^{14}C-dating method, first proposed by Libby and his associates,[61] is suit-able for accurate age estimates up to 50,000 years, but extension to 70,000 years is possible.

Naturally occurring ^{14}C is formed by solar radiation in the atmosphere. The cosmic radiation produces neutrons, some of which are captured by N with the production of ^{14}C, according to the equation:

$$^{14}N + neutron \rightarrow {}^{14}C + H$$

The ^{14}C combines with atmospheric O_2 to form $^{14}CO_2$, which is carried to the earth's surface by convection and turbulent mixing in the troposphere. Through photosynthesis, plants acquire small amounts of this ^{14}C. When the plant (or the animal that consumes the plant) dies and CO_2 assimilation from the atmosphere ceases, the ^{14}C content of the material (the $^{14}C/^{12}C$ ratio) diminishes with time because of radioactive decay. The basis for age estimate is the quantitative determination of the charged beta particles released during radioactive decay. The number of atoms decaying per unit time is proportional to the ^{14}C content of the material under investigation.

The ^{14}C radioactivity of a given sample decreases in a predictable manner with time, the half-life being 5,730 years. After this period, only one-half of the original ^{14}C activity will be left. After twice this period, or 11,460 years, only one-fourth will be left, and after another 5,730 years, or a total of 17,190 years, only one-eighth will be left, and so on. The formula relating age to ^{14}C activity is:

$$\log A = \log A_0 - \frac{\log 2}{t_{1/2}} t$$

where A is the number of radioactive nuclei remaining after time interval t, A_0 is the number of radioactive nuclei present at zero time, t is time or age since zero time, and $t_{1/2}$ is the half-life of the radioactive nuclide. The steady-state specific activity of ^{14}C (A_0) has been determined to be 15 disintegrations per minute per gram of C (d/m/g). A wood fragment found to have a specific activity of 5 d/m/g would have an estimated age of 9100 years.

The fundamental assumption underlying the ^{14}C dating method is that production of ^{14}C by cosmic radiation has been constant for at least 70,000 years and that C exchange among the reservoirs (atmosphere, biosphere, and oceans) has occurred at the same rate. These conditions appear to have been fulfilled, although deviations are known to have occurred. For example, since about 1870 the distribution of ^{14}C has been altered by burning of fossil fuel of very low ^{14}C content (crude oil and coal are several million years old and contain almost no ^{14}C), which has diluted the ^{14}C in the atmosphere by release of $^{12}C-CO_2$. More recently, the explosion of thermonuclear bombs has added large amounts of ^{14}C to the atmosphere; in 1964, the ^{14}C level of the atmosphere was twice the pre-1950's level. At present, the ^{14}C content of the atmosphere is still much higher than the previous normal level, but is declining as the result of transfer of $^{14}C-CO_2$ into plant biomass and equilibration with marine pools of CO_2/HCO_3^-.

Mean Residence Time (MRT) of Soil Organic Matter

Because of the very low levels of naturally occurring ^{14}C, special preparation techniques are required for age determinations and care must be taken to prevent contamination of the older organic matter with newer organics, such as decomposition products of plant residues.[62] The problem is particularly acute with very old organic matter, where contamination with as little as 0.1% (w/w) of C younger than the indigenous organic matter can severely affect the accuracy of the age determination.[63]

In order for the level of radioactivity to be increased, organic matter is usually extracted from the soil's mineral matrix, converted to CO_2 or elemental C (charcoal), and then synthesized into compounds of high solubility in a liquid scintillation fluid. One method of sample preparation involves conversion of the CO_2 into benzene, followed by determination of ^{14}C activity with a liquid scintillation counter.[64] The reactions are:

$$\text{Organic matter} \xrightarrow{+O_2} 6\ CO_2,\ \text{or}$$

$$\text{Organic matter} \xrightarrow{+Ar} 6\ C^0$$

$$6\ CO_2\ \text{or}\ 6\ C^0 \xrightarrow{Li,\,heat} 3\ Li_2C_2$$

$$3\ Li_2C_2 \xrightarrow{H_2O} 3\ C_2H_2\ \text{(acetylene)}$$

$$3\ C_2H_2 \xrightarrow{Cr\ or\ V\ catalyst} C_6H_6\ \text{(benzene)}$$

Methods based on the production of acetylene rather than benzene have also been used.

Representative MRTs for organic C in the surface horizon of soils from western Canada and the United States are recorded in Table 1.5.[65-67] While considerable variation in mean ages has been reported (210 to 1900 years), the findings attest to the high resistance of humus to microbial attack. In general, the humus of Mollisols is more stable than that of other mineral soils. As one might expect, the MRT of organic matter increases with soil depth and can be more than 10,000 years in buried soils in which there is little microbial activity.[62] Results of MRT measurements show that much of the humus in present-day soils of North America is derived from the vegetation (grasses, forests) that existed on them long before the land was colonized by Europeans after 1500 A.D.

Campbell et al.[66] used a complex fractionation scheme similar to that of Martel and Paul[67] to isolate organic matter components from a Chernozem (Mollisol) and a Gray-Wooded Podzolic (Ultisol) soil. The various fractions of the Chernozem soil were dated with MRTs that ranged from 25 to 1400

TABLE 1.5   Mean Residence Time (MRT) of Organic Matter in Some Typical Soils

Sample Description and Location	MRT	Reference
Mollisol		
a. Saskatchewan	1000	Paul et al.[65]
Saskatchewan	870 ± 50	Campbell et al.[66]
Saskatchewan	545	Martel and Paul[65]
b. North Dakota		
Virgin soil	1175 ± 100	Paul et al.[65]
Clean cultivated	1900 ± 120	Paul et al.[65]
Manured orchard	880 ± 74	Paul et al.[65]
Bridgeport loam, Wyoming		
a. Surface sod layer	3280	Paul et al.[65]
b. Continuous wheat plot	1815	Paul et al.[65]
Alfisol, Saskatchewan	250 ± 60	Paul et al.[65]

years. The main findings are shown in Table 1.6. The MRT of the unfractionated soil was 870 ± 50 years. Stability of the major humus fractions followed the order: humic acid > humin > fulvic acid.

The material hydrolyzable in 6 N HCl, consisting of amino acids, carbohydrates, and other biomass components, had the lowest MRT. For the Ultisol (data not shown), fractions were obtained with MRTs of from 0 to 485 years; MRT for the unfractionated soil was 250 ± 60 years. Humic fractions from the Mollisol were more stable than their counterparts from the Ultisol.

Results obtained for MRTs of individual humus fractions suggest that within the organic pool of the soil some components are more stable than others. As noted earlier for ^{14}C-labeled substrates, MRTs for humus and its components have been used to simulate the dynamics of soil organic matter and the cycling of nutrients in the soil–plant system (see Chapter 3).

Absolute Ages of Buried Soils

The humus component is seldom suitable for absolute age estimations of buried soils and this analysis is usually employed only as a last resort. More suitable materials for ^{14}C dating are charcoal, fossil wood, uncontaminated peaty layers, and well-preserved shells.[64] Contamination of buried soils with more recent C, by processes such as by downward movement of soluble organics in percolating water and penetration of the buried soil by plant roots, is always a problem. Visible contaminants (e.g., root hairs) can be physically removed. The problem centers on elimination of younger humus material. Principles involved in dating pedogenic events in the Quaternary are discussed by Ruhe.[68]

TABLE 1.6   Mean Residence Times (MRT) for Different Humus Fractions of a "Chernozemic" (Mollisol) Soil (Melfort silt loam)[a]

Component	MRT, years[b]
Unfractionated soil	870
Acid extract of soil	325
Fulvic acid	495
Humic acid	
Total sample	1425
Acid hydrolysate	25
Nonhydrolyzable	1400
Humin	
Total sample	1140
Acid hydrolysate	465
Nonhydrolyzable	1230

[a]From Campbell et al.[66]
[b]MRT = mean residence time. Ages are accurate to ± 60 years.

Recognition of contamination may be difficult, particularly when no reason exists to suspect that the data obtained are in error. The likelihood of foreign C contamination is often indicated by widely divergent dates from the same stratigraphic unit or by the occurrence of inversions (i.e., higher ^{14}C content of material taken at a deeper depth).

Ruhe[68] cites examples where buried soils were dated and a comparison was made with dates of the NaOH-soluble (mobile) organic fraction. In one case, the insoluble component had an age of 20,500 ± 400 years, whereas the mobile fraction had an age of 13,400 ± 600 years, indicating that contamination had occurred. Problems in obtaining reliable ages for buried soils are discussed by Goh[64] and Orlova and Panychev.[62]

In a study of radiocarbon dates obtained from soil-buried charcoals, Goh and Molloy[69] observed that the larger charcoal particles yielded a significantly older age than the finer particles. This was attributed to the presence of easily extractable fulvic acid in the smaller charcoal particles.

As an adjunct to ^{14}C-dating, sediments can be dated based on the racemization of amino acids. With the exception of a few D-amino acids in bacterial cell walls, living organisms synthesize amino acids that are largely in the L-stereoisomeric configuration. After burial, these compounds undergo slow conversion to the D-configuration. The degree of racemization provides an estimate of age and has been used for dating sediments and fossil calcareous shells.[70]

IN SITU LABELING OF SOIL ORGANIC C FROM BOMB-DERIVED ^{14}C–CO$_2$

An aftermath of nuclear bomb testing has been an increase in the ^{14}C–CO$_2$ content of the atmosphere (and hence the ^{14}C content of the plant and organic matter derived therefrom). The increase can be regarded as an *in situ* labeling of soil organic C and permits estimates to be made for the MRT's and turnover rates of both the older and younger organic C.[71–75] A theoretical discussion of the approach is given by Hsieh.[73]

Values obtained for ^{14}C can conveniently be expressed in terms of the per mil excess ^{14}C, or delta ^{14}C (δ^{14}C). The equation is:

$$\delta^{14}C\permil = \left(\frac{R_{sample}}{R_{standard}} - 1 \right) \times 1000$$

where R is the ratio of ^{14}C/^{12}C.

A δ^{14}C of +10 indicates that the experimental sample is enriched by 1% in atom % ^{14}C as compared to the standard.

Data for δ^{14}C and the distribution of C and N among various organic matter fractions are given in Table 1.7. The soil was hydrolyzed and fractionated as shown in Fig. 1.18.[67] In this study, organic-C was distributed about 60:40

TABLE 1.7   Distribution of C and N among Organic Matter Fractions of Various Ages in a Brown Chernozemic Soil[a] (A flow diagram of the fractionation procedure is shown in Fig. 1.18.)

Soil Fraction	MRT Years B.P.	$\delta^{14}C$	—% of total—		
			^{14}C	Soil C	Soil N
Light material	Modern	+243	8	6	3
0.5 N HCl hydrolysate	Modern	+7	37	35	41
6 N HCl hydrolysate	Modern	+61	24	22	43
NaOH extract	1,190	−212	22	27	9
Water extract	1,790	−200	3	3	2
Humin	1,330	−153	6	7	2

[a]From Martel and Paul.[67]

between relatively young and relatively old organic compounds. The light fraction, with a $\delta^{14}C$ value of +243‰, illustrates the impact of nuclear bomb testing on increasing the ^{14}C–CO_2 content of the atmosphere (and hence the ^{14}C content of the plant and organic matter derived therefrom). The positive δ‰ ^{14}C values for the hydrolysates show that much of the C in these fractions was of recent origin, being derived from ^{14}C–CO_2 since the early 1960s. From Table 1.7 it can also be seen that there were differences in N-content between the older and younger materials, with the younger materials containing most of the N. The relative age (degradability) of the N-containing fractions has a major influence on availability of N to plants, as described in Chapter 4.

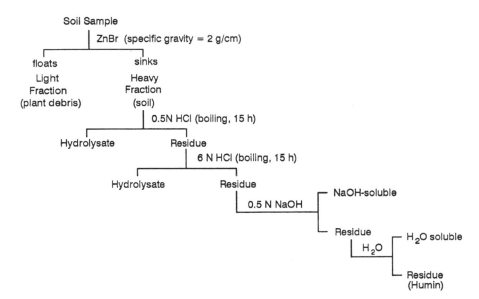

Fig. 1.18 Fractionation scheme for ^{14}C-dating of soil organic constituents described in Table 1.7. Adapted from Martel and Paul.[67]

On the basis of ^{14}C measurements, Hsieh[73] determined mean ages for the "active" and stable pools of organic matter in the Morrow Plots at the University of Illinois and the Sanborn Field at the University of Missouri. Mean ages for the stable pool were 2973 and 853 years, respectively. As shown in Fig. 1.19, the percentage of the total organic C as "active" organic C for the Morrow Plot soils ranged from less than 20% to over 60%, depending upon cropping history.

USE OF ^{13}C AND THE ^{13}C/^{12}C RATIO

Relatively little use has been made of the stable isotope ^{13}C in soil organic matter studies, even though the isotope is safer to use than the radioactive isotope ^{14}C and thus more suitable for field trials where restrictions in the use of ^{14}C apply. For detailed laboratory studies, ^{14}C has been the isotope of choice, primarily because of the ease and high sensitivity of radiochemical methods for isotope analysis (see previous section).

The ^{13}C content of organic materials or CO_2 is determined by mass spectrometry, a method that separates ^{12}C and ^{13}C atoms on the basis of the small difference in their atomic weight.[76] Major limitations to the method are the high instrumentation cost and, to some extent, the limited number of commercially available ^{13}C-substrates.

Measurements for the ^{13}C/^{12}C ratios of soil organic C have provided useful information on MRTs and turnover rates.[77-82] The results are usually expressed in terms of the per mil excess ^{13}C, or delta ^{13}C (δ^{13}C). The equation is:

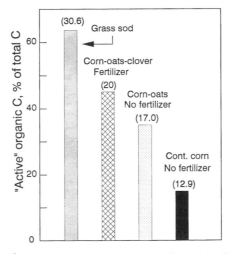

Fig. 1.19 Influence of management system on total organic-C (in brackets, g C/kg soil) and % of total organic-C in the "active" organic-C fraction as estimated from assimilation of ^{14}C "bomb" C. From data of Hsieh.[73]

$$\delta^{13}C\text{\textperthousand} = \left(\frac{R_{\text{sample}}}{R_{\text{standard}}} - 1\right) \times 1000$$

A $\delta^{13}C$ of $+10$ indicates that the experimental sample is enriched by 1% in atom % ^{13}C as compared to the standard.

The $\delta^{13}C$ value varies widely in nature, depending upon the source of the C and the plant species that assimilated the C during photosynthesis.[83] In general, plants using the C3 pathway for C-assimilation discriminate against ^{13}C and, as a result, the $\delta^{13}C$ for the plant tissues is lower than atmospheric CO_2. On the other hand, plants using the C4 pathway do not discriminate and the $^{13}C/^{12}C$ ratio of the plant is similar to that of atmospheric CO_2. Hence, addition of residues of C4 plants to soils which developed under C3 native vegetation (or under long-term cultivation with C3 plants) can be regarded as an *in situ* labeling of the organic matter with products from the added residues. Accordingly, formation of new humus can be followed by tracing the flow of ^{13}C from the added residues to the older organic matter. [77-82]

In the study by Balesdent et al.,[77] two cases of continuous corn cultivation (*Zea mays*) were examined, in one case, with a silty clay soil, and a sandy loam in the other case. In both locations, after 13 years, 22% of the total organic C had turned over. Physical fractionations showed that stability of organic matter was greater for the silt fraction than for the clay fractions, a finding in agreement with other studies using the technique of ^{14}C–NMR (see Chapter 3).

TRANSFORMATIONS IN WATERLOGGED SOILS AND SEDIMENTS

Soils and sediments that are water-saturated have substantially different physical and chemical properties and biota than aerobic surface soils.[84-90] Accordingly, many of the concepts developed for well-drained soils do not apply to waterlogged environments such as flooded soils, peat bogs, marshes, clay- and silt-based lake and stream bottoms, and similar situations. Not only do wet sediments support a different microflora, but decay of organic materials differs qualitatively and quantitatively from that of aerobic sediments.[84-86]

Because of the practical difficulties associated with working with anaerobic sediments and associated organisms, our knowledge of anaerobic transformations is very limited. There is much current interest in restoring natural wetlands or creating artificial wetlands as effective, low-technology systems for pollution abatement and pollutant degradation (see Moshiri[91]).

Strictly anaerobic or facultative bacteria are the predominant degraders of organic residues in saturated environments. Since most fungi and actinomycetes are aerobic organisms, their activities are limited in the nearly O_2-free situation that is typical of organic-rich flooded soils. Degradation of plant residues is relatively slow, and some compounds such as lignin and long-chain hydrocarbons, degrade slowly if at all. Accumulation of organic matter and hydrocarbon pollutants in sediments is a practical consequence of incomplete degradation of these materials.

The degradation of plant components (e.g., cellulose) by fermentative organisms results in the production of end-products such as short-chain organic acids and alcohols, followed by their further degradation by specific organisms (e.g., denitrifying and sulfate-reducing bacteria and methane producers). The most common organic acids found in recently flooded soils are acetic and formic acids, with smaller amounts of propionic acid. In natural substrata subjected to long-term flooding, organic acids do not accumulate, due to their degradation by denitrifying, sulfate-reducing, or methanogenic organisms. Over time, any water-soluble compounds that are not degraded (or adsorbed to the solid phase) will diffuse from the anaerobic zone into the overlying aerobic layer, where further degradation will occur.

The basic metabolism of fermentative bacteria differs from that of aerobic organisms in two main ways. First, the end-product of glycolysis, pyruvate, is not oxidized by enzymes of the tricarboxylic acid (Krebs) cycle, as is the case for aerobes. Instead, pyruvate and closely related metabolites are used as acceptors for electrons generated during glycolytic and other reactions (Fig. 1.20), thereby forming characteristic end-products of fermentation such as acetate, lactate, and formate.

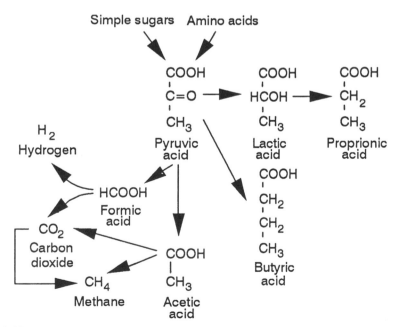

Fig. 1.20 Reactions of pyruvate in anaerobic systems, showing major organic acid and CH_4 formation. Carbon dioxide produced during fermentation serves as an electron (hydrogen) acceptor, with formation of CH_4 (see text). Adapted from Yoshida.[86]

A second difference is that denitrifying and sulfate-reducing bacteria utilize nitrate (NO_3^-), nitrite (NO_2^-), sulfite (SO_3^{2-}), or sulfate (SO_4^{2-}) as terminal electron acceptors for their respiratory (cytochrome) systems instead of O_2. The principal substrates for these organisms are the organic acids and alcohols that are final products of fermentation. Methane-generating microorganisms oxidize H_2 to produce electrons needed for energy generation and, in the process, use either CO_2 as the terminal electron acceptor or carry out an internal oxidation–reduction of the two carbons of acetate to yield CO_2 and methane (CH_4). The various reactions are:

$$2 H_2 \rightarrow 4 H^+ + 4e^-$$

$$\underline{HCO_3^- + 4e^- + 6 H^+ \rightarrow CH_4 + 3 H_2O}$$

$$(net) \ 2 H_2 + HCO_3^- + 6 H^+ \rightarrow CH_4 + \ H_2O$$

or

$$*CH_3COO^- + H^+ \rightarrow {}^*CH_4 + CO_2$$

Net energy yield for many anaerobes is less than for aerobes; accordingly, biomass production for an equivalent amount of energy (substrate) is less. A consequence of the reduction in biomass is a decrease in the assimilation of N, P, and S, with the result that the inorganic N content (primarily NH_4^+) of sediments can be quite high as compared to that of aerobic soils.

Thus, in contrast to aerobic systems, where CO_2, H_2O, and microbial cells are the main end-products of degradation, anaerobic systems are characterized by formation of organic acids and alcohols, CH_4, CO_2, N_2, and hydrogen sulfide (H_2S). Secondary products, some formed biologically and others by chemical reactions, are often present. They include mercaptans, aldehydes, ketones, and primary, secondary, and tertiary amines, all of which contribute to the malodorous condition of putrefying organic matter. Many of the compounds listed above are toxic to plants and animals. For example, butyric acid is believed to contribute to the poor plant growth often observed in rice paddy fields.[92,93] Ethylene in toxic amounts has been reported to be produced by decay of organic residues under anaerobic soil conditions.[94-96] According to Kilham and Alexander,[97] accumulations of organic matter in some flooded soils are due to the inhibitory effect of organic acids and H_2S on microorganisms.

A list of compounds peculiar to wet sediments is given in Table 1.8. A brief discussion of individual items follows.

Protein Degradation

Protein degradation in anaerobic environments yields organic acids, NH_4^+, and a wide range of foul-smelling amines, mercaptans, H_2S, and partial degradation products of amino acids, whose very names are indicative of their

TABLE 1.8   Organic Compounds Peculiar to Wet Sediments

Class	Comments
Fermentation products	Incomplete oxidation leads to production of CH_4, organic acids, amines, mercaptans, aldehydes, and ketones.
Modified or partially modified remains of plants	In addition to slightly altered lignins, long-chain aliphatic hydrocarbons, carotenoids, sterols, and porphyrins of chlorophyll origin are preserved.
Synthetic organic chemicals[a]	Many man-made chemicals (e.g., DDT) degrade very slowly under anaerobic conditions and can thus persist for long periods.[a]
Carcinogenic compounds[a]	Synthesis of methylmercury, dimethylarsine, dimethylselenide, and nitrosamines of various types.[a]

[a]Environmental aspects of the C cycle are discussed in Chapter 4.

odorous nature; skatole, cadaverine, and putrescine are typical examples. These products are not generally found in aerobic soils, where the end-products of metabolism are CO_2, NO_3^-, SO_4^{2-}, and H_2O.

Cellulose Degradation

Many microorganisms are capable of degrading cellulose in the absence of O_2, the most common being species of *Clostridium*. Some *Clostridium* species are specific in that they require cellulose as an energy source. The main end-products of cellulose decomposition are CO_2, H_2, ethanol, and various organic acids. These products are used, in turn, by methanogens and other bacteria to yield a wide range of end-products.

Lignin Modification

Mention is made above of the significance of fungi and actinomycetes in lignin degradation. Since the activity of these organisms is restricted in anaerobic aquatic sediments, lignin is degraded slowly or not at all[98] and (in a modified form) tends to accumulate.[99] The main modification of lignin appears to be enzymatic cleavage of methoxyl ($-OCH_3$) groups, presumably by anaerobic bacteria. The fate of the remainder of the lignin molecule is uncertain.

Preservation of Other Plant Components

Various lipid components appear to be less subject to degradation in anaerobic systems than in aerobic soils. Carotenoids, certain sterols, and porphyrins derived from chlorophyll are generally prevalent in wet sediments, whereas in well-drained soils they occur in trace amounts only.

Oxidation–Reduction Reactions

Depletion of dissolved O_2 resuls in a shift from aerobic to anaerobic metabolism in wet sediments with an accompanying change in the oxidation–reduction potential, E_h. The absolute value for E_h serves as an indication of the tendency of organic and inorganic substances to gain electrons (become reduced) or lose electrons (become oxidized). Thus, a high E_h indicates an oxidizing environment; a low or negative value suggests a reducing environment. Negative E_h values are common in rice paddy fields and aqueous sediments containing decaying organic materials.

A typical sequence in transiently flooded soils is that microbes reduce NO_3^- before SO_4^{2-}; both are reduced before CH_4 formation is observed. The ability of inorganic substances to act as electron acceptors—to be reduced— would be expected to follow the thermodynamic sequence shown in Table 1.9. However, the kinetics and kind of reduction products are determined by a variety of factors, including amount of organic matter, pH, temperature, duration of flooding, and kind and concentration of inorganic electron acceptors.

Changes in Soils and Sediments During Submergence

Substantial changes occur when soils or sediments are submerged, whether flooded by natural or artificial means. The major changes can be summarized as follows:

TABLE 1.9   Thermodynamic Sequence for Reduction of Inorganic Substances

Reaction	E_h^a
Disappearance of O_2	
$O_2 + 4H^+ + 4e \rightarrow 2H_2O$	0.816 V
Disappearance of NO_3^-	
$NO_3^- + 2H^+ + 2e \rightarrow NO_2^- + H_2O$	0.421 V
Formation of Mn^{2+}	
$MnO_2 + 4H^+ + 2e^- \rightleftharpoons Mn^{2+} + 2H_2O$	0.396 V
Reduction of Fe^{3+} to Fe^{2+}	
$Fe(OH)_3 + 3H^+ + e^- \rightleftharpoons Fe^{2+} + 3H_2O$	−0.182 V
Formation of H_2S	
$SO_4^{2-} + 10H^+ + 8e^- \rightarrow H_2S + 4H_2O$	−0.215 V
Formation of CH_4	
$CO_2 + 8H^+ + 8e^- \rightarrow CH_4 + 2H_2O$	−0.244 V

[a] At pH 7.0.

1. The utilization of organic matter by microorganisms results in a marked drop in dissolved O_2 and an increase in CO_2 content. The net result is a concurrent, but often temporary, drop in E_h. Microorganisms cause the change in E_h by consuming O_2 and liberating reduced products.
2. There may be changes in pH as well as in electrolyte concentration in general.
3. Denitrification may occur with loss of $NO_3^- -N$ as N_2 and/or N_2O. On the other hand, NH_3 (as NH_4^+) may accumulate.
4. Intermediate organic matter decomposition products such as organic acids may accumulate.
5. Sulfate is reduced to H_2S.
6. Manganese and Fe in the form of sparingly soluble oxides may be reduced and their solubilities enhanced by generation of water-soluble Mn^{2+} and Fe^{2+}. These transformations are controlled to a considerable extent by microorganisms (change in E_h accompanying depletion of O_2 leads to changes in oxidation state). Various other inorganic constituents, such as phosphate, may become more soluble.

Taken as a group, these changes account for the typical behavior of pond waters, where periodic bubbling is often observed (formation of CO_2 and CH_4), where the bottom sediment is black (from formation of FeS; i.e., $Fe^{2+} + S^{2-} \rightarrow FeS$), and where foul odors are emitted when the bottom sediment is disturbed (from organic end-products and H_2S).

Nitrogen Transformations

All aspects of the N cycle must be viewed in a new context when considered in terms of anaerobic aquatic systems (see Chapters 5 and 6). For example, many plants that form N_2-fixing symbiotic associations in aerobic soils do not grow in flooded soils. In water-saturated environments, fixation of N_2 is due primarily to cyanobacteria and certain green and purple photosynthetic bacteria. A schematic representation of N transformations in a flooded soil ecosystem is shown in Fig. 1.21.

The recurrent cycles of N mineralization–immobilization observed in aerobic soils are also modified in zones where O_2 is absent. Mineralization produces NH_3, but further oxidation to NO_2^- and NO_3^- by nitrifying bacteria is restricted because O_2 is required. Oxidation of NH_4^+ may occur, however, at the water–sediment interface or in the water above sediments, where O_2 is present. The ammonification process (conversion of organic N to NH_3) may also be curtailed or otherwise modified. Since fermentation provides little energy for microbial biosynthesis in comparison to aerobic processes, larger amounts of substrate must be oxidized per unit of C assimilated, with the result that there is less assimilation of mineral N by microbial cells than observed under aerobic conditions. Accordingly, competition of microorga-

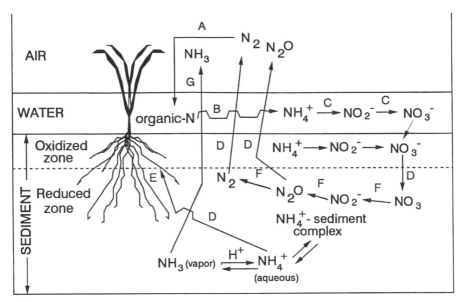

Fig. 1.21 Schematic representation of N transformations in a flooded soil ecosystem. Processes are: A, biological nitrogen fixation; B, mineralization; C, nitrification; D, diffusion; E, plant uptake; F, dentrification; G, volatilization. Adapted from Reddy.[85]

nisms with plants for available N may be less severe in aquatic environments than in aerobic soils.

Nitrate, introduced into aqueous sediments in surface or ground waters, applied as fertilizer, or formed at the oxidized layer, is subject to gaseous loss through denitrification. A quantitative measure of denitrification losses is considered basic to an understanding of the N balance in lakes and streams and for the proper utilization of fertilizer N in rice fields. The subject of denitrification is considered in greater detail in Chapter 5.

SUMMARY

The decay of dead plant and animal materials in soil is a fundamental biological process that is carried out by bacteria, actinomycetes, and fungi. Part of the C is utilized for the synthesis of microbial cell components (the soil biomass); part is incorporated into stable humus. Biochemical transformations of organic matter result largely from the enzymatic activities of proliferating microorganisms active in the decay process.

Humic substances represent a complex mixture of molecules having various molecular weights and geometry, but no completely satisfactory scheme has been developed for their isolation, purification, and fractionation. There is some evidence to indicate that, in any given soil, the various fractions obtained on the basis of solubility characteristics (e.g., humic acid, fulvic acid, and others) represent a system of polymers whose chemical properties (elemental composition, functional group content, etc.) change systematically with increasing molecular weight.

Considerable information has been obtained on degradation processes in recent years. Based on short-term and long-term studies with ^{14}C-labeled substrates, the following conclusions can be drawn.

1. The concept of soil organic matter as an "inert" biologically resistant material has not been substantiated.
2. Addition of fresh organic residues to the soil may result in a small priming action on the native soil organic matter.
3. Humus can be synthesized by microorganisms utilizing nonlignin C sources. Freshly incorporated C first enters into microbial cells (soil biomass), then "labile" fractions of soil organic matter, and finally into complex humic polymers during advanced stages of humification.

The MRT of freshly-added C in soil is initially short but approaches that of the native humus C within a relatively short time. The MRT of stable humus varies from several hundred to somewhat over 1000 years.

The conditions in wet sediments considerably alter the processes whereby organic residues undergo decay in soils. Plant remains are not completely metabolized (notably, lignin) and intermediate decomposition products accumulate, some of which are toxic to plants. In addition, hazardous carcinogenic and toxic compounds such as methylmercury and dimethylarsine are produced (see Chapter 4).

REFERENCES

1. E. A. Paul and J. A. van Veen, *Trans. 11th Intern. Congr. Soil Sci.,* **3,** 61 (1978).
2. D. S. Jenkinson, *Soil Sci.,* **111,** 64 (1971).
3. B. Bolin, *Science,* **196,** 613 (1977).
4. W. B. McGill and C. V. Cole, *Geoderma,* **26,** 267 (1981).
5. H. L. Bohn, *Soil Sci. Soc. Amer. J.,* **40,** 468 (1976).
6. B. Bolin and R. B. Cook, Eds., *The Major Biogeochemical Cycles and Their Interactions,* Wiley, New York, 1983.
7. R. M. Garrels, F. T. MacKenzie, and C. Hunt, *Chemical Cycles and the Global Environment,* W. Kauffmann, Los Altos, California, 1975.
8. B. Bolin, E. T. Degens, S. Kempe, and P. Ketner, *The Global Carbon Cycle,* in *SCOPE Report 13,* Wiley, New York, 1979.
9. F. Andreux, "Humus in World Soils," in A. Piccolo, Ed., *Humic Substances in Terrestrial Ecosystems,* Elsevier, Amsterdam, 1996, pp. 45–100.

10. K. Paustian, E. T. Elliott, H. P. Collins, C. V. Cole, and E. A. Paul, *Aust. J. Experimental Agr.*, **35**, 929 (1995).

11. H. W. Scharpenseel, M. Schomaker, and A. Ayoub, Eds., *Soils on a Warmer Earth,* Elsevier, Amsterdam, 1990.

12. B. A. Kimball, Ed., *Impact of Carbon Dioxide, Trace Gases, and Climatic Change on Global Agriculture,* SSSA Special Publication 53, Soil Science Society of America, Madison, Wisconsin, 1990.

13. R. Lal, J. M. Kimble, E. Levine, and B. A. Stewart, Eds., *Soils and Global Change,* Lewis Boca Raton, Forida, 1995.

14. L. A. Harper et al., Eds., *Agricultural Ecosystem Effects on Trace Gases and Global Climatic Change,* Special Publication 55, America Society of Agronomy, Madison, Wisconsin, 1993.

15. F. J. Stevenson, *Humus Chemistry: Genesis, Composition, Reactions,* 2nd ed., Wiley, New York, 1994.

16. H. Dinel, M. Schnitzer, and R. R. Mehuys, "Soil Lipids: Origin, Nature, Content, Decomposition, and Effect on Soil Physical Properties," in J.-M. Bollag and G. Stotsky, Eds., *Soil Biochemistry,* Vol. 6, Marcel Dekker, New York, 1990.

17. E. B. Rogertson, S. Sarig, and M. K. Firestone, *Soil Sci. Soc. Amer. J.,* **55**, 734 (1987).

18. M. B. Molope, I. C. Grieve, and E. R. Page, *J. Soil Sci.,* **38**, 71 (1987).

19. P. Vandevivere and P. Baveye, *Soil Sci. Soc. Amer. J.,* **56**, 1 (1987).

20. A. Szalay, *Geochim. Cosmochim. Acta,* **22**, 1605 (1964).

21. M. M. Kononova, *Soil Organic Matter,* Pergamon Press, Oxford, 1966.

22. M. S. A. Leisolo and S. Garcia, "The Mechanism of Lignin Biodegradation," in M. P. Coughian, Ed., *Enzyme Systems for Lignocellulose Degradation,* Elsevier, Amsterdam, 1990, pp. 89–99.

23. K. Haider, "Advances in the Basic Research of the Biochemistry of Humic Substances," in N. Senesi and T. M. Miano, Eds., *Humic Substances in the Global Environment and Implications on Human Health,* Elsevier, Amsterdam, 1994, pp. 91–107.

24. S. M. Shevchenko and G. W. Bailey, *Critical Rev. Environ. Sci. Tech.,* **26**, 95 (1996).

25. K. Haider and J. P. Martin, *Soil Sci. Soc. Amer. Proc.,* **39**, 657 (1975).

26. R. M. Pengra, M. A. Cole, and M. Alexander, *J. Bacteriol.,* **97**, 1056 (1969).

27. K. Haider and J. P. Martin, *Soil Biol. Biochem.,* **2**, 145 (1970).

28. J. P. Martin and K. Haider, *Soil Sci.,* **107**, 260 (1969).

29. J. P. Martin and K. Haider, *Soil Sci.,* **111**, 54 (1971).

30. J. P. Martin, K. Haider, and D. Wolf, *Soil Sci. Soc. Amer. Proc.,* **36**, 311 (1972).

31. N. Senesi and T. M. Miano, Eds., *Humic Substances in the Global Environment and Implications on Human Health,* Elsevier, Amsterdam, 1994.

32. J. Dec and J.-M. Bollag, "Dehalogenation of Chlorinated Phenols During Binding to Humus," in T. A. Anderson and J. R. Coats, Eds., *Bioremediation through Rhizosphere Technology,* American Chemical Society, Washington, D.C., 1994, pp. 102–111.

33. C. W. Bingeman, J. E. Varner, and W. P. Martin, *Soil Sci. Soc. Amer. Proc.,* **17,** 34 (1953).

34. M. J. Hallam and W. V. Bartholomew, *Soil Sci. Soc. Amer. Proc.,* **17,** 365 (1953).

35. B. J. O'Brien and J. D. Stout, *Soil Biol. Biochem.,* **10,** 309 (1978).

36. International Atomic Energy Agency (IAEA), *Isotopes and Radiation in Soil Organic Matter Studies,* FAO/IAEA, Vienna, 1968, pp. 3–11, 57–66, 197–205, 241–250, 351–361.

37. Y.-P. Hsieh, "Radiocarbon Methods for Dynamics of Soil Organic Matter," in N. Senesi and T. M. Miano, Eds., *Humic Substances in the Global Environment and Implications on Human Health,* Elsevier, Amsterdam, 1994, pp. 305–311.

38. D. C. Wolf, J. O. Legg and T. W. Boutton, "Isotopic Methods for the Study of Soil Organic Matter Dynamics," in R. W. Weaver et al., Eds., *Methods of Soil Analysis: Part 2, Microbiological and Biochemical Properties,* Soil Science Society of America, Madison, Wisconsin, 1994, pp. 865–906.

39. D. L. Crawford, R. L. Crawford, and A. L. Pometto, *Appl. Environ, Microbiol.,* **33,** 1247 (1977).

40. M. A. Cole, unpublished data.

41. L. H. Sørensen and E. A. Paul, *Soil Biol. Biochem.,* **3,** 173 (1971).

42. K. D. Ivarson and I. L. Stevenson, *Can. J. Microbiol.,* **10,** 677 (1964).

43. J. A. Shields, E. A. Paul, and W. E. Lowe, *Soil Biol. Biochem.,* **6,** 31 (1974).

44. E. A. Paul and A. D. McLaren, "Biochemistry of the Soil Subsystem," in E. A. Paul and A. D. McLaren, Eds., *Soil Biochemistry, Vol. 3,* Marcel Dekker, New York, 1975, pp. 1–36.

45. W. J. Payne, *Ann. Rev. Microbiol.,* **24,** 17 (1970).

46. A. G. Moat and J. W. Foster, "Carbohydrate Metabolism and Energy Production," in *Microbial Physiology,* Wiley-Liss, New York, 1995, Chapter 7, pp. 305–409.

47. J. W. Foster and S. A. Waksman, *J. Bacteriol.,* **37,** 599 (1939).

48. D. S. Jenkinson, *J. Soil Sci.,* **16,** 104 (1965).

49. D. S. Jenkinson, *J. Soil Sci.,* **17,** 280 (1966).

50. J. A. Shields and E. A. Paul, *Can. J. Soil Sci.,* **53,** 297 (1973).

51. J. W. Nyhan, *Soil Sci. Soc. Amer. Proc.,* **39,** 643 (1975).

52. D. S. Jenkinson and J. H. Rayner, *Soil Sci.,* **123,** 298 (1977).

53. F. E. Clark and E. A. Paul, *Adv. Agron.,* **22,** 375 (1970).

54. J. N. Ladd, *Plant Soil,* **58,** 401 (1981).

55. J. N. Ladd, J. M. Oades, and M. Amato, *Soil Biol. Biochem.* **13,** 119 (1981).

56. H. Sørensen, *Soil Sci.,* **95,** 45 (1963).

57. F. Löhnis, *Soil Sci.,* **22,** 355 (1926).

58. F. E. Broadbent and W. V. Bartholomew, *Soil Sci. Soc. Amer. Proc.,* **13,** 271 (1948).

59. F. E. Broadbent and A. G. Norman, *Soil Sci. Soc. Amer. Proc.,* **11,** 264 (1946).

60. M. J. Hallam and W. V. Bartholomew, *Soil Sci. Soc. Amer. Proc.,* **17,** 365 (1953).

61. W. Libby, *Radio Carbon Dating,* University of Chicago Press, Chicago, 1955.

62. L. A. Orlova and V. A. Panychev, *Radiocarbon,* **35,** 369 (1993).

63. K. A. Arslanov, T. V. Tertychnaya, and S. B. Chernov, *Radiocarbon,* **35,** 393 (1993).

64. K. M. Goh, "Carbon Dating," in D. C. Coleman and B. Fry, Eds., *Carbon Isotope Techniques,* Academic Press, New York, 1991, pp. 125–145.

65. E. A. Paul, C. A. Campbell, C. A. Rennie and K. J. McCallum, *Trans. 8th. Intern. Congr. Soil Sci.,* **3,** 201 (1964).

66. C. A. Campbell, E. A. Paul, D. A. Rennie and K, J. McCallum, *Soil Sci.,* **104,** 81, 217 (1967).

67. Y. A. Martel and E. A. Paul, *Soil Sci. Soc. Amer. Proc.,* **38,** 501 (1974).

68. R. V. Ruhe, *Soil Sci.,* **107,** 398 (1969).

69. K. M. Goh and B. P. J. Molloy, *J. Soil Sci.,* **29,** 567 (1978).

70. G. Dungworth, *Chem. Geol.,* **17,** 135 (1976).

71. B. J. O'Brien and J. D. Stout, *Soil Biol. Biochem.,* **10,** 309 (1978).

72. B. J. O'Brien, *Soil Biol. Biochem.,* **16,** 115 (1984).

73. Y.-P. Hsieh, *Soil Sci. Soc. Amer. J.,* **56,** 460 (1992).

74. Y.-P. Hsieh, *Soil Sci. Soc. Amer. J.,* **60,** 1117 (1996).

75. Y.-P. Hsieh, "Radiocarbon Methods for Dynamics of Soil Organic Matter," in N. Senesi and T. M. Mianao, Eds., *Humic Substances in the Global Environment and Implications to Human Health,* Elsevier, Amsterdam, 1994, pp. 305–311.

76. T. W. Boutton, "Stable Carbon Isotope Ratios of Natural Materials: 1. Sample Preparation and Mass Spectrometric Analysis," in D. C. Coleman and B. Fry, Eds., *Carbon Isotope Techniques,* Academic Press, San Diego, California, 1991, pp. 155–171.

77. J. Balesdent, A. Mariotti, and B. Guillet, *Soil Biol. Biochem.,* **19,** 25 (1987).

78. J. Balesdent, G. H. Wagner, and A. Mariotti, *Soil Sci. Soc. Amer. J.,* **52,** 118 (1988).

79. T. A. Bonde, B. T. Christensen, and C. C. Cerri, *Soil Biol. Biochem.,* **24,** 275 (1992).

80. A. Martin, A. Mariotti, J. Balesdent, P. Lavelle, and V. Vuattoux, *Soil Biol. Biochem.,* **22,** 517 (1990).

81. J. O. Skjemstad, P. P. Le Feuvre, and R. E. Prebble, *Aust. J. Soil Res.,* **28,** 267 (1990).

82. J. D. Jastrow, T. W. Boutton, and R. M. Miller, *Soil Sci. Soc. Amer. J.,* **60,** 801 (1996).

83. B. N. Smith and S. Epstein, *Plant Physiol.,* **47,** 380 (1971).

84. N. K. Savant and S. K. De Datta, *Adv. Agron.,* **35,** 241, 1982.

85. K. R. Reddy, *Plant Soil,* **67,** 209 (1982).

86. T. Yoshida, "Microbial Metabolism of Submerged Soils," in E. A. Paul and A. D. McClaren, Eds., *Soil Biochemistry,* Vol. 3, Marcel Dekker, New York, 1975, pp. 449–465.

87. D. Gunnison, R. M. Engler, and W. H. Patrick, Jr., "Chemistry and Microbiology of Newly Flooded Soils: Relationship to Reservoir Water Quality," in D. Gunnison, Ed., *Microbial Processes in Reservoirs,* Dr. W. Junk, Dordrecht, 1985, pp. 39–57.

88. B. B. Jorgensen, "Processes at the Sediment–Water Interface," in B. Bolin and R. B. Cook, *The Major Biogeochemical Cycles and Their Interactions (Scope 21),* Wiley, New York, 1983, pp. 477–508.

89. W. M. Lewis (Chair), *Wetland Creation and Restoration,* Island Press, Washington, DC, 1990.

90. W. H. Patrick, "Nitrogen Transformations in Submerged Soils," in F. J. Stevenson, Ed., *Nitrogen in Agricultural Soils,* American Society of Agronomy, Madison, Wisconsin, 1982, pp. 449–465.

91. G. A. Moshiri, Ed., *Constructed Wetlands for Water Quality Improvement,* Lewis, Boca Raton, Florida, 1993.

92. Y. Takai, *J. Sci. Soil Manure,* **28,** 40, 138 (1967).

93. Y. Takijima, *Soil Sci. Plant Nutr.,* **10,** 7, 14 (1964).

94. G. Goodlass and K. A. Smith, *Soil Biol. Biochem.,* **10,** 201 (1964).

95. K. A. Smith and S. W. F. Restall, *J. Soil Sci.,* **22,** 430 (1971).

96. K. A. Smith and R. S. Russell, *Nature,* **222,** 769 (1969).

97. O. W. Kilham and M. Alexander, *Soil Sci.,* **137,** 419 (1984).

98. J. G. Zeikus, "Fate of Lignin and Related Aromatic Substrates in Anaerobic Environments," in T. K. Kirk, T. Higuchi, and H. Chang, Eds., *Lignin Biodegradation: Microbiology, Chemistry, and Potential Applications:* Vol. 1, CRC Press, Boca Raton, Florida, 1980, pp. 101–109.

99. E. Fustec, E. Chauvet, and G. Gas, *Appl. Environ Microbiol.,* **55,** 922 (1989).

2

SOIL CARBON BUDGETS AND ROLE OF ORGANIC MATTER IN SOIL FERTILITY

All soils contain carbon (C) in the form of organic matter, or humus, the two terms being used synonymously. Quantities of C vary widely, from under 1% (by weight) in coarse-textured soils (sands) to 3.5% in grassland soils (e.g., Mollisols). Poorly drained soils (Aquepts) often have C contents of 10% or more. Many tropical soils, such as the Oxisols, are very low in organic C.

Most N in surface soil is in organic forms, and therefore the organic C content and total-N content of soil are closely related. The N-content is easier to determine than C-content and is frequently used as an index of organic matter or C content. The C/N ratio of the soil is usually between 10:1 to 12:1.

The factor 1.724 is normally used to convert organic C to organic matter. It is based on an assumed C content of 58% for the organic matter. However, organic C content can vary considerably from this value, and the percentage usually decreases with depth in the soil profile.

Topics discussed in this chapter include:

1. Factors affecting organic C content of soil, including effects of cropping and management systems
2. Role and function of humus
3. Organic matter and sustainable agriculture, and
4. Value of commercial humates as soil amendments.

For additional details on factors affecting organic matter levels in soil, the work of Jenny[1-4] is recommended. Reviews are also available on C (and N) storage in soils,[5-8] on the role and function of organic matter,[9] and on crop residue management for erosion control and protection of the environment.[10-12]

FACTORS AFFECTING LEVELS OF ORGANIC C IN SOILS

As early as 1900, British soil scientists had developed a qualitative perception of the relationship among organic matter content, soil development, and the productivity of cropped soils, as illustrated by the following quotation:

> It is well known that the fertility of "virgin" soils is due to the accumulation of the debris of a natural vegetation which has been in occupation of the soil for a long epoch previously. Only when the climate and rainfall are suitable to the growth of the plants and the partial preservation of their residues does a virgin soil of any richness arise; on the one hand, virgin soil may be as poverty stricken as the most worn-out European field because it has never carried any vegetation; on the other hand, as in the tropics, the debris of an extensive vegetation may decay with such rapidity that no reserve of fertility accumulates. In temperate climates, and with a particular distribution of the annual rainfall, occur the grassy treeless prairies and steppes which provide the ideal conditions for the accumulation of fertility. But that fertility does increase when land is in the state of permanent grass has long been an axiom in our farming. . . .   (Hall[13])

Later work showed that organic C content does not increase indefinitely in well-drained soils, and with time an equilibrium level is attained that is governed by the soil-forming factors of climate, topography or relief, vegetation and organisms, parent material, and time.[1-3] The numerous combinations under which the various factors operate accounts for the great variability in the organic C content of soils, even within a localized area.

Jenny and coworkers[1-4] attempted to evaluate the importance of soil-forming factors on the N content of the soil by treating each factor as an independent variable. However, altering any one factor produces changes in one or more of the remaining factors. Despite this limitation, his studies have contributed substantially to our understanding of the factors influencing N and organic C content of soil. According to Jenny,[1] the relative importance of soil-forming factors which determine N content of loamy soils within the United States as a whole is: climate > vegetation > topography = parent material > age.

Figure 2.1 depicts differences in organic matter content within a soil association as influenced by such factors as topography, drainage, parent material, and vegetation.[14] In this case, parent material and climate are constant for all soils, but variations in topography have greatly influenced drainage, plant species, and biomass production, with the result that soils having substantially different properties and organic matter contents have developed over time.

The Time Factor of Soil Formation

The influence of time on soil organic matter (C and N) levels has been examined by evaluating time sequences (or chronosequences) on mud flows,

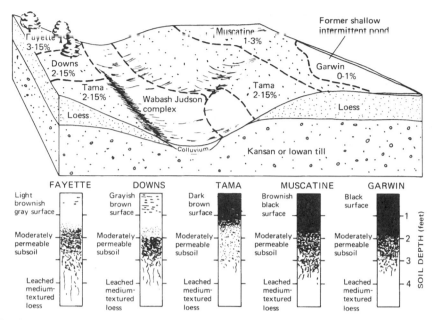

Fig. 2.1   Variability in soil organic matter content within a soil association as influenced by such factors as topography, drainage, parent material, and vegetation. Two of the soils are Alfisols (Fayette and Downs), while three are Mollisols (Tama, Muscatine, and Garwin). From Brady,[14] reproduced by permission of Macmillan Publishing Co.

spoil banks, sand dunes, road cuts, and moraines of receding glaciers and in ecosystems during the transition from annual crops to perennial grasses or following abandonment and revegetation of depleted soils. This work[6] has shown that organic matter levels increase rather rapidly during the first few years of soil formation, the rate subsequently slows down, and an equilibrium level characteristic of the environment under which the soil was formed is attained. On a geological time scale, this process can be relatively rapid. For example, equilibrium levels of C and N were attained in moraines of the Alaskan glaciers in about 110 years.[15] Similar results have been attained during soil development on other landscapes, including strip mine spoils.[16,17]

Much longer periods may be required to attain equilibrium C levels under drier conditions. Syers et al.[18] found that organic C was still increasing after 10,000 years of soil formation on wind-blown sand in New Zealand. On the scale of geological time, C content can change because of climatic variation or alteration in soil composition through pedogenic processes such as leaching and mineral deposition.

An idealized diagram showing the effect of time on the N (C) content of loam soils of the Indo-Gangetic Divide in India[4] is shown in Fig. 2.2. In this case, steady-state levels of N during soil development on sediments deposited

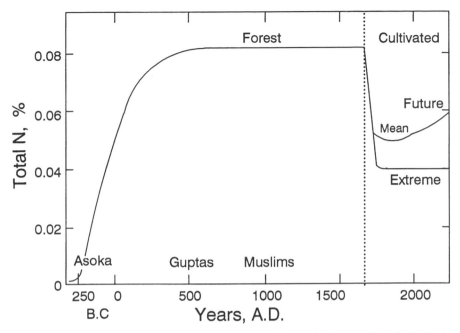

Fig. 2.2 Idealized time function for N accumulations in loam soils of the Indo-Gangetic Divide in India as envisioned by Jenny and Raychaudhuri.[4]

by flood waters during the Asoka period (250 B.C.) were attained near the end of the Gupta Dynasty (500 A.D.) and remained in this condition for a millennium until the reign of Shah Jahn (of Taj Mahal fame, about 1650 A.D.), at which time the area was converted to agricultural use. Subsequent decreases in soil N were the result of man's activities, with the establishment of new steady-state levels of organic N and C.

Although several reasons have been given for the phenomenon of equilibrium levels of organic matter, none has been entirely satisfactory. Among these explanations are:

1. Organic colloids (e.g., humic acids) are produced that resist attack by microorganisms. In this case, the chemical structure of organic matter is the prime determinant of residence time in soil.

2. Humus is protected from decay through its interaction with mineral matter (e.g., polyvalent cations and clay) or because it is occluded in pore spaces that are unoccupied or inaccessible to microorganisms and/or degrading enzymes. In this case, the organic matter may be biodegradable, but inaccessible to microorganisms.

3. A particular soil type has a finite number of protective sites where organic matter can be present in a nonaccessible form. When this ca-

pacity is exceeded, further increments of stabilized organic matter are not possible.

Soil processes can promote migration of organic matter into the lower soil horizons, often in association with clay or metal ions. Worm and root channels and ped surfaces become coated with dark-colored mixtures of humus and clay with substantially higher organic matter content than the surrounding soil.[19] In some soils, streaks or tongues result from the downward seepage of humus. Illuvial humus also appears as coatings on sand and silt particles. Localized accumulations of sesquioxides (Fe and Al) and humus are common. In many soils, a secondary maximum in humus content coincides with an accumulation of clay in the B horizon. This humus is probably transported from the surface and immobilized by adsorption to the clay. Transfer of organic matter into the lower soil horizons can continue for some time after equilibrium levels are reached in the surface layer; eventually, the total quantity in the profile stabilizes and remains essentially constant over time.

Organic matter can be transported downward by soil animals. Earthworms, for example, can completely mix soil to depths of a meter or so, transferring organic matter downward in the process. Burrowing animals move soil material low in organic matter from the deeper horizons to the surface and vice versa.[20] Soil micro- and macrofaunal organisms can have significant effects on soil development by affecting litter decay, nutrient dynamics, and microbial ecology.[21–24]

Release of organic compounds from the rhizosphere, along with death and degradation of deeply penetrating roots, can also result in substantial deposition of organic materials in the subsurface. Martin and Merckx[25] found that 26% and 33% of the soil organic C formed from ^{14}C-labeled wheat root exudates consisted of humic acid and humin, respectively.

As noted below, root systems of native prairie plants extend to depths of 1.5 to 2.5 m in many cases. Roots typically contain a higher percentage of lignin and other aromatic polymers than do aerial portions of the same plant; these compounds degrade slowly in soil and yield precursors for humus synthesis (see Figs. 1.8 and 1.9). Taken collectively, these observations may explain the exceptionally deep and organic-rich prairie soils (Mollisols) of the American Midwest.

Effect of Climate

Climate is the most important single factor that determines the array of plant species at any given location, the quantity of plant material produced, and the intensity of microbial activity in the soil; consequently, this factor plays a prominent role in determining organic matter levels. Considering climate in its entirety, a consistently humid climate leads to forest associations and the development of Spodosols and Alfisols; a semiarid or seasonally dry climate leads to grassland associations and the development of Mollisols. Grassland soils exceed all other well-aerated soils in organic matter content; desert, semidesert, and certain tropical soils are the lowest. The profile distribution of C in soils representative of three great soil groups is shown in Fig. 2.3.

Soils formed under restricted drainage (Histosols and Inceptisols) do not follow a climatic pattern. In these soils, O_2 deficiency—and in some cases, low pH—prevent the complete destruction of plant remains by microorganisms. When sufficient water is available to keep the surface layer saturated throughout the year, Histosols and Inceptisols will form over a wide range of temperature and parent materials.

Jenny et al.[1–4] studied the effect of climate on N levels in soil by analyzing north-to-south transects of the semiarid, semihumid, and humid regions of the United States (Fig. 2.4.). In each case, N content of the soil decreased two- to three-fold for each 10°C rise in mean annual temperature. Whereas relationships derived for U.S. soils cannot be extrapolated directly to other areas, many soils of warmer climatic zones have very low N (and C) contents. However, some tropical soils have organic matter contents that compare favorably with those of temperate regions,[7] which is attributed to low biodegradability due to strong binding of organic matter by mineral matter (e.g., allophane). Sanchez[26] indicates that many tropical soils that would be expected to be fertile based on their C and N contents actually have limited productivity, and for only a short cropping duration, because the organic matter is protected from attack by microorganisms.

Several explanations have been given for the decrease in soil N (and C) levels with an increase in mean annual temperature. Jenny[1,2,4] suggests that this relationship was due to the effect of temperature on soil microbial activity,

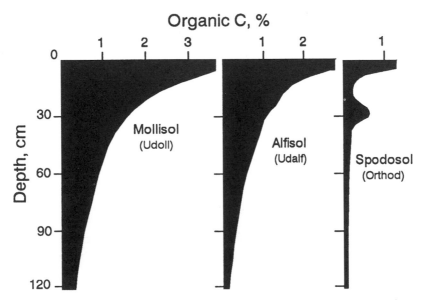

Fig. 2.3 Profile distribution of organic C in soils representative of three major great soil groups of the North Central Region of the United States. Adapted from Stevenson.[6]

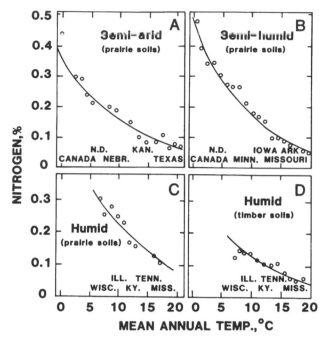

Fig. 2.4 Average total N content as related to mean annual temperature for soils along north to south transects of the semiarid, semihumid, and humid regions of the central United States,[6] reproduced by permission of the American Society of Agronomy.

a theory that fails to account for: (1) temperature effects on photosynthesis (production of raw material for humus synthesis), (2) differences in organic matter availability among different soil types, and (3) variations in the duration of the growing season. In cold climates, for example, the soil is frozen for many months and full-year biological activity is low in comparison to warmer climates, with longer periods of microbial and plant activity.

Enders[27] presented the striking concept that the best soil conditions for the synthesis and preservation of humic substances are frequent and abrupt changes in the environment (e.g., humidity and temperature); consequently, soils formed in harsh continental climates should have high C and N contents. Harmsen[28] used this same theory to explain the greater synthesis of humic substances in grassland soils, as compared to arable land, claiming that in the former the combination of organic substrates in the surface soils and frequent and sharp fluctuations in temperature, moisture, and irradiation lead to greater synthesis of humic substances. According to Harmsen,[28] the extreme surface of the soil (upper few millimeters) is the site of synthesis of humic substances and fixation of N.

Increasing rainfall (moisture component of climate) promotes greater plant growth and the production of larger quantities of raw material for humus synthesis. Plant productivity and organic materials added to soil can vary widely. However, total rainfall is not a satisfactory index of available soil moisture, because of great variations in evaporation and plant transpiration. As an index of available moisture, Jenny[1,2] used what is known as the NS quotient, which is the ratio of precipitation (in mm) to the absolute saturation deficit of the air (in mm of Hg).

For grassland soils along a west to east transect of central United States, a definite correlation exists between the depth of root penetration (and thickness of the grass cover) and the amount of N contained in the surface layer,[29] as can be seen by comparing Fig. 2.5 with Fig. 2.6.

Vegetation

It is well known that plant species have a profound effect on the organic matter content of the soil. Other factors being constant, the C content of grassland soils (e.g., Mollisols) is substantially higher than that of forest soils (e.g., Alfisols). Some of the reasons given for this difference are as follows:

1. Larger quantities of raw material for humus synthesis are produced under grass.
2. Nitrification is inhibited in grassland soils, thereby leading to the preservation of N in a low mobility form (NH_4^+) that cannot be converted to N_2 by denitrification. Conservation of N may lead to further synthesis of soil organic matter.
3. Humus synthesis occurs in the rhizosphere, which is more extensive under grass than under forest vegetation.
4. Inadequate aeration occurs under grass, thereby contributing to organic matter preservation.

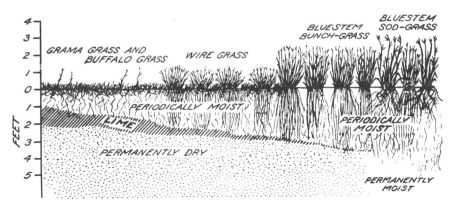

Fig. 2.5 Relationship between vegetative growth and moisture supply along a west-to-east transect of the Great Plains. Adapted from Shantz.[29]

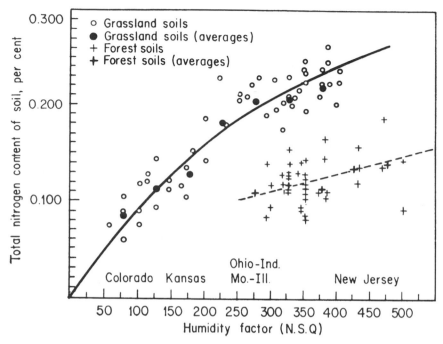

Fig. 2.6 Relationship between N content and humidity factor for soils along a west to east transect of the Central United States. From Jenny.[1]

A combination of several factors is probably involved, with item 3 being of major importance. In a study involving transformation of ^{14}C-labeled organic substrates, Reid and Goss[30] found that decay rates for applied organic matter were reduced in the rhizosphere of corn and perennial rye. Prairie grasses also have a significantly greater portion of their total biomass as roots when compared to woody forest species; as much as 90% of total plant biomass in mesic prairies occurs as roots, in contrast to forests, where only 20 to 50% of the total biomass is subterranean.

In the case of forest soils, differences in the profile distribution of C and N occur by virtue of the manner in which the leaf litter becomes mixed with mineral matter. In soils formed under deciduous forests on sites that are well drained and well supplied with Ca (Alfisols), the litter becomes well mixed with the mineral layer by earthworms and other fauna. In this case, the top 10–15 cm of soil become coated with humus. On the other hand, on sites low in available Ca (Spodosols), the leaf litter is not mixed with the mineral layer but forms a mat on the soil surface. An organic-rich layer of acid (mor) humus accumulates at the soil surface, and humus accumulates only in the top few centimeters of soil.

Parent Material

Parent material affects the C content of the soil through its influence on texture and adsorptive properties. The fixation of humic substances as organic-mineral complexes serves to preserve organic matter. Thus, heavy-textured soils with a high clay and silt content contain more organic-C than coarser-textured (sandy) soils.[31] A positive correlation between organic matter content and particles less than 0.005 mm is commonly found (Fig. 2.7).[32] Parton et al.[33] concluded that texture is a major factor affecting organic matter levels of grassland soils of the Great Plains, United States.

Soil organic matter, irrespective of soil type, has several characteristics (e.g., resistance to attack by microorganisms and to removal by chemical extractants) that suggest that it occurs in intimate association with mineral matter. Retention may also be affected by the type of clay mineral present. Montmorillonitic clays, which have high adsorption capacities for organic molecules, are particularly effective in protecting nitrogenous constituents against attack by microorganisms.

Topography

Local variations in topography, or relief, such as knolls, slopes, and depressions, modify the microclimate (defined by Aandahl[34] as the climate in the immediate vicinity of the soil profile) rather dramatically, and thereby influence soil formation. Substantial temperature and moisture gradients can exist between the tops and bottoms of sloped areas. Lower areas, for example, are generally cooler because cold air is more dense than warm air and flows (like water) to the lowest place in the landscape. Variations in water content are the result of differential rates of runoff, evaporation, and transpiration. Soils occurring in depressions, where the climate is "locally humid" and cooler than in the knolls, have higher C contents than those occurring on the knolls, where the climate is "locally arid" and warmer.

Organic matter accumulation is especially evident in Histosols (peats and mucks), but many swamps are also rich in organic matter. A serious economic problem is encountered when Histosols are drained for agricultural use, due to loss of soil through subsidence. The major cause of subsidence is enhanced microbial oxidation of the soil organic matter, with accompanying conversion of soil organic matter to CO_2.

Continuously moist and poorly drained soils are typically found at the lowest elevation in the landscape, and it is in these localized areas that small pockets of organic-rich soils are often found. In the northern Great Plains (United States), one often finds spots rich in organic matter, the so-called "prairie potholes," even though evaporation exceeds precipitation by 30 to 40 cm per year and adjacent uplands have plant species characteristic of semiarid grassland.[35] These soils are very productive, and large areas of the

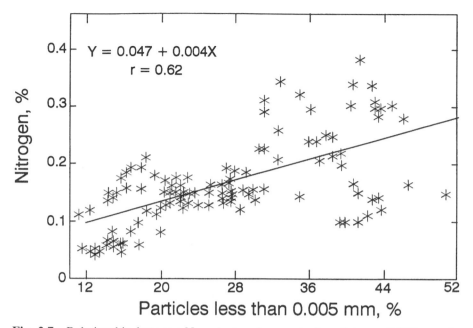

Fig. 2.7   Relationship between N content and percent of particles < 0.005 mm for the surface soils of virgin sod from 13 locations in the Great Plains Region of the United States. From Haas et al.[32]

wetlands have been drained for agricultural use, thereby diminishing available nesting and feeding areas for wildlife.

EFFECT OF CROPPING ON SOIL ORGANIC MATTER LEVELS

Marked changes are brought about in the C (and N) content of the soil through the activities of man. Usually, but not always, organic C levels decline when soils are first placed under cultivation, as documented elsewhere.[6]

Historical Aspects

The initial cultivation of virgin soils has long been known to lead to losses of organic matter and has often been believed to create other problems, as illustrated by the following description of pioneer times in west-central Indiana (United States):

> In 1822 the pioneers saw a land rich indeed in natural wealth, but, from its very fertility, presenting great difficulties. In all the most fertile spots the pea vine grew in tangled masses, cropped by the cattle, which often fattened upon this food alone. . . . But mud and malaria were the evils of the times. The latter did

not manifest itself till after the land clearing began, but the mud they had from November till June. . . ."

The pioneers now devoted all their energies to clearing, and soon every house was the center of a cultivated tract of from ten to forty acres. The first consequences were frightful. When the virgin soil, with a thousand years' accumulation of moldering leaves, with the natural earthy damp and decaying roots, was suddenly turned to the sun, there went up therefrom a literal savor of death. When the shaded brooks were suddenly deprived of their leafy covering, and the hot sun drank their sluggish waters, they became as open sewers running with malaria.—-Travelers tell of riding ten miles along Raccoon [Creek] without finding a house free from sickness, and often whole families were so prostrated that not one could aid another. This reached its climax usually two or three years after settlement. . . .   (Beadle[36])

The water pollution and disease problems that were perceived to have resulted when the forests were cleared can be attributed to leaching of organic matter decomposition products and nutrients, with an increase in microbial activity in the stream water, a subsequent lowering of oxygen supply, and creation of an environment (e.g., stagnant waters) conducive to proliferation of mosquitoes and disease-causing microorganisms.

The rates of N (and C) loss upon the clearing of forested areas in temperate climate regions are provided by the work of Likens et al.,[37] who cleared the vegetation from a forested watershed and prevented regrowth by periodic herbicide applications. For a three-year period, N losses were 120 kg N/ha/ yr in the devegetated areas, but negligible in adjacent uncut areas. At this rate, only a few years would be required to significantly reduce soil organic matter content and productivity, which was typical of results observed in New England in colonial times.[38]

Not until the 1920s were quantitative studies of the effects of management practices on soil organic matter and N content common in the United States. Studies such as those of Salter and Green[39] for rotation experiments at the Ohio Agricultural Experiment Station showed that losses of soil N were least with crop rotations containing a legume and greatest under continuous row cropping. Salter and Green[39] observed that losses of soil N (and C) were more or less linear over the 31-year period of the experiment. Therefore, unless a change occurred, organic matter content would ultimately approach zero. They described N loss with the following equation:

$$N = N_0 e^{-rt}, \quad \text{or} \quad dN/dt = -rN \tag{1}$$

where N_0 is the initial N content of the soil and r is the fraction of the N remaining after a single year's cropping.

The conclusion that virtually all of the organic matter (and N) might eventually be lost from soil by cropping caused considerable alarm in agronomic circles. It was feared that unless drastic measures were taken to maintain organic matter reserves, many soils would become unproductive, as had been found historically:

> The most important material problem of the United States is to maintain the fertility of the soil, and no extensive agricultural country has ever solved this problem. The frequent periods of famine and starvation in the great agricultural countries of China, India, and Russia, and the depleted lands and abandoned farms of our own eastern United States are facts that serve as a constant proof that the common practice of agriculture reduces the productive power of land. (Hopkins[40])

Jenny's[1,2] early findings for losses of soil N over a 60-year period under average farming conditions in the Corn Belt section of the United States agreed generally with the results of Salter and Green, as well as historical records, in indicating that cropping caused a decline in organic matter levels. However, Jenny's work suggested that destruction of organic matter would not be complete but that new equilibrium levels would be attained. For Corn Belt soils, about 25% of the N was found to be lost in the first 20 years, 10% in the second 20 years, and 7% in the third 20 years. The data suggested that new equilibrium levels would be attained within a 60- to 100-year time period. As shown earlier, a similar time period was required to attain equilibrium levels of organic matter during soil formation.

The validity of Jenny's equilibrium concept has been authenticated in several long-time cropping systems. Results obtained for shortgrass prairie soils in western Kansas[41] and for the Morrow Plots at the University of Illinois (originally tallgrass prairie soils) are given in Figs. 2.8 and 2.9, respectively. In both cases, organic-N losses during the early years were much more rapid than in later years. In the Morrow plots, losses were greatest with continuous

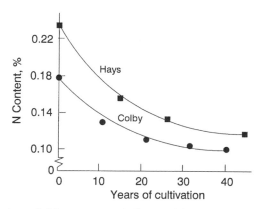

Fig. 2.8   Decline in soil N content as influenced by years of cropping in western Kansas. Adapted from Hobbs and Brown.[41]

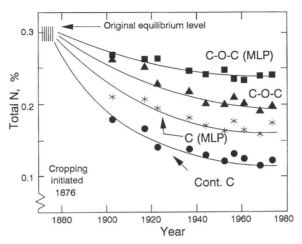

Fig. 2.9 Effect of long-time rotations on the N content of select soils from the Morrow plots at the University of Illinois. C = corn; O = oats; Cl = clover; MLP = manure-lime-phosphate. Adapted from Stevenson.[6]

corn and no fertilization and least with a corn–oats–clover rotation amended with manure, lime, and phosphate.

Like the Morrow Plot results, the Kansas studies indicated that organic amendments could substantially decrease the rate of organic C and N loss associated with cropping. Interpretation of the Kansas data is somewhat difficult because the area had been subjected to extreme drought and wind-driven soil erosion during the "Dust Bowl" period (1933–1940); accordingly, some of the decline in C (and N) content may have been due to loss of topsoil by wind erosion, rather than to enhanced microbial oxidation attributable to cropping.

Under humid tropical conditions, the rate of change in soil N reserves (and productivity) may be dramatically faster than for temperate zone soils, as illustrated in Fig. 2.10 (note the sharp drop in N content upon cultivation and increase upon revegetation with a secondary forest).[26] The precipitous decline in N content of many tropical soils is invariably accompanied by dramatic yield decreases following forest clearing,[42,43] as shown in Fig. 2.11. The loss of food-producing capacity in less-developed countries where fertilizers are not readily available (or are unaffordable) is of serious concern to the international community. It should be noted, however, that soil productivity in some cases has been maintained for centuries by using traditional cultivation systems that included growing of legumes or application of organic fertilizers (e.g., night soil, *Azolla*). If such methods, or the concepts underlying them, could be applied to some of the newly emerging countries, improvement in soil quality and productivity would be substantial.

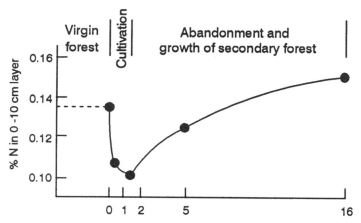

Fig. 2.10   Nitrogen content of surface soil (0–10 cm) in a virgin tropical forest (Columbia, South America) prior to clearing, following cultivation, and after abandonment and regrowth. Adapted from Sánchez.[26]

Jenny attempted to correct the inadequacy of Salter and Green's equation by including a factor for the annual return of N to the soil. Thus, Eq. (1) becomes:

$$dN/dt = -rN + A \tag{2}$$

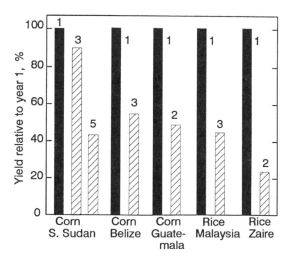

Fig. 2.11   Decrease in yields of corn and rice in tropical forest environments through loss of organic matter and nutrients after forest clearing. Numbers indicate years after clearing. Adapted from Norman.[43]

where A is the annual rate of N addition. A limitation of this approach is that account is not given to N losses (leaching and denitrification) or to the fate of N (and C) from stabilized materials (e.g., composts and sludges) from which N (and C) release is relatively low.

This equation was transformed by Bartholomew and Kirkham to:[44]

$$N = A/r - (A/r - N_0)e^{-rt} \tag{3}$$

A plot of N vs e^{-rt} should, therefore, yield a straight line in which the y intercept (A/r) would describe the expected equilibrium value and $(N_0-A/r)e^{-rt}$ the change process. The change in the magnitude of the latter with time provides a measure of the rate of establishment of a new equilibrium level.

Bartholomew and Kirkham[44] used graphical methods to obtain the constants A and r for the experimental plots of several long-time rotation experiments. Like Jenny, they concluded that new equilibrium levels can be expected to be attained within a 60 to 100 year period of cultivation.

Crop yields have a potential effect on organic N levels through a feedback effect (greater return of plant residues as yields increase). Russell[45] used computer-based numerical methods to predict long-term effects of increased yields on N levels for the Morrow Plots at the University of Illinois and the Sanborn Field at the University of Missouri. The basic equation was:

$$dN/dt = -K_1(t)N + K_2 + K_3(t)Y(t) \tag{4}$$

where $Y(t)$ is plant yield at time t, $K_1(t)$ is the decomposition coefficient, K_2 represents addition to soil organic matter from noncrop sources (e.g., manures), and $K_3(t)$ is a coefficient related to the specific crop at time t. This equation permits estimates to be made of the effect of crop yield within a rotation on soil N levels.

For the Morrow Plots, increasing corn yields in a continuous corn system had negligible effects on soil N levels, but strong positive effects were noted for oats and clover. All crops in the Sanborn Field had some feedback on soil N levels.

The decline in the organic matter content of the soil when land is cultivated cannot be attributed entirely to a reduction in the quantity of plant residues available for humus synthesis. A temporary increase in respiration rate occurs each time an air-dried soil is wetted,[46] and since considerable amounts of fresh soil are subjected to repeated wetting and drying through cultivation, losses of organic matter by this process could be appreciable. Microbial activity may also be stimulated by exposure of previously inaccessible organic matter by cultivation.

Several agricultural practices may accelerate mineralization of soil humus, including:

Cultivation, which improves aeration and moisture and rearranges soil aggregates, thereby increasing microbial activity and the release of soluble organic compounds.

Irrigation, which improves the moisture status of the soil, with enhancement of microbial activity and plant productivity.

Liming, which increases the activities of earthworms and other fauna; encourages actinomycetes, which may be more effective degraders of humic materials than fungi or other bacteria; facilitates precipitation of metallic cations that are effective in stabilizing humic substances.

Green manuring (e.g., plowing under a cover crop), which greatly increases supplies of readily degradable organic materials, whose decay supports a larger microbial population and thereby increases the rate of oxidation of soil organic matter. The effect of this practice is multifold, however, since the green manure also contributes C that becomes new soil organic matter.

Addition of supplemental organic materials (e.g., manures, composts).

Maintenance of Soil Organic Matter (C)

For many cultivated soils, particularly Mollisols, organic matter can be maintained only at a level approaching that of the native uncropped soil by inclusion of a sod crop in the rotation, use of perennial plants, or frequent and heavy applications of manures and crop residues. Increases in soil N (and C) levels have been observed by returning previously tilled soil to a grass sod,[47–50] typical results being shown in Fig. 2.12. Introduction of legumes to soils initially low in N has been reported to lead to increased levels of soil

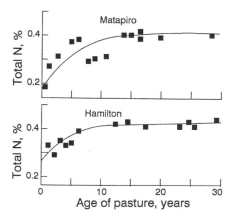

Fig. 2.12   Changes in the N content of the surface soil (0–7.6 cm) vs. time after establishment of permanent pasture on two New Zealand soils. Adapted from Jackman.[49]

N and to prolong the productive period of tropical soils.[26,42] Increases in N and C content in the top few centimeters of soil have been also observed through zero tillage,[51-53] a management system in which crop residues are not plowed into the soil.

In contrast to research that demonstrated a decreasing rate of loss of soil N with time after a change in cultural practice, linear changes were observed by Rasmussen et al.[54] for a wheat-fallow cropping sequence on a Pacific Northwest semiarid soil (45-year cropping period). As shown in Fig. 2.13, linear declines in soil N occurred with or without supplemental additions of N (34 to 45 kg/ha/yr) or pea vines at a rate of 2.24 kg/ha/year. A linear increase was observed by addition of strawy manure at a high rate (22.4 metric tons ha/yr). The greatest loss of N occurred when the in-field wheat straw residue was burned rather than returned to the soil. Supplemental addition of 5000 kg crop residue/ha/yr (about 2000 kg C/ha/yr) was regarded as necessary to maintain the initial soil organic matter content. For a wheat-fallow system in a cooler climate, a smaller quantity of plant residues may be needed.[55]

Jenkinson and Johnson[56] demonstrated a pronounced increase in soil N content through long-time applications of manure (35 tons/ha/y) to soil under permanent barley at the Rothamsted Experimental Station in England (Fig. 2.14). The plots were established in 1852, and equilibrium levels had still not been attained in the manured plot by the time they were sampled 123 years later (in 1975). Only minor changes in soil N levels occurred in the unmanured plot, while a slow but continuous decline occurred in a plot receiving farmyard manure annually between 1852 and 1871, but not thereafter. Anderson and Peterson[57] observed increases in soil N content by long-time applications of manure (27 metric tons/ha/yr) to a Nebraskan soil.

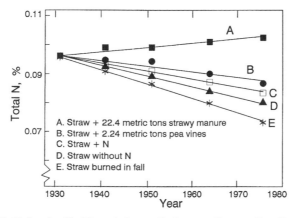

Fig. 2.13 Soil N levels (0–30 cm) in a silt loam soil near Pendleton, Oregon, as related to crop residue treatment in a wheat–fallow rotation system. Adapted from Rasmussen et al.[54]

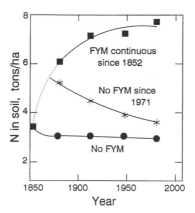

Fig. 2.14   Changes in soil N levels (0–23 cm) over time as influenced by manure (FYM) additions to soil under continuous barley at the Rothamsted Experimental Station. Adapted from Jenkinson and Johnson.[56]

   Results obtained by Larson et al.[58] indicated that application of plant residues at a rate of 6 tons/ha/yr was needed for maintenance of organic matter in a Typic Haplaquoll soil in Iowa (Marshall silty clay loam); higher application rates led to increases in soil N. Hobbs and Brown[41] concluded that 56 to 67 metric tons of farmyard (ruminant) manure applied every three years were necessary to minimize loss of organic matter from some prairie soils of Kansas. In other work, Swafakab[59] observed that residues rich in polyphenols promote the synthesis of large amounts of "true humic matter" with a high *N* content.
   The effectiveness of organic residues in maintaining soil organic matter reserves depends upon such factors as rate of application, kind of residue and its *N* content (*C/N* ratio), manner of incorporation into the soil, soil characteristics, and seasonal variations in temperature and moisture. Select references on soil management factors affecting organic matter levels in soils from widely different climatic regions of the earth can be found in several reviews.[4,60] Factors affecting the fate of organic wastes when applied at high disposal rates to soils of the various climatic regions of the United States (arid, cool subhumid and humid, hot humid) are discussed by Elliott and Stevenson[61] and in European agroecosystems by Hansen and Henrickson.[62]

PALEOHUMUS

Remnants of plant and animal life, as well as products produced from them through humification, have been observed in buried soils (paleosols) of all ages.[63–65] This organic matter, or paleohumus, is of interest to geologists and pedologists because of its importance as a stratigraphic marker and as a clue to the environment of the geologic past. Thus, the occurrence of dark-colored humus zones, when used in conjunction with other pedological observations, has served as a basis for the identification of buried soils, from which it has been possible to draw conclusions about climate, vegetative patterns, and the morphology of former land surfaces.[63]

Humus is the fraction of the soil most susceptible to change. Thus, in most instances, the humus found in buried soils represents only a small fraction of the amount initially present. The degree to which humus has been lost will depend upon such factors as the change in environment which preceded the new cycle of sedimentation, the circumstances under which the soil was buried, and activities (if any) of living organisms in the buried sediment.

In Fig. 2.15, the organic C contents of two paleosols are compared with those of their modern counterparts. In both cases, much of the humus had been lost, particularly from the surface layer.

ROLE AND FUNCTION OF SOIL HUMUS

Organic matter (humus) has a substantial influence on crop production and soil properties. It is the presence of organic matter that distinguishes the soil from an inert mass of weathered rock products or sand and that gives life to the soil by providing a favorable habitat for micro- and macrofaunal organisms and carbon and energy sources for microorganisms.

Other factors being constant (climate, drainage, topography, parent material, etc.), soils rich in humus are generally more productive than those poor

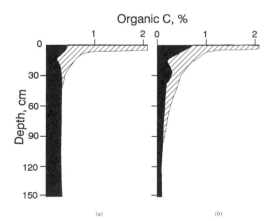

Fig. 2.15 Comparison of organic C contents of two paleosols with their supposed modern counterparts. Solid areas represent paleosols, hatched areas are modern soils. From Stevenson[63] as prepared from literature data. (*a*) Farmdale "Gray-brown Podzolic." From Hogan and Beatty.[65] (*b*) Yarmouth "Gray-brown Podzolic." From Simonson.[64]

in humus. The high productivity of the dark-colored soils of the Corn Belt section of the United States is usually attributed to their high content of stable humus.

Humus contributes to the fertility, or productivity, of the soil through its positive effects on the chemical, physical, and biological properties of the soil. It has a nutritional function in that it serves as a reservoir of *N*, *P*, and *S* for plant growth, a physical function in that it promotes good soil structure, and a biological function in that it serves as a source of C and energy for soil organisms.

Nutrient Availability

Three interrelated organic matter functions must be taken into account in considering nutrient–organic matter interactions in soil:[9] (1) plant and animal residues, whose degradation provides N, P, and S for plant growth, (2) the microbial biomass, which serves as a temporary storage unit for nutrients (MRT of several months to a few years), and (3) the resistant humus fraction, which has a long MRT (250 to 1000+ years). Under conditions where steady-state levels of organic matter have been attained, mineralization of native humus is compensated for by synthesis of new humus.

Conversion of organic N, P, and S to available mineral forms (NH_4^+, PO_4^{3-}, SO_4^{2-}, HS^-) occurs through the activity of microorganisms, and the rate of conversion is influenced by factors affecting microbial activity such as temperature, moisture, and pH, as well as by the N, P, and S content of soil and decaying plant residues. The process is referred to as *mineralization* and is nearly always accompanied by conversion of mineral forms of the nutrient to organic forms, or *immobilization*.

$$\text{Organic N, P, S} \xrightleftharpoons[\text{immobilization}]{\text{mineralization}} NH_4^+, PO_4^{3-}, SO_4^{2-}, HS^-$$

The assumption is often made that 1 to 3% of the soil organic matter is mineralized during the course of the growing season, with net release of corresponding percentages of the N, P, and S. This statement must be accepted with reservations, since the humus content of most soils is in a state of quasi-equilibrium. A net annual release of nutrients occurs only when organic matter levels are declining, a condition that is to be avoided because such exploitation generally leads to a reduction in the productive capacity of the soil. Any soil under constant management attains a balance between gains and losses of N, P, and S. Thereafter, nutrients continue to be liberated, but the amounts released are often compensated for by incorporation of equal amounts into newly formed humus. On a sustained basis, plant nutrition cannot be improved by adoption of cultivation practices that are designed to accelerate degradation of soil humus, since loss of humus results in the loss of organic pools of required nutrients and can result in loss of soil structure, with accompanying compaction and decreased root growth.

In developed countries, adequate quantities of externally provided plant nutrients (N, P, and K fertilizers) are available to the farmer at reasonable cost. Undoubtedly, this has changed our concept of the importance of humus in supplying nutrients for crop growth. Nevertheless, humus should be regarded as a positive asset to high crop production. Soil scientists generally agree that practices that deplete humus reserves lower the yield potential, even when soils are supplied with nutrients from external sources.

The C/N/P/S Ratio

While considerable variation is found in the C/N/organic P/total S ratio for individual soils, the mean for soils from different regions of the world is remarkably similar (see Chapter 9). As an average, the proportion of C/N/P/S in soil humus is approximately 140:10:1.3:1.3. Ratios recorded for C and P are somewhat more variable than for C and N, or C and S. One explanation for the wider range of C/P ratios is that, unlike N and S, P is not a structural constituent of humic and fulvic acids.

The mineralization of organic pools of N, P, or S can display a number of patterns, the more common ones being:

1. Initial net immobilization followed by net mineralization in later stages
2. Net mineralization with time
3. Initial net mineralization followed by net immobilization

Additions of plant residues of moderately high C/N, C/P, or C/S ratios favor pattern 1, while residues with low ratios (or removal of straw by burning) favor pattern 2. Pattern 3 is observed when residue additions vary widely from year to year. The pattern of nutrient release is not related to any given soil property, including the N, P, or S content of the soil.

If soil organic matter were a relatively homogeneous material, one would predict that the relative rates of N, P, and S mineralization would be similar and that they would be released in the same ratios in which they occur in soil organic matter. However, this has not always been the case. Differences in the relative amounts of N, P, and S released from soil organic matter may be due to a number of factors.

1. Nitrogen, P, and S occur in different organic compounds and organic matter fractions; consequently, they are not released in the same ratios because the degradability of the organic compounds containing these different elements is different.
2. The application of plant and animal residues results in differential mineralization–immobilization of nutrients. The N, P, and S contents of applied plant residues, as reflected by the C/N, C/P, and C/S ratios, play a major role in regulating the amounts of these nutrients that accumulate in available mineral forms at any one time (see Chapter 6).

As a general rule, an initial net mineralization will occur at C/N ratios of less than 20:1 and at C/P and C/S ratios of less than 200:1. In contrast, an initial net immobilization of N will occur at C/N ratios of 30:1, of P at C/P ratios over 300:1, and of S at C/S ratios over 400:1. Intermediate ratios lead to neither a gain nor a loss of the nutrient. In the use of these ratios, it is assumed that the C, N, P, and S are in organic forms with a similar degradation rate, which is seldom the case. If, for example, most of the C is in slowly degraded organic forms (e.g., lignin), whereas the N is in rapidly degraded forms (e.g., protein), net mineralization will occur even though the C/N ratio would suggest otherwise. Sewage sludges often have low C/organic-N ratios, on the order of 6:1, but little organic N will be mineralized, due to the recalcitrant nature of the organic N. As a result of these uncertainties, the ratio of C to other elements in soil cannot be regarded as a reliable guide to element availability.

3. The actual amount of NH_4^+ formed by mineralization is obscured by fixation reactions between NH_4^+ and soil minerals, NH_3 volatilization, and denitrification and/or leaching of NO_3^-.

4. The actual release of PO_4^{3-} and SO_4^{2-} is masked by formation of insoluble salts (e.g., Ca-, Al-, or Fe-phosphates and $CaSO_4$).

Indirect Nutrient Effects

In addition to serving as sources of N, P, and S, organic matter influences nutrient supply in other ways. For example, denitrification is affected by the supply of degradable organic matter. Several investigators have demonstrated a direct correlation between denitrification rate and the content of soluble C in the soil solution (see Chapter 5).

The availability of phosphate in soil is often limited by fixation reactions, which convert the monophosphate ion ($H_2PO_4^-$) to various insoluble forms. However, additions of organic residues to the soil often enhance the available of the native soil P to higher plants.

Aluminum toxicity is a major problem in many acid soils. However, acid soils rich in native organic matter, or amended with large quantities of organic residues, give low Al^{3+} concentrations in the soil solution and permit good growth of crops under conditions where toxicities would otherwise occur. The role of organic matter in alleviating toxicity effects of heavy metals is discussed in Chapter 11.

Source of Energy for Soil Organisms

Humus serves as a source of energy for both macro- and microorganisms. The numbers of bacteria, actinomycetes, and fungi in the soil are related in a general way to humus content; macrofaunal organisms are similarly affected because these organisms feed on microbes or partially degraded plant materials.

The role of soil fauna has not been completely elaborated, but the functions they perform are multiple and varied. For instance, earthworms may be important agents in producing good soil structure. They construct extensive channels through the soil that serve not only to loosen the soil but also to improve drainage and aeration. Earthworms can flourish only in soils with good physical structure (a feature that is enhanced by residue additions) and that are well provided with organic matter.

Growth of Higher Plants

In addition to the indirect effects noted above, organic substances have a direct physiological effect on plant growth. Varanini and Pinton[66] have recently reviewed this subject. Some compounds, such as phenolic and short-chain fatty acids, have phytotoxic properties; others, such as the auxins, enhance the growth of higher plants. The condition of "soil sickness" has frequently been connected with the accumulation of organic toxins. Phytotoxic substances are believed to be responsible for the low yields of wheat often observed under the stubble-mulch system of farming in semiarid regions of the world.

Numerous studies have shown that under certain circumstances humic and fulvic acids can stimulate plant growth, the usual explanation being that they act as growth hormones (auxins) by serving as activators of O_2 during photosynthesis. Favorable effects of humic substances under laboratory conditions include (1) greater length and fresh and/or dry weights of shoots and roots, (2) increased root initiation and number of lateral roots, (3) more rapid seedling growth after germination, and (4) improved flowering. This work has been reviewed elsewhere.[67,68] Organic-rich composts produced from a variety of waste materials enhance plant growth as well.[69]

Soil Physical Condition

Humus has a profound effect on the structure of most soils. The deterioration of structure that accompanies intensive tillage is usually less severe in soils adequately supplied with humus (e.g., see Olmstead[70]). When humus is lost, soils tend to become hard, compact, and cloddy. Seedbed preparation and tillage operations are easier to carry out and are more effective when humus levels are high.

Aeration, water-holding capacity, and permeability are favorably affected by humus. The frequent addition of easily decomposable organic residues leads to the synthesis of polysaccharides and other complex organic compounds that bind soil particles into structural units called aggregates. These aggregates help to maintain a loose, open, granular condition. Water is then better able to enter and percolate downward through the soil. The relationship between structure and organic matter storage in soil has been covered in detail elsewhere.[8]

Soil Erosion

Humus definitely increases the ability of the soil to resist erosion and enables the soil to hold more water. Even more important is its effect in promoting soil granulation (aggregation) and maintaining large pores through which water can enter rapidly and percolate downward. In a granular soil, individual particles are not easily carried along by moving water. The combination of increased water penetration and formation of erosion-resistant soil particles dramatically reduces erosion and improves water retention for later use by plants or regeneration of groundwater reserves. Loss of pore space in long-cultivated soils is mainly a loss of the larger pore spaces. As a result, the soil becomes dense and compact, soil particle size is reduced in comparison to a well-aggregated soil, water enters slowly, and surface runoff increases. Increased quantity and velocity of surface runoff accelerates erosion, deprives the soil of water reserves, and aggravates downstream flooding.

Buffering and Exchange Capacity

From 20 to 70% of the exchange capacity of the soil is due to colloidal humic substances. Exchange acidities of isolated fractions of humus range from 300 to 1400 cmoles/kg. Humus exhibits buffering over a wide pH range.

Health Effects

Since ancient times, peat treatments and mud baths have been thought to promote wound healing and to cure such human ailments as rheumatoid arthritis. These beneficial effects have generally been attributed to humic acids and related substances. This work has been reviewed by Klöcking.[71]

Miscellaneous Effects

In recent years, information relative to a number of additional effects of soil humus has come forth. These newly recognized features have implications with respect to both the yields of crop plants and the quality of the environment. They include (1) interaction of humus with organic pesticides and environmental contaminants[72] and (2) sorption of N and S oxides from the atmosphere (see Chapter 6). Industrial applications of humic acids and related substances are reviewed by MacCarthy and Rice.[73]

Organic matter imparts a dark color to the soil which can enhance soil warming and thereby promote earlier plant growth.

ORGANIC MATTER AND SUSTAINABLE AGRICULTURE

Sustainable agriculture is a method of land management designed to maintain soil productivity by adoption of management practices that simulate native ecosystems, such as forests and prairies.[74,75] From minimizing tillage operations, and thereby maximizing the return of organic residues to the soil, benefits accrue due to better soil tilth, reduced erosion, preservation of stable humus, and improved nutrient cycling through conservation of nutrients that would otherwise be lost through leaching.

Under conventional tillage, mixing of the surface soil is carried out by plowing and additional tillage operations are carried out for seedbed preparation and weed control. In contrast, under no-till, the soil left in an undisturbed state and crop residues are allowed to remain on the surface as a mulch, thereby decreasing erosion. The difference in residue placement leads to changes in the soil's physical, chemical, and biological properties, including activities of enzymes.[76,77] Studies on microbial and biochemical changes associated with reduced tillage have been carried out by Doran.[78]

The statement has been made that proper management of organic matter is the heart of sustainable agriculture,[11,79] a concept that has been promoted by organic farmers for years.[80] A particularly important role is played by the so-called "active" fraction (discussed in Chapter 3), consisting of litter, the light fraction (partially degraded plant residues), and the microbial biomass. A stimulating discussion of the significance of organic matter in sustainable agriculture is provided by Weil.[79]

COMMERCIAL HUMATES AS SOIL AMENDMENTS

From time to time, lignites or products derived from them have been marketed for use as organic soil amendments. In some cases, rather extravagant claims have been made for their beneficial effects on the physical, chemical, and biological properties of the soil. The main focus of this section will be on the potential value of such products for use on *normally productive agricultural soils*.[81] The conclusions reached may or may not apply to the use of commercial humates as amendments for lawns, turf, or certain problem soils, such as sands.

Origin and Nature of Commercial Humates

Commercial humates are defined as oxidized lignites, or products divided from them. Oxidized lignite is an earthy, medium-brown, coal-like substance associated with lignitic outcrops. The material typically occurs at shallow depths, overlying or grading into the harder and more compact lignite, a type of soft coal. A unique feature of oxidized lignites is their unusually high content of humic acids, of the order of 30 to 60% of the material as mined.

Extensive deposits of oxidized lignites occur in North Dakota, Texas, New Mexico, Idaho, and elsewhere. The deposit in North Dakota, often referred to as Leonardite, is about 1.8 m thick and lies 2.4 to 10.7 m beneath the soil surface.

Oxidized lignites are undesirable as fuel because of a low caloric content. While normally discarded during mining, they have been utilized in a limited way in industry, such as for viscosity control in oil well drilling muds. The use of "coal humates" as soil additives dates back at least five decades.

Chemical Properties and Agricultural Value of Lignite Humic Acids

Mined lignites have low water solubilities, the insolubility being due to the tie-up of humic acids with mineral matter. Some oxidized lignites are essentially salts of humic acids mixed with mineral matter, such as gypsum, silica, and clay.

Two main types of commercial humate preparations have been marketed for agricultural purposes:

Mined lignites: The original mined product is crushed, pulverized, and usually fortified with commercial fertilizer. The material itself does not contain significant amounts of N, P, or K, but is assumed to have a favorable effect on soil properties. Inclusion of mined lignite in the fertilizer mixture is sometimes said to increase the efficiency with which plants utilize the applied nutrients.

Ammonium humate fertilizer: This is a soluble product containing available N, P, and K. In contrast to oxidized lignites, which have low solubility, NH_4^+-humates are easily dispersed in water.

Because of manufacturing and transportation costs, commercial humates are relatively expensive. Their use can be justified only if it results in extra yields from the humates they contain.

Promoters of coal-derived humates often give the impression that humates from oxidized lignites have biological and chemical properties similar to those of the humus in soil. In reality, the composition and properties of lignite humic acids are substantially different from those of soil humic substances. Unlike soil humus, coal humates are essentially free of such biologically important compounds as proteins and polysaccharides; furthermore, they contain few if any fulvic acids. In comparison to soil humic acids, lignite humic acids have higher C contents, which indicates that they will be less soluble (see Chapter 1).

The virtual absence of proteins and other nitrogenous biochemicals indicates that commercial humates are not good N sources for plant growth. Mucopolysaccharide gums, the constituents of soil humus that are the most effective in forming stable aggregates, are absent from commercial humates. Due to adsorption processes, the application of humates as their Na^+, K^+, or NH_4^+ salts will not ensure that they will persist in soluble forms, which means that they will have little opportunity to directly influence plant growth by acting as growth hormones.

Other benefits sometimes attributed to commercial humates are as follows:

1. Exchange and buffering capacities of the soil are increased.

Because commercial humates resist microbial attack, some increase in organic matter content is to be expected, with accompanying increases in cation-exchange and buffering capacities. However, at the rates at which commercial humates are normally applied (200–400 kg/ha or less), any increase in organic matter content will be small, since the soil already contains much larger amounts of soil organic matter. Prolonged and costly applications will be required to bring about a noticeable increase in the organic matter content of most soils.

Even so, the cation-exchange and buffering capacities of the soil may not be significantly increased. The potential cation-exchange capacity of commercial humates (value obtained in the laboratory) is undoubtedly much higher than will be attained in the soil. To a large extent, mined humates are inert owing to their combination with mineral matter.

2. Lignites function as controlled release N fertilizers.

Although commercial humates *per se* are known to have little fertilizer value, claims persist that they provide a continuous supply of N for plant growth in subsequent seasons following application. While this claim may be true, the residual value for any one year would be negligible.

Assuming that availability of N in oxidized humates approaches that of native humus, only 2 to 3% of the N will become available to plants during any given growing season. From a single application of 500 kg of mined lignite per hectare (N content of 1.5%), a maximum of 0.225 kg of N would become available during any given season ($500 \times 0.015 \times 0.03 = 0.225$), which is equivalent to 0.3 kg of anhydrous NH_3 or 0.6 kg of NH_4NO_3. A build-up of 20,000 kg of oxidized lignite per hectare would result in the release of only 9 kg of N per hectare.

Commercial humates applied to *normally productive agricultural soils* at rates recommended by their promoters would not appear to contain sufficient quantities of the necessary ingredients to produce the desired, and sometimes claimed, beneficial effects.[81] Yield increases, if any, from the use of such products would appear to be insufficient to offset increased production costs to the farmer.

SUMMARY

Soils vary widely in their organic matter (C) contents. In undisturbed (un-cultivated) soils, the amount present is governed by the soil-forming factors of age (time), climate, vegetation, parent material, and topography. During soil development, considerable migration of organic matter occurs into the lower soil horizons, often in association with clay and polyvalent metal cations.

Organic matter is usually lost when soils are first placed under cultivation, and a new equilibrium is reached which is characteristic of cultural practices and soil type. Increases in plant yield brought about by improved varieties, more widespread use of fertilizers, or adoption of better management practices would be expected to have a positive effect on equilibrium levels of organic matter in soil through return of larger quantities of plant residues. However, the increase will generally be slight. For most soils, organic matter can only be maintained at high levels by inclusion of a sod crop in the rotation, by frequent addition of large quantities of organic residues (e.g., animal manures, composts), or by adoption of perennial cropping systems such as permanent pasture or woodland.

Humus serves as a reservoir of N, P, and S for higher plants; improves structure, drainage, and aeration; increases water-holding, buffering, and exchange capacity; and serves as a source of energy for the growth and development of microorganisms. Commercial humates rich in humic acids have been marketed for agricultural use, but, as applied to normally productive agricultural soils, their cost would appear to be prohibitive when considered in terms of expected yield increases.

REFERENCES

1. H. Jenny, *Missouri Agr. Exp. Sta. Res. Bull.,* **152,** 1 (1930).
2. H. Jenny, *Missouri Agr. Exp. Sta. Res. Bull.,* **765,** 1 (1960).
3. H. Jenny, *The Soil Resource,* Springer-Verlag, New York, 1980.
4. H. Jenny and S. P. Raychaudhuri, "Effect of Climate and Cultivation on Nitrogen and Organic Matter Reserves in Indian Soils," *Indian Council Agric. Res.,* New Delhi, 1960.
5. P. A. Sánchez, M. P. Gichuru, and L. B. Katz, *Trans. 12th. Intern. Congr. Soil Sci., Symposium Papers,* **1,** 99 (1982).
6. F. J. Stevenson, "Origin and Distribution of Nitrogen in Soil," in F. J. Stevenson, Ed., *Nitrogen in Agricultural Soils,* American Society of Agronomy, Madison, Wisconsin, 1986, pp. 1–42.
7. D. C. Coleman, J. M. Oades, and G. Uehara, Eds., *Dynamics of Soil Organic Matter in Tropical Ecosystems,* University of Hawaii Press, Honolulu, 1989.
8. M. R. Carter and B. A. Stewart, Eds., *Structure and Organic Matter Storage in Agricultural Soils,* Lewis, Boca Raton, Florida, 1995.

9. F. J. Stevenson, *Trans. 12th. Intern. Congr. Soil Sci., Symposium Papers,* **1,** 137 (1982).

10. J. L. Hatfield and B. A. Stewart, Eds., *Crop Residue Management,* Lewis, Boca Raton, Florida, 1994.

11. F. Magdoff, *Building Soils for Better Crops: Organic Matter Management,* University of Nebraska Press, Lincoln, Nebraska, 1992.

12. P. W. Unger, Ed., *Managing Agricultural Residues,* Lewis, Boca Raton, Florida, 1994.

13. A. D. Hall, *J. Agr. Sci.,* **1,** 241 (1905).

14. N. C. Brady, *The Nature and Properties of Soils,* MacMillan, New York, 1974.

15. R. L. Crocker and B. A. Dickson, *J. Ecol.,* **45,** 169 (1957).

16. G. R. Hallberg, N. C. Wollenhaupt, and G. E. Miller, *Soil Sci. Soc. Amer. J.,* **42,** 339 (1978).

17. R. M. Smith, E. H. Tyron, and E. H Tyner, *West Virginia Agric. Exp. Sta. Bull.,* **604T,** 1 (1971).

18. J. K. Syers, J. A. Adams, and T. W. Walker, *J. Soil Sci.,* **21,** 146 (1970).

19. S. W. Buol and F. D. Hole, *Soil Sci. Soc. Amer. Proc.,* **23,** 239 (1959).

20. J. M. Anderson, M. A. Leonard, and P. Ineson, "Lysimeters with and without Roots for Investigating the Role of Microfauna in Forest Soils," in A. F. Harrison, P. Ineson, and O. W. Heal, Eds., *Nutrient Cycling in Terrestrial Ecosystems,* Elsevier, London, 1989, pp. 347–355.

21. J. N. Klironomos, P. Widden, and I. Desandes, *Soil Biol. Biochem.,* **24,** 685 (1992).

22. P. Lavelle and A. Martin, *Soil Biol. Biochem.,* **24,** 1491 (1992).

23. B. E. Ruz-Jerez, P. R. Ball, and R. W. Tillman, *Soil Biol. Biochem.,* **24,** 1529 (1992).

24. G. W. Yeates and D. C. Coleman, "Role of Nematodes in Decomposition," in D. W. Freckman, Ed., *Nematodes in Soil Ecosystems,* University of Texas Press, Austin, 1982, pp. 55–80.

25. J. K. Martin and R. Merckx, "The Partitioning of Root-derived Carbon within the Rhizosphere of Arable Crops," in K. Mulongoy and R. Merckx, Eds., *Soil Organic Matter Dynamics and Sustainability of Tropical Agriculture,* West Sussex, England, 1993, pp. 101–107.

26. P. A. Sánchez, *Plant Soil,* **67,** 91 (1982).

27. C. Enders, *Biochem. Z.,* **315,** 259, 352 (1943).

28. G. W. Harmsen, *Plant Soil,* **3,** 110 (1951).

29. H. L. Shantz, *Ann. Assoc. Amer. Geog.,* **13,** 81 (1923).

30. J. B. Reid and M. J. Goss, *Soil Biol. Biochem.,* **15,** 687 (1983).

31. C. Feller, "Organic Inputs, Soil Organic Matter and Functional Soil Organic Compartments in Low-activity Soils of Tropical Zones," in K. Mulongoy and R. Merckx, Eds., *Soil Organic Matter Dynamics and Sustainability of Tropical Agriculture,* West Sussex, England, 1993, pp. 77–88.

32. H. J. Haas, C. E. Evans, and M. L. Miles, USDA Tech. Bull., **1164,** 1 (1957).

33. W. J. Parton, D. S. Schimel, C. V. Cole, and D. S. Ojima, *Soil Sci. Soc. Amer. J.,* **51,** 1173 (1987).

34. A. R. Aandahl, *Soil Sci. Soc. Amer. Proc.,* **13,** 449 (1948).

35. A. van der Valk, Ed., *Northern Prairie Wetlands,* Iowa State University Press, Ames, Iowa, 1989.

36. J. H. Beadle, "History of Parke County," in H. W. Beckwith, Ed., *History of Vigo and Parke Counties (Indiana),* Hill and N. Iddings, Chicago, 1880, pp. 25–26.

37. G. E. Likens, F. H. Bormann, M. Johnson, D. W. Fisher, and R. S. Pierce, *Ecol. Monogr.,* **40,** 23 (1970).

38. W. Cronin, *Changes in the Land,* Hill & Wang, New York, 1983.

39. R. M. Salter and T. C. Green, *J. Amer. Soc. Agron.,* **23,** 622 (1933).

40. C. G. Hopkins, *Soil Fertility and Permanent Agriculture,* Ginn, Boston, 1910.

41. J. A. Hobbs and P. L. Brown, *Kansas Agric. Exp. Sta. Tech. Bull.,* **144,** 1, 1965.

42. P. A. Sánchez, "Nitrogen in Shifting Cultivation Systems of Latin America," in G. P. Robertson, R. Herrara, and T. Rosswell, *Nitrogen Cycling in Ecosystems of Latin America and the Caribbean,* Martinus Nijhoff, The Hague, 1982, pp. 91–103.

43. J. T. Norman, *Annual Cropping Systems in the Tropics: An Introduction,* University Press of Florida, Gainesville, Florida, 1979.

44. W. V. Bartholomew and D. Kirkham, *Trans. 7th Intern. Congr. Soil Sci.,* **2,** 471 (1960).

45. J. S. Russell, *Soil Sci.,* **120,** 370 (1975).

46. H. F. Birch, *Plant Soil,* **11,** 262 (1959).

47. C. R. Clement and T. E. Williams, *J. Agr. Sci.,* **69,** 133 (1967).

48. J. Giddens, W. E. Adams, and R. N. Dawson, *Agron. J.,* **63,** 133 (1967).

49. R. H. Jackman, *New Zealand J. Agric. Res.,* **7,** 445 (1964).

50. E. M. White, C. R. Krueger, and R. A. Moore, *Agron. J.,* **68,** 581 (1976).

51. A. L. Azevedo, *Ann. Inst. Super. Agron. Univ. Tec. Lisboa,* **34,** 63 (1973).

52. A. L. Azevedo and M. L. V. Fernandes, *Ann. Inst. Super. Agron. Univ. Tec. Lisbon,* **34,** 115 (1973).

53. H. Fleige and K. Baeumer, *Agro-Ecosystems,* **1,** 19 (1974).

54. P. E. Rasmussen, R. R. Allmaras, C. R. Rohde, and N. C. Roager, Jr., *Soil Sci. Soc. Amer. J.,* **44,** 596 (1980).

55. A. L. Black, *Soil Sci. Soc. Amer. Proc.,* **37,** 943 (1973).

56. D. S. Jenkinson and A. E. Johnson, "Soil Organic Matter in the Hoosfield Continuous Barley Experiment," in *Rothamsted Exp. Sta. Report for 1976, Part 2,* Harpenden, Herts., England, 1977, pp. 81–101.

57. F. N. Anderson and G. A. Peterson, *Agron. J.,* **65,** 697 (1973).

58. W. E. Larsen, C. E. Clapp, W. H. Pierre, and Y. B. Morachan, *Agron. J.,* **64,** 204 (1972).

59. K. Swafakab, *Soil Biol. Biochem.,* **14,** 309 (1982).

60. K. Mulongoy and R. Merckx, Eds., *Soil Organic Matter Dynamics and Sustainability of Tropical Agriculture,* Wiley, New York, 1993.

61. L. F. Elliott and F. J. Stevenson, Eds., *Soils for Management of Organic Wastes and Waste Waters,* American Society of Agronomy, Madison, Wisconsin, 1977.

62. J. A. Hansen and K. Hendricksen, *Nitrogen in Organic Wastes Applied to Soils,* Academic Press, London, 1989.

63. F. J. Stevenson, *Soil Sci.,* **107,** 470 (1969).
64. R. W. Simonson, *Amer. J. Sci.,* **252,** 705 (1954).
65. J. D. Hogan and M. T. Beatty, *Soil Sci. Soc. Amer. Proc.,* **27,** 345 (1963).
66. Z. Varanini and R. Pinton, *Prog. Bot.,* **56,** 97, (1995).
67. Y. Chen and T. Avid, "Effects of Humic Substances on Plant Growth," in P. MacCarthy et al., Eds., *Humic Substances in Soil and Crop Sciences,* American Society of Agronomy, Madison, Wisconsin, 1990, pp. 161–186.
68. D. Vaughn and R. E. Malcolm, "Influence of Humic Substances on Growth and Biochemical Processes," in D. Vaughn and R. E. Malcolm, Eds., *Soil Organic Matter and Biological Activity,* Martinus Nijhoff/Dr. W. Junk, Dordrecht, 1985, pp. 37–75.
69. W. A. Dick and E. L. McCoy, "Enhancing Soil Fertility by Addition of Compost," in H. A. J. Hoitink and H. M. Keener, Eds., *Science and Engineering of Composting,* Renaissance, Worthington, Ohio, 1993, pp. 622–644.
70. L. B. Olmstead, *Soil Sci. Sci. Amer. Proc.,* **11,** 89 (1946).
71. R. Klöcking, "Humic Substances as Potential Therapeutics," in N. Senesi and T. M. Miano, Eds., *Humic Substances in the Global Environment and Implications on Human Health,* Elsevier, Amsterdam, 1994, pp. 1245–1257.
72. W. F. Guerin and S. A. Boyd, "Bioavailability of Sorbed Naphthalene to Bacteria: Influence of Contaminant Aging and Soil Organic Carbon Content," in D. M. Linn, T. H. Carski, M. L. Brusseau, and F.-H. Chang, Eds., *Sorption and Degradation of Pesticides and Organic Chemicals in Soil,* Soil Science Society of America, Madison, Wisconsin, 1993, pp. 197–208.
73. P. MacCarthy and J. A. Rice, "Industrial Applications of Humus: An Overview," in N. Senesi and T. M. Miano, Eds., *Humic Substances in the Global Environment and Implications on Human Health,* Elsevier, Amsterdam, 1994, pp. 1209–1223.
74. W. Jackson, "A Search for the Unifying Concept for Sustainable Agriculture," in W. Jackson, *Altars of Unhewn Stone,* North Point Press, San Francisco, 1987, 119–146.
75. J. D. Soule and J. K. Piper, *Farming in Nature's Image,* Island Press, Washington, D.C., 1992.
76. W. A. Dick, *Soil Sci. Soc. Amer. J.,* **48,** 569 (1984).
77. S. P. Deng and M. A. Tabatabai, *Biol. Fert. Soils,* **22,** 202, 208 (1996).
78. J. W. Doran, *Soil Sci. Soc. Amer. J.,* **44,** 518, 765 (1980).
79. R. R. Weil, "Inside the Heart of Sustainable Farming," *The New Farm,* Jan. 1992, pp. 43–48.
80. J. I. Rodale, *Pay Dirt,* Rodale Press, Emmaus, Pennsylvania, 1945.
81. F. J. Stevenson, *Crops and Soils,* **31,** 14 (1979).

3

SOIL ORGANIC MATTER QUALITY AND CHARACTERIZATION

It is generally accepted that astute management of organic matter is the key to sustainable agriculture. Optimizing the return of crop residues and external organic amendments (e.g., manures, composts) and reducing tillage operations causes benefits to accrue due to improved soil tilth, more efficient cycling of nutrients, reduced soil erosion, and preservation of stable humus.

A stimulating discussion of the role of organic matter and soil quality in sustainable farming is given by Weil;[1] the multiple benefits of sustainable agricultural systems are described by Acton and Gregorich[2] and Soule and Piper.[3] Cole[4] discusses the topic of soil quality from a broad environmental perspective. More detailed deliberations on specific aspects of organic matter management and sustainable agriculture are also available.[5-10] Descriptive aspects of soil quality, with emphasis on biological indicators, are discussed in Doran et al.[11]

This chapter focuses on pools of organic matter in soil, their function, dynamics, and biological significance, the location of organic matter within the soil (i.e., physical fractionation), and the applications of ^{13}C–NMR and analytical pyrolysis for the determination of soil organic matter quality.

POOLS OF ORGANIC MATTER

On a conceptual basis, organic materials in soils can be partitioned into the following forms or pools:[12-14]

> *Litter:* Macroorganic matter (e.g., crop residues) that lies *on* the soil surface
>
> *Light Fraction:* Plant residues and their partial decomposition products that reside *within* the soil proper

Microbial biomass: Cells of living microorganisms, notably bacteria, actinomycetes, and fungi

Faunal biomass: Tissues of animals (primarily invertebrates)

Belowground plant constituents: Primarily roots with lesser amounts of dead roots and exudates

Water-soluble organics: Organic substances dissolved in the soil solution

Stable humus: Humified remains of plant and animal tissues that have become stabilized by microbial and chemical transformations and/or by association with inorganic soil components

The size of individual pools varies seasonally and is dependent on features such as soil type, vegetation and climate. The biological significance of the different pools is discussed in Coleman et al.[6]

The concept of pools of organic matter that differ in their susceptibilities to microbial decomposition and their longevity in soil has provided a basis for understanding the dynamic nature of soil organic matter and how nutrient availability (e.g., N, P, S, trace elements) is influenced by management practices and changes in the soil environment. If this concept is to be studied or used for practical purposes, the pools should be measurable, so that fluxes of nutrients among them can be determined. Although the pools concept is old, experimental methods to define and quantify the various pools are a relatively recent advance, with most of the research being done during the last decade.

Enzymes constitute still another important component of soil organic matter and are also covered in this chapter.

Litter

This pool has long been known to be important in the cycling of nutrients in forest soils, natural grasslands, and crop production agriculture. Increasing attention is now being given to cropping systems where residues are allowed to remain on the soil surface following harvest, such as in minimal tillage or no-till. The literature is extensive, as recent reviews indicate.[7,8,12–14] An interesting historical account of the successful renovation of a Chilean soil through no-till is given by Lal et al.[15]

Herbaceous litter is normally determined by collecting the macroorganic matter from measured plot areas of the ecosystem under study. Litter quantities are expressed in terms of g/m^2 or t/ha. Anderson and Ingram[9] provide guidelines for quantification of litter inputs.

In intensively cultivated soils, crop residues have traditionally been incorporated into the soil well in advance of planting, in which case the litter becomes part of the light fraction. With greater acceptance of cultural practices designed to reduce erosion (e.g., minimal tillage or no-till), greater amounts of crop residues will be left on the soil surface and for a longer time, ultimately to be mixed with the surface soil through the activities of macrofaunal organisms or through leaching of partial decomposition products.

Light Fraction

The light fraction consists of organic residues in various stages of decomposition that exist within the soil proper (i.e, not as surface litter). Root residues, long ignored in considering C cycling in soil, often make a major contribution to this fraction, particularly in perennial systems.

The light fraction has been used as an indicator of changes in labile organic matter as affected by tillage, cropping practice, and environmental factors affecting microbial activity.[16–20] Since the material has not been subjected to extensive microbial degradation, the light fraction has a rapid turnover rate and thus serves as a ready source of nutrients for plant growth.

Liquids with densities of from 1.6 to 2.0 are commonly used to separate the light fraction. Practically speaking, the "light fraction" includes materials of lower density than the solvent, and therefore float on the surface of the liquid. Other organic fractions are associated with dense mineral components of the soil and sink in the dense liquids. The liquid can be inorganic (e.g., NaI, $ZnBr_2$, CsCl) or organic (e.g., halogenated solvents such as tetrabromomethane). Organic liquids are toxic to microorganisms and can cause problems when the isolated material is to be used in biochemical studies; some inorganic solutions are also toxic.

The size of the light fraction is somewhat less than for stable humus but can constitute as much as 30% of the organic matter in some soils.[6] Janzen et al.[19] found that the light fraction accounted for from 2.0 to 17.5% of the organic matter in the soils of some long-term crop rotations in Saskatchewan, Canada. Differences in the light fraction among sites and treatments were attributed to variable residue inputs and rates of substrate decay.

The amount of light fraction material in any given soil varies with season and is greatest immediately following incorporation of crop residues through tillage. Factors affecting the amount of light fraction include the quantity of plant litter introduced into the soil and extent of decomposition as affected by temperature, texture, moisture, and soil pH. The chemical composition of the light fraction, as determined by ^{13}C–NMR spectroscopy (discussed below), has been found to be comparable to that of plant litter.[16] Hence, the light fraction is primarily very small pieces of relatively undegraded debris formed by physical fragmentation and the activities of chewing invertebrates.

For modeling purposes, the C of plant residues has been further separated into several subpools (as many as four: soluble, polysaccharides, protein, lignin), which vary in susceptibility to decay. In the simplest model, the residue C is subdivided into two main pools with initial turnover times of a few months for a *labile* fraction to about five years for a *stable* fraction (see section on Simulation Models below).

Microbial Biomass

Interest in the microbial biomass ensues because of its importance in nutrient cycling as well as soil aggregation.[21,22] As Jenkinson[23] elegantly states: "The soil biomass is the eye of the needle through which all natural organic materials that enter the soil must pass, often more than once, as they are degraded to inorganic compounds."

Microbes play a dual role in the soil: first, as agents for the degradation of plant residues, with concurrent release of nutrients and CO_2 formation, and second, as a labile pool of nutrients. The size of the microbial biomass can be appreciable and may account for as much as 2% of the total soil C (e.g., see Anderson and Domsch[24]). Reviews concerning the biomass include those of Jenkinson and Ladd,[25] Paul and Voroney,[26] and Insam.[27]

Theng et al.[28] subdivided the living component of soil organic matter into three compartments: microorganisms, 60–80%; macroorganisms or fauna, 15–30%; and plant roots, 5–10%. Faunal organisms are particularly abundant in the humus layers of forest soils, where populations of microinvertebrates may be 40,000 to 50,000/m^2.[29] Roots may have a greater influence on soil processes than their pool size suggests, due in part to microorganisms that live in the rhizosphere.

Methods for biomass C include counts for total microorganisms and measurements for metabolic activities. Included with the latter are O_2 consumption, enzymatic activity, heat production, adenosine triphosphate (ATP) content, and respiratory measurements for CO_2 production. Methods for the biochemical determination of the biomass are discussed by Tunlid and White[30] and by Horwath and Paul.[31]

Estimates from Microbial Counts   Biomass estimates based on microbial counts require a knowledge of cell numbers and volume, which can be obtained from microscopic observations. Biomass is calculated as:

$$\text{Biomass} = \text{number of cells} \times \text{volume} \times \text{density}$$

From data for numbers of soil microorganisms (Table 3.1), as recorded by Martin and Focht,[32] along with an assumed density for microbial cells (1.5 g/cm^3), the total live weight of bacteria in soil is on the order of 0.15 to 1.5 g/kg of soil, or about 340 to 3400 kg/ha to plow depth. From a similar approach, the fungal biomass in a kilogram of soil ranges from 540 to 5400 kg/ha to plow depth of soil.

Smith and Paul[33] summarized values for microbial (bacterial + fungal) biomass ranging from 110 to 2240 kg/ha. The difference in the values reported by the different authors is, to some extent, an indication of the uncertainty associated with biomass estimations of soil organisms. Although numbers of bacteria greatly exceed those of fungi in most soils, the total mass

TABLE 3.1   Approximate Number of Organisms Commonly Found in Soils[a]

Organism[b]	Estimated Numbers/g
Bacteria	3,000,000 to 500,000,000
Actinomycetes	1,000,000 to 20,000,000
Fungi	5,000 to 900,000
Yeasts	1,000 to 100,000
Algae	1,000 to 500,000
Protozoa	1,000 to 500,000
Nematodes	50 to 200

[a] From Martin and Focht[32]
[b] Numbers for bacteria, actinomycetes, fungi, and yeasts are based on plate counts. Other organisms found in soil include viruses, arthropods, and earthworms.

of the latter is larger due to the larger individual cell size. Numbers of microorganisms in soil, and thus the biomass, are subject to daily and seasonal variations, as dictated by such factors as moisture, temperature, pH, and energy supply.

Chemical Methods   Attempts have been made to use chemical methods to determine the amount of certain cellular components such as muramic acid for bacteria and chitin for fungi and thereby to estimate biomass.[34,35] Assays for nucleic acid (DNA and RNA)[36,37] and adenosine triphosphate (ATP) have also been suggested.[38,39] This idea is not straightforward in theory or in practice, because numerous problems are encountered using these methods. A universal chemical technique for estimating the active microbial biomass should meet the following requirements.[40]

1. The component to be measured must be present in all living cells and absent from dead cells (as well as nonliving soil components), and decay rapidly upon death of the organism.
2. It must be present in fairly uniform concentration in all cells, regardless of cell age, growth conditions, and environmental stresses.
3. The technique employed must be relatively rapid and easy to use for large-scale programs involving numerous samples.

Conditions 1 and 3 are met reasonably well by adenosine triphosphate (ATP). This compound is very unstable outside living cells, and quantification by the bioluminescence technique is relatively easy. The method is based on measurement of the light emitted from the reaction of extracted ATP with luciferin, luciferase, and atmospheric O_2, as detailed in Chapter 7. Condition 2 is not easily met, since ATP content of cells is variable among species and it is influenced by substrate availability, environmental conditions and metabolic state of the organism.

Fumigation Method   A popular approach currently used to estimate the magnitude of the biomass is to measure the increase (flush) in CO_2 (and release of inorganic nutrients) upon incubation of fumigated soil when compared to a nonfumigated soil.[41-46] The flush in CO_2 evolution is believed to result from the degradation of microbial cells killed by fumigation, thereby providing an indication of the size of the biomass. The assumption is made that all of the extra CO_2 released by incubation of the fumigated soil is derived from cells of microorganisms killed by fumigation and subsequently degraded by recolonizing microbes.

Biomass C is calculated from the relationship:

$$\text{Biomass C} = F_c/k_c$$

where F_c is the difference between the amount of CO_2–C released by incubation of the fumigated and unfumigated soil and k_c is the fraction of the biomass C mineralized to CO_2 during the course of the incubation period (usually 10 to 14 days).

The value chosen for k_c is of some importance and is not precisely the same for all soils. However, k_c values obtained for pure cultures of bacteria and fungi span a rather narrow range, with most values being within the range of 0.43 to 0.50.[32] For biomass C calculations, a k_c value of 0.45 is commonly used.[44]

Assumptions underlying the fumigation method include the following:

1. All microorganisms are killed during fumigation, and C in the dead organisms is more rapidly mineralized than that in living organisms; furthermore, microbial degradation of native humus occurs at the same rate in both fumigated and unfumigated soil.
2. The only effect of fumigation is to kill microorganisms—death of organisms in the unfumigated (control) soil is negligible compared to that in the fumigated soil.
3. The fraction of the dead biomass C mineralized over a given time period is similar for all soils under investigation.

The last assumption is not entirely valid, notably for soils receiving large inputs of fresh substrates,[44,47] nor is it valid for soils with high clay contents, in which the released biomass constituents can adsorb to clay surfaces and decay relatively slowly. An evaluation of methods for measuring the microbial biomass in soils following incorporation of plant residues is given by Ocio and Brookes.[48]

Extraction Method   An extraction method has been described for the direct determination of biomass C in fumigated soil.[49,50] The basis for this method is that the amounts of organic C in extracts of fumigated soils are reasonably well correlated with the corresponding amounts recovered as CO_2 by incubation. The reagent used for extraction is $0.5M$ K_2SO_4.

This method has a great advantage over the fumigation procedure in that there is no need for complete removal of fumigant or for extended incubation of the soil under carefully controlled conditions. The soil is fumigated and extracted and organic C is analyzed immediately by a simple chemical procedure. Results can be obtained in one day rather than two weeks and with much less analytical effort.

Conversion factors for estimating biomass C from extracted organic C are highly variable; for mineral soils, a provisional value of 0.33 has been recommended. This variability is a serious constraint on utility of the method, since *a priori* the appropriate conversion factor for a previously unstudied soil is unknown.

Amounts and turnover rates of C and N in the microbial biomass for two temperate zone soils (England, Canada) and a tropical zone soil (Brazil) are given in Table 3.2.[51] Biomass values were lower in the tropical soil even though C inputs were higher. This result was attributed to accelerated decay of organic matter in the tropical soil, as indicated by a shorter microbial turnover rate (see last column of the table). The turnover rate for the biomass is considerably higher than for stable humus.

Figure 3.1 shows the relationship between biomass C and total organic C for a number of soils from temperate and tropical regions, as reported by Theng et. al.[28] Ratios for biomass C/soil organic C (lines marked A and B on the diagram) show that, for most soils, from 1 to 3% of the soil organic C is biomass C. In other work, the size of the microbial biomass has been found to be related to aggregate sizes.[21,22]

TABLE 3.2   Amount and Turnover Rates of C and N in the Microbial Biomass for Cultivated Soils at Three Locations[a,b]

Climate and Location	Microbial C	Microbial N	C Inputs	Microbial Turnover Time
	kg/ha	kg/ha	mg/ha/yr	yr
Temperate				
England	570	95	1.2	25
Canada	1600	300	1.6	6.8
Tropical				
Brazil	460	84	13.0	0.24

[a]Adapted from Duxbury.[51]
[b]Calculated from simulation modeling.

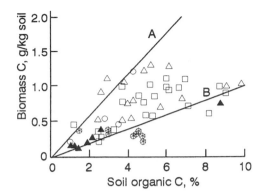

Fig. 3.1 Relationship between microbial biomass C and total soil organic C for a range of temperate and tropical soils. Included are four New Zealand soils and one each from England, Scotland, Nigeria, and India. Values outside the two demarcation lines are for soils in which pH, moisture, and other soil factors affected the results. From Theng et al.[28]

Faunal Biomass

In general, soil fauna, particularly soil microfauna such as nematodes and microscopic insects and mites, have not been studied as intensively or extensively as soil microorganisms. Earthworms are an exception to this statement; they are undoubtedly the most widely studied soil invertebrate.[52,53] In many cases, soil fauna have been studied on a population basis rather than a biomass basis, and hence it is difficult to make direct biomass comparisons with other groups of organisms. Populations of this group are affected by management practices and season of the year.[54] The faunal biomass comprises but a small percentage of the total biomass or soil organic matter in grassland systems.[55] Faunal organisms, however, are of major importance to soil organic matter dynamics because they convert macroscopic (visible) plant residues in soil to microscopic material ("light fraction"), and they also are major agents for redistribution of litter materials from the soil surface to deeper regions of the profile. Belowground food webs involving these organisms are complex,[56] but they are difficult to study because the organisms are distributed very erratically and their density (organisms per volume of soil) is often low. Methods for extraction and enumeration of soil fauna are described by Ingham[57] and Moldenke.[58]

Belowground Plant Constituents

Most ecological and agricultural research has been directed to the aboveground plant biomass and fauna, with much less emphasis being given to belowground biotic components such as roots and soil fauna. The difference in the amount of research is primarily a matter of practicality: collection of material from the mineral portion of the soil is much more difficult than from aboveground, disruptive sampling is required, and the subsurface biomass is not distributed as uniformly as aboveground biomass.[59] As a consequence of these difficulties, a large and important component of the soil ecosystem is neglected in many studies. Depending on the plant species and growing conditions, roots comprise between 18 and 83% of total biomass of perennial herbaceous plants and up to 35% of total biomass of woody plants.[14,60] The data below and in Table 3.3[55] provide examples of the magnitude of aboveground and belowground biomass and C pools. Note that in both examples soil organic matter and belowground biomass are substantial fractions of the total organic pool of the system. In addition to root biomass, most plants lose a substantial amount of photosynthate as root exudates, by death of small roots and as sloughed-off root material. Under field conditions, these components are unquantifiable; in laboratory studies, plants have been shown to lose between 20 and 40% of the total photosynthate as root exudates and sloughed-off root material.[61,62] Hence, large (but unmeasured) contributions are made to the overall C balance of soil during the growing season. Most small roots and root exudates decay rapidly and enter other C pools in the soil.

Distribution of organic materials in a 55-year-old hardwood forest was shown by Bormann and Likens[14] to be:

Component	Amount (kg/ha)
Aboveground	
Live plant	132,810
Dead wood	4,400
Belowground	
Live roots	28,260
Dead wood	29,000
Litter layer	48,000
Soil organic matter	173,000
Total	415,470

Water-Soluble Organics

Considerable attention has been given in recent years to organics contained in the soil solution, from the standpoint of both plant nutrition (source of nutrients) and the environment (alleviation of metal ion toxicities through chelation, acidifying effects on natural waters, and carriers of xenobiotics). Zsolnay reviews the extensive literature on dissolved organic matter in soils.[63]

Recovery of the aqueous phase from the soil *in situ* can be achieved by applying suction to a porous collector inserted into the soil. The sampler (i.e., ceramic or fritted glass plate) is installed to the desired depth, a vacuum is

TABLE 3.3   Distribution of Organic Materials in Two Prairie Soils[a]

Component	Tallgrass Prairie	Shortgrass Prairie
	kg/ha	kg/ha
Aboveground		
Leaves and stems	1520	1007
Dead plant	6490	2910
Invertebrates	2	2
Vertebrates + birds	<1	<1
Total	**8012**	**3919**
Belowground		
Live roots and crowns	136301	2840
Invertebrates	8	1
Microorganisms	6350	670
Soil organic matter[b]	403000	179200
Total	**422988**	**192711**

[a] Adapted from data in French et al.[55]
[b] Based on 6% organic matter to 0.5 m soil depth (tallgrass) and 4% organic matter to
0.3 m soil depth (shortgrass).

applied, and pore water is drawn into the sample chamber through the porous section.

The soil solution can also be obtained from field-moist soil in the laboratory. Methods include column miscible displacement, centrifugation with and without an immiscible liquid, and use of ceramic or plastic filters.[64,65]

Stable Humus

In most soils, the bulk of the organic matter occurs as stable humus. Both chemical and physical fractionation procedures have been used in attempts to separate the various components of humus and to ascertain their location within the soil matrix. Chemical fractionation methods are described in Chapter 1.

Soil Enzymes

Soil enzymes are a quantitatively minute, but very important, fraction of soil organic matter because all biochemical action is dependent upon, or related to, them. Because of the complexity and variety of substrates that serve as energy sources for microorganisms, the soil would be expected to contain a wide array of enzymes.[38–40] Each soil may have its own characteristic pattern of specific enzymes, as stated by Kuprevich and Shcherbakova:[39]

> The differences in level of enzymatic activity are caused primarily by the fact that every soil type, depending on its origin and developmental conditions, is distinct from every other in its content of organic matter, in the composition and

activity of living organisms inhabiting it, and consequently, in the intensity of biological processes. Obviously, it is probable that each type of soil has its own inherent level of enzymatic activity.

A partial list of enzymes found in soils is given in Table 3.4. It should be noted that enzymes in soil are determined not by direct analysis as the number of molecules of enzyme, but indirectly through their ability to transform a given amount organic substrate into known products in a specified time. For example, urease activity is estimated from the conversion of added urea to NH_3 and CO_2 (urea $\rightarrow$ 2 NH_3 + CO_2). Methods for the determination of soil enzymes are described by Tabatabai[40] and Burns.[66]

Certain enzymes are ubiquitous; that is, they are found in virtually all soils. Typical examples are urease, catalase, phosphatase, and peptidases of various types. Other enzymes may be produced in soil only under special circumstances. Dehydrogenase activity appears to be related to the quantity of decomposable organic matter and is closely related to the soil biomass.

Soil enzymes are not readily recovered from soil; only rarely have they been isolated in a pure form. McLaren[67] concluded that most enzymes are complexed, or immobilized, with clay minerals and/or humus. Attempts have been made to correlate enzyme activities to overall metabolic activity of the

TABLE 3.4   Partial List of Enzymes in Soil

Enzyme System	Reaction Catalyzed
α- and β-Amylase	Hydrolysis of α-1, 4-glucan bonds
Arylsulfatases	$R\text{-}SO_3^- + H_2O \rightarrow ROH + H^+ + SO_4^{2-}$
Asparaginase	Asparagine + $H_2O \rightarrow$ aspartate + NH_3
Catalase	$2\ H_2O_2 \rightarrow 2\ H_2O + O_2$
Cellulase	Hydrolysis of β-1, 4-glucan bonds
Chitinase	Hydrolysis of β-1, 4-aminoglucan bonds
Deamidase	Carboxylic acid amide + $H_2O \rightarrow$ carboxylic acid + NH_3
Dehydrogenases	$X\text{-}2e^- +$ acceptor (dye) $\rightarrow X +$ acceptor-$2e^-$
α- and β-Galactosidase	Galactoside + $H_2O \rightarrow ROH +$ galactose
α- and β-Glucosidase	Glucoside + $H_2O \rightarrow ROH +$ glucose
Lichenase	Hydrolysis of β-1,3-cellotriose bonds
Ligninase	Free-radical scission of bonds in lignin
Lipase	Triglyceride + 3 $H_2O \rightarrow$ glycerol + 3 fatty acids
Nucleotidases	Dephosphorylation of nucleotides
Peptidases	Peptides $\rightarrow$ amino acids
Phenoloxidases	Diphenol + 1/2 $O_2 \rightarrow$ quinone + H_2O
Phosphatase	Phosphate ester + $H_2O \rightarrow ROH + PO_4^{3-}$
Phytases	Inositol hexaphosphate + $6H_2O \rightarrow$ inositol + 6 PO_4^{3-}
Proteases	Proteins $\rightarrow$ peptides and amino acids
Pyrophosphatase	Pyrophosphate + $H_2O \rightarrow 2PO_4^{3-}$
Urease	Urea $\rightarrow 2NH_3 + CO_2$

soil, but these efforts have not been very successful. The frequently poor correlation between overall metabolic activity and activity of a particular enzyme is probably the result of stabilization of extracellular enzymes by association with soil organic matter and/or clay surfaces. In this case, the cell that produced the enzyme in the past may be dead or inactive, but the enzyme that it produced and released into the soil environment remains active. The best results have been obtained with dehydrogenase because this enzyme is *intracellular* and probably does not exist in soil in the free state.[68,69] The substrate (a dye) for dehydrogenase captures electrons generated during metabolism. Ordinarily, these electrons would be transferred to molecular O_2, using carriers such as nicotinamide dinucleotide (NAD) or flavin adenine dinucleotide (FAD). This activity is not the result of a substrate-specific enzyme (in contrast to many of the enzymes listed in Table 3.4, which are substrate specific). Instead, many enzymes with electron-transfer capabilities will transfer electrons to the dye.

Like microbial populations, enzymatic activity is not static but fluctuates with biotic and abiotic conditions. Activities would be expected to be particularly high in productive soils rich in organic matter. Such factors as aeration, temperature, moisture, soil structure, organic matter content, seasonal changes, and soil treatment all have an influence on the presence and abundance of enzymes. A marked change occurs in the kinds and amounts of enzymes when virgin lands are first placed under cultivation.

SIMULATION MODELS

A major aim of research on biologically significant organic matter pools in soil is to organize the data into simulation models whereby predictions can be made regarding nutrient cycling (N, P, S), organic matter storage, and the impact of residue management practices on environmental quality and soil productivity. A key component of most models is the position of organic matter, or the nutrients contained therein, into fractions (pools) that have distinct turnover times or mean residence times (MRT). For this purpose, and on a conceptional basis, the organic matter has also been partitioned into *labile* (active) and *stable* compartments or fractions. It is generally accepted that the labile components decay within a few weeks or months, while stable components are resistant to decomposition and persist in soils for a few years, decades or longer.

Included in the labile fraction are plant litter, light fraction, microbial biomass, dead fauna and their wastes, and nonhumic substances not bound to mineral constituents. The stable fraction (i.e., passive humus) functions as a *reservoir* of nutrients and is important to the long-term C balance of the soil.

The labile fraction serves as a ready source of nutrients for plant growth and microbial activity and is particularly important for maintaining the productivity of the soil under sustainable agriculture, where maximum use is made of organic residues under conditions of minimum tillage and reduced fertilizer inputs. As one might expect, the size of the labile fraction is related to residue inputs and is influenced by degradation patterns as affected by microbial activity.

Many current models have evolved from the five-pool model of Jenkinson and Rayner,[70] which includes the microbial biomass, two forms of stabilized organic matter (physically protected and chemically protected), and the plant residue component, partitioned into two subpools (readily decomposable and resistant). Other C (and N) models,[12,13,71–73] while being more complex and with different terminologies, have somewhat the same general format. For example, in the organic N model of Paul and Juma,[73] the following pools are recognized:

Pool Description	Half-life
Microbial biomass	24 weeks
Active nonmicrobial biomass	77 weeks
Stabilized organic N	27 years
Old organic N	600 years

A similar approach has been used to simulate the dynamics of soil organic matter in natural grasslands and their associated agroecosystems. In the CENTURY model,[74–76] three soil organic matter fractions are identified and incoming plant residues (from roots and shoots) are divided into metabolic and structural pools as a function of the lignin to N ratio of the residues. Mean residence times for the model (illustrated in Fig. 3.2) are as follows:[75]

Pool	MRT
Metabolic C of plant residues	0.1–1 years
Structural C of plant residues	1–5 years
Active soil C	1–5 years
Slow soil C	20–40 years
Passive soil C	200–1500 years

The CENTURY model has also been used for the modeling of tropical soil organic matter,[75] and submodels have been devised for N, P, and S.[74,76]

Limitations of the discrete pool approach to describe organic matter quality and nutrient availability are as follows:[71]

1. Labile components are in a continuous state of flux, and the size of the active fraction does not define an absolute amount of a specific form of

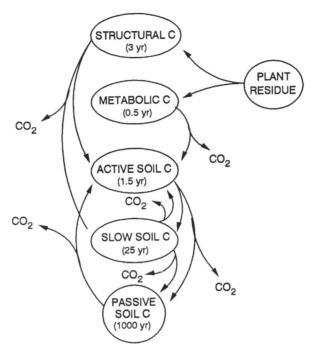

Fig. 3.2 Flow diagram for the CENTURY model. Values in parentheses are estimated turnover times of the pools. Adapted from Parton et al.[74,75]

the nutrient but rather provides a relative index of nutrient availability under a given set of circumstances.

2. Not all fractions (pools) can be experimentally measured (note item 1).

3. Under conditions of net immobilization, the biomass functions as a sink for nutrients and results in greater pressure on plants in terms of nutrient acquisition. The mineralization-immobilization of nutrients may depend on the state of the system.

4. The concept of pools, and of models based on them, will never completely match with biological, chemical, or physical reality.

Grant and coworkers[77] used the discrete pool approach to develop a sophisticated supercomputer-based model for C and N dynamics in soil by using a combination of published literature and experimentally obtained data at specific sites. In terms of scope and complexity, this model provides a dramatic example of the complexity of organic matter and N dynamics in soil.

A simpler model has been developed by Gilmour et al.[78] to estimate N released during decay of soil-applied wastes such as sewage sludge. In this model, laboratory data were obtained with a specific soil–sludge combination at various temperatures and moisture contents and used to predict decay of the organic fraction of the waste under field conditions.

The modeling of soil organic matter is a recent and complex addition to the field of soil science. As computational methods and analytic techniques for determination of soil organic fractions improve, computer-based methods may become a practical addition or substitute for time- and labor-intensive field and laboratory studies. The reader is referred to the review of Parton et al.[75] and to Swartzman and Kaluzny[79] for additional information.

PHYSICAL FRACTIONATION OF ORGANIC MATTER

Interest in the "physical fractionation" of organic matter arose from the observation that nutrient turnover depends not only on the kind (and amount) of organic matter in soil but on its *location* within the soil; furthermore, location of organic matter has an impact on soil physical properties. Physical fractionation methods are reviewed by Christensen[80] and Stevenson and Elliott.[81]

Physical fractionation procedures have been used in the following types of studies:

1. To recover the "light fraction," consisting largely of undecomposed plant residues and their partial decomposition products
2. To establish the nature and biological significance of organic matter in the various size fractions of the soil[82-87]
3. To determine the types of organic matter involved in the formation of water-stable aggregates[88]

A typical fractionation procedure is shown in Fig. 3.3. The light fraction is separated using a liquid of high density (mentioned earlier), following which further separations are made by sedimentation. Common fractions are sand, course silt, fine silt, coarse clay, and fine clay, although more complete separations are often made. The various fractions have been subjected to chemical,[82-84] microbiological,[85] and microscopic examination.[86]

There is a large body of information supporting the quantitative importance and resistant nature of organic matter associated with the fine silt/coarse clay fraction. Organic matter of the fine clay fraction has been shown to have a faster turnover rate than that of the silt and coarse clay fraction during decades of cultivation.[86]

Results of incubation studies on mineralization rates of organic N, P, and S have shown that differences exist in the recalcitrance of organic matter in the various size fractions. Apparently, C and N of the silt fraction is more recalcitrant than that of the clay fractions.[87]

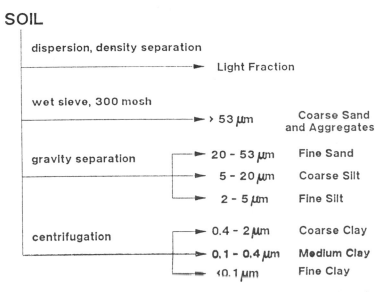

Fig. 3.3 Scheme for the physical fractionation of soil organic matter. From Stevenson and Elliott,[81] reproduced by permission of the University of Hawaii Press.

INSTRUMENTAL METHODS FOR DETERMINING ORGANIC MATTER QUALITY

Two techniques, nuclear magnetic resonance spectroscopy (NMR) and analytical pyrolysis, can be used for direct analysis of the forms of organic matter in soil and are thus valuable tools for the determination of organic matter quality.

Of the two approaches, ^{13}C–NMR has been the most frequently used and the most widely accepted. Pyrolysis techniques have been used only sparingly because of high instrument costs and the scarcity of equipment.

Nuclear Magnetic Resonance Spectroscopy (NMR)

Like other spectroscopic methods, NMR spectroscopy depends upon the interaction of electromagnetic radiation with nuclear, atomic, or molecular species. By this technique, estimates can be obtained for the major types of organic C in soils, namely, aliphatic C of alkanes and fatty acids, N–alkyl (proteinaceous substances) + methoxyl, carbohydrates, aromatic C, and –COOH. The method has been used for examination of mineral soils and their size and density fractions,[89–95] forest litter,[96–104] particulate organic matter,[105–107] composts,[108–109] and organic soils.[110] The technique has also been used to determine cultivation effects on soil organic matter.[111]

The literature on application of the method to soil organic matter and its fractions is extensive. Additional references can be found in comprehensive reviews on the subject.[99,112,113]

Brief Theory of NMR Spectroscopy   All nuclei carry a charge. In some nuclei (e.g., ^{13}C), this charge spins about the nuclear axis and thereby generates a magnetic dipole along the axis. Under the influence of an external magnetic field, H_o, the nuclei precess about the magnetic field in much the same manner as a gyroscope precesses under the influence of gravity.

In NMR spectroscopy, electromagnetic radiation is applied and an alternating field, H_1, is generated at right angles to the external magnetic field. The molecule absorbs energy and, at the point where the frequency matches the precession frequency, the nuclei flip over or resonate, thereby inducing a voltage change (resonance signal), which is amplified and recorded. The theory and practice of NMR spectroscopy are covered by Wilson[112] and Silverstein et al.[114]

The fundamental equation relating the frequency of electromagnetic radiation, ν, to the magnetic field strength, H_o, is given by:

$$\nu = \frac{\gamma H_o}{2\pi}$$

where γ, the gyromagnetic ratio, is constant for a particular nuclear type.

To facilitate comparisons, results of an NMR experiment are expressed in terms of "chemical shifts" with respect to a reference standard (usually tetramethylsilane, $Si(CH_3)_4$). The chemical shift, δ, is given by:

$$\delta = \frac{\nu_{sample} - \nu_{reference}}{\nu_{reference}} \times 10^6$$

Chemical shift, δ, is the principal parameter from which structural information is obtained. The basis for the analysis is that the frequency at which nuclei resonate (e.g., ^{13}C) is governed by the chemical environment of the nuclei in the experimental sample. To use one example, the chemical shift δ of ^{13}C in an aromatic ring is different from that of a –COOH group; thus, the two can be distinguished by ^{13}C–NMR spectroscopy.

The utility of solid-state ^{13}C–NMR spectroscopy has been greatly enhanced in recent years by the development of techniques which reduce line broadening and improve spectral quality. They include:

High-power proton decoupling: Reduces line broadening by eliminating or minimizing the magnetic influence of a neighboring proton nucleus (^1H).

Cross-polarization (CP): Ibid., but by transfer of net magnetization from the abundant ^1H spins to the less abundant ^{13}C spins.

Magic-angle spinning (MAS): Elimination of the remaining vestiges of dipolar $^{13}C-^1H$ interactions and chemical shift anisotropy effects by rapidly rotating the sample at the "magic-angle" of 54.7° with respect to the applied magnetic field.

The symbol "CPMAS" is used to designate application of cross-polarization and magic-angle spinning in the solid-state analysis of organics (i.e., *"CPMAS $^{13}C–NMR$ spectroscopy"*)

Assignment of Chemical Shift Zones   Chemical shifts δ of ^{13}C in structures of importance in the analysis of soil organic matter are given below and diagrammatically in Fig. 3.4.

Unsubstituted aliphatic C (e.g., alkanes, fatty acids)	0–50 ppm
N–Alkyl (e.g., amino acid-, peptide-, and protein C) + methoxyl C	50–60 ppm
Aliphatic C–O (notably carbohydrates)	60–110 ppm
Aromatic C, consisting of unsubstituted and alkylsubstituted aromatic C (110–150 ppm) and phenolic C (150–160 ppm).	110–160 ppm
Carboxyl C (includes the carboxylate ion, COO⁻)	160–190 ppm
Ketonic C=O of esters and amides	190–200 ppm

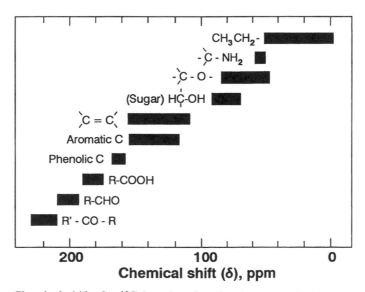

Fig. 3.4   Chemical shifts for ^{13}C in select functional groups relative to tetramethylsilane. From Stevenson,[113] reproduced by permission of John Wiley & Sons, Inc.

In some studies, the first two chemical shift zones (i.e., from 0–60 ppm) have been combined and labeled as alkyl C; in other cases, the 50–60 ppm zone has been assigned exclusively to methoxyl groups. The region corresponding to 60–110 ppm (C–O resonance) is normally attributed to carbohydrates, but other compounds may contribute to resonance in this region.

The aromatic region (110–160 ppm) consists of two main groups of resonances, one centered near 130 ppm and the other near 150 ppm. The latter is characteristic of C next to an oxygen atom and has been assigned to phenolic C. It should be noted that alkenes (aliphatic compounds containing the double bond) may also contribute to resonance in the 110–160 ppm region.

As applied to soils, results of ^{13}C–NMR spectroscopy must be interpreted with caution. While it is usual to divide the spectrum into regions based on results obtained for known compounds, these "chemical-shift" regions are not completely exclusive or specific. In the case of humic substances, complex chemical structures may be present that are no longer amenable to conventional interpretations. In this respect, it is instructive that results acquired by ^{13}C–NMR spectroscopy have not been in complete agreement with results obtained by chemical approaches, which may be due to failure of ^{13}C in the various groups to follow precisely the chemical shifts obtained for well-defined organic molecules. To some extent, these limitations are within accepable limits for comparative studies.

Limitations for Direct Analysis of Soils   There are several limitations in the application of CPMAS ^{13}C–NMR spectroscopy for the analysis of soils, including the following:

1. The magnitude of the signal and the quality of the resulting spectra are dependent upon the amount of ^{13}C in the soil. The low C content of most mineral soils, along with the low natural abundance of ^{13}C (1.1% of the total organic C), make the acquisition of high-quality spectra extremely difficult unless excessively long scan periods are used. Physical fractionations have been used to concentrate C by removal of the coarser particle size fractions, and chemical procedures have been used to selectively remove inorganic components.[89,115] The addition of ^{13}C-labeled plant residues to soil, followed by a period of time without further additions (to allow the ^{13}C to move into all of the soil organic matter fractions), has been used to circumvent the problem of low natural abundance of ^{13}C.[116]

2. The presence of inorganic paramagnetic species can reduce the efficiency of signal acquisition. The major paramagnetic component in mineral soils is Fe^{3+}, but other paramagnetic species may also be present. Reduction of Fe^{3+} to the nonparamagnetic Fe^{2+} has been used to improve signal resolution.[89]

3. A solid-state ^{13}C–NMR spectrum represents the sum of major organic structures in the sample, but it does not distinguish between humic

substances that vary in molecular weight or of structures that are associated with inorganic soil components. According to Wilson,[112] there is no reason to believe that *all* of the C in a soil is being observed by CPMAS ^{13}C–NMR.

A CPMAS ^{13}C–NMR spectrum for a Mapourika soil from New Zealand is shown in Fig. 3.5. As commonly seen with many other soils,[89–94] the most dominant peaks are the aliphatic C of alkanes (30 ppm) and the C—O of carbohydrates and related substances (73 ppm). Variations in the magnitude of the aromatic C signals (110 to 160 ppm) have been used to demonstrate differences in the aromaticity of the organic matter in soils. As a rule, from 15 to 20% of the organic C in mineral soils occurs as aromatic C. Zech et al.[94] found that cultivation led to an increase in the aromaticity of humus in some Vertisol soils of Mexico.

Results of ^{13}C–NMR studies indicate a higher proportion of O–alkyl structures (carbohydrate C) in soil organic matter than can be accounted for by conventional analyses for monosaccharides by acid hydrolysis,[91] which suggests that C—O structures other than those of carbohydrates contribute to resonance in the 60–110 region of the spectrum. It should also be noted that many spectra indicate a higher proportion of alkyl or lipid-like C (resonance peak at 30 ppm) than can be accounted for by extraction with an organic solvent.

The relative proportion of each type of C in the size fractions of some Australian soils is given in Fig. 3.6. A unique feature of the results is the relatively high content of aliphatic (alkyl) C in the fine clay fraction, particularly for the Urrbrae soil. The high content of aliphatic C in the fine clay

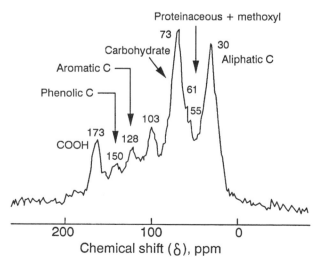

Fig. 3.5 CPMAS ^{13}C–spectrum of a Mapourika soil. Adapted from Wilson.[112]

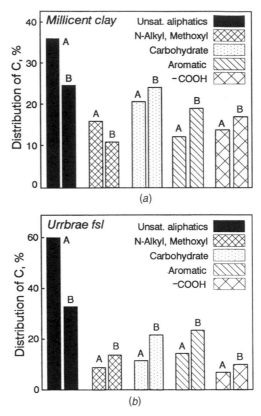

Fig. 3.6 Composition of organic C in the fine clay (A) and coarse clay (B) fraction of two Australian soils as determined by CPMAS ^{13}C–NMR. From Stevenson,[112] as adapted from Baldock et al.[90] Reproduced by permission of John Wiley & Sons, Inc.

was thought to be due to accumulation of recalcitrant plant waxes strongly associated with soil clays.[90]

The technique of CPMAS ^{13}C–NMR spectroscopy has been used to follow the humification process in composts[108,109] and in the organic layers of forest soils.[99] Among the changes noted for forest litter are the following:

1. Polysaccharides of plant litter are extensively decomposed and substituted by microbial polysaccharides.
2. Lignin is partly decomposed and the remnant molecule is transformed through side-chain oxidation and ring cleavage.
3. Further modifications of aromatic C lead to an increase of C-substituted aromatic C and a concomitant loss of phenolic structures in the humic acid fraction during humification.

4. The refractory alkyl–C moieties in humified forest soil organic matter do not result from selective preservation of plant-derived biomacro-molecules.

5. Aliphatic biopolymers (i.e., cutin and suberin) do not accumulate, but there may be an increase in cross-linking of lipid- and/or cutin-type and suberin-type materials.

Results of CPMAS ^{13}C–NMR studies have shown that forest soil organic matter contains from 20 to 30% aromatic C.[96]

ANALYTICAL PYROLYSIS

Pyrolysis is the controlled thermal decomposition of a sample. When used in tandem with gas–liquid chromatography (Py–GC), mass spectrometry (Py–MS), or a combination of the two (Py–GC–MS) to separate and identify partial degradation products arising from the sample, this method has emerged as a viable approach for characterization of soil organic matter without prior extraction from the soil. Advantages of Py–MS, in which the pyrolysis products are passed directly into the mass spectrometer, include better reproducibility, fast analysis time, and amenability to multivariate or other chemometric methods of pattern analysis. However, Py–Ms does not distinguish between compounds of similar molecular weights and fragmentation patterns and thus does not provide positive identification of all pyrolysis products. This can best be done by an initial separation of degradation products by gas chromatography, in which case the column effluent is allowed to flow directly into the mass spectrometer (i.e., Py–GC–MS). Identification of pyrolysis products is then made from the mass spectrometric data in combination with a knowledge of retention times for standard compounds on the gas chromatograph.

In analytical pyrolysis, a pulse of thermal energy is applied to the sample, thereby fracturing weaker linkages of macromolecules, with release of products characteristic of their structures. Two techniques have been used for pyrolysis: (1) quasi-instantaneous heating, or Curie-point pyrolysis, and (2) controlled temperature programming. A criticism of 2 is that secondary reactions during pyrolysis can lead to the formation of compounds unrelated to the material being pyrolyzed. Curie-point methods are highly reproducible, and they can be applied to small samples. Ratios obtained for pyrolysis products derived from polysaccharides, polypeptides, lignin, and so on can serve as a semiquantitative index of soil humus type and "degree of humification."

Essentially, a mass spectrometer bombards the substance under investigation (i.e., pyrolysis products) with an electron beam and records the results as a spectrum of positive ion fragments. Mass spectra are scanned continuously during passage of the pulse of products through the source and are "averaged" to obtain a representative spectrum (i.e., plot of relative abundances versus mass of the ion fragments divided by charge, m/z). A computer is generally used for instrument control and data collection.

It should also be noted that, unlike ^{13}C–NMR spectroscopy, pyrolysis does not provide quantitative data for structural units or components of humic substances, but only identification of the types of chemical moieties in the sample. Another limitation is that some of the products may be artifacts of the degradation process. The production of artifacts can be minimized by carrying out pyrolysis under vacuum or in a stream of inert gas.

The literature on the application of pyrolysis techniques to soil organic matter is extensive and has been covered in several reviews.[90,117,118] Pyrolysis methods have been used to distinguish between the organic matter in soils of different origins,[119–121] characterize organic matter in size and density fractions of the soil,[90] delineate the raw humus layers of forest soils,[122] follow the humification process in peat,[123–124] examine litter decomposition and humification in forest soils,[99,104,124] and to establish horizon differentiation in Histosols.[117]

Results of pyrolysis studies indicate that during humification, complex polysaccharides of crop residues (e.g., cellulose and xylans) are lost and lignin is transformed through loss of methoxyl groups and oxidation of C-3 side chains. As decomposition proceeds, a new set of products is produced through synthesis by microorganisms, including secondary or pseudo-polysaccharides. A portion of the aromatic rings of lignin remains behind and contributes to the substituted benzenes and phenols obtained on pyrolysis.

SUMMARY

Pools of organic matter in soil that can be measured, albeit with difficulty in some cases, include litter (labile and resistant components), microbial biomass, "light" fraction, and labile and resistant components of soil humus. A major goal in the study of soil organic matter is to relate information regarding the size and composition of the pools to soil quality and productivity. A need exists to identify "labile" components and the rates at which the nutrients contained therein are released to available mineral forms. Results of physical fractionations indicate that nutrient turnover depends not only on the kind (and amount) of organic matter but on its location within the soil. The analytical techniques of ^{13}C–NMR and analytical pyrolysis have the potential for determining the main classes of organic matter in the intact soil (and size fractions) and thus organic matter quality.

REFERENCES

1. R. R. Weil, *The New Farm*, Jan. 1992, pp. 43–48.
2. D. F. Acton and L. J. Gregorich, Eds., *The Health of Our Soils: Toward Sustainable Agriculture in Canada*, Centre for Land and Biological Resources Research, Ottawa, Canada, 1995.

3. J. D. Soule and J. K. Piper, *Farming in Nature's Image: An Ecological Approach to Agriculture*, Island Press, Washington, D.C., 1992.

4. M. A. Cole, "Soil Quality as a Component of Environmental Quality," in C. R. Cothern and N. P. Ross, Eds., *Environmental Statistics, Assessment, and Forecasting*, Lewis, Boca Raton, Florida, 1993, pp. 223–237.

5. M. R. Carter and B. A. Stewart, Eds., *Structure and Organic Matter Storage in Agricultural Soils*, Lewis, Boca Raton, Florida, 1994.

6. D. C. Coleman, J. Malcolm Oades, and G. Uehara, Eds., *Dynamics of Soil Organic Matter in Tropical Ecosystems*, University of Hawaii Press, Honolulu, 1989.

7. J. L. Hatfield and B. A. Stewart, Eds., *Residue Management*, Lewis, Boca Raton, Florida, 1994.

8. P. W. Unger, *Managing Agricultural Residues*, Lewis, Boca Raton, Florida, 1994.

9. J. M. Anderson and J. S. I. Ingram, Eds., *Tropical Soil Biology and Fertility Handbook of Methods*, Commonwealth Agricultural Bureau, Wallingford, England, 1989.

10. J. Hassink, F. J. Matus, C. Chenu, and J. W. Dalenberg, "Interactions Between Soil Biota, Soil Organic Matter, and Soil Structure," in L. Brussard and R. Ferrera-Cerrato, Eds., *Soil Ecology in Sustainable Agricultural Systems*, Lewis, Boca Raton, Florida, 1967, pp. 15–35.

11. J. W. Doran, D. C. Coleman, D. F. Bezdicek, and B. A. Stewart, *Defining Soil Quality for a Sustainable Environment*, SSSA Special Publication 35, Soil Science Society of America, Madison, Wisconsin, 1994.

12. O. L. Smith, *Soil Microbiology: A Model of Decomposition and Nutrient Cycling*, CRC Press, Boca Raton, Florida, 1992.

13. G. S. Innis, "Objectives and Structure for a Grassland Simulation Model," in G. S. Innis, Ed., *Grassland Simulation Model*, Springer-Verlag, New York, 1978, pp. 1–21.

14. F. H. Bormann and G. E. Likens, *Pattern and Process in a Forested Ecosystem*, Springer-Verlag, New York, 1970.

15. R. Lal, M. McMahon, R. Papendick, and G. Thomas, Eds., *Stubble over the Soil: The Vital Role of Plant Residues in Soil Management to Improve Soil Quality*, Soil Science Society of America, Madison, Wisconsin, 1996.

16. J. O. Skjemstad, R. C. Dalal, and P. F. Barron, *Soil Sci. Soc. Amer. J.*, **50,** 354 (1986).

17. R. C. Dalal and R. J. Mayer, *Aust. J. Soil Res.*, **24,** 293 (1986).

18. H. H. Janzen, *Can. J. Soil Sci.*, **67,** 845 (1987).

19. H. H. Janzen, C. A. Campbell, S. A. Brandt, G. P. Lafond, and L. Townley-Smith, *Soil Sci. Soc. Amer. J.*, **56,** 1799 (1992).

20. G. Spycher, P. Sollins, and S. Rose, *Soil Sci.*, **135,** 79 (1983).

21. D. L. Egerton, J. A. Harris, P. Birch, and P. Bullock, *Soil Biol. Biochem.*, **27,** 1499 (1995).

22. V. V. S. R. Gupta and J. J. Germida, *Soil Biol. Biochem.*, **20,** 777 (1988).

23. D. S. Jenkinson, *New Zealand Soil News*, **25,** 213 (1977).

24. T. H. Anderson and K. H. Domsch, *Soil Biol. Biochem.*, **21,** 471 (1989).

25. D. S. Jenkinson and J. N. Ladd, "Microbial Biomass in Soil, Measurement and Turnover," in E. A. Paul and J. N. Ladd, Eds., *Soil Biochemistry*, Vol. 5, Marcel Dekker, New York, 1981, pp. 415–472.

26. E. A. Paul and R. P. Voroney, "Field Interpretation of Microbial Biomass Activity Measurements," in M. J. Klug and C. A. Reddy, Eds., *Current Perspectives in Microbiology Ecology*, American Society for Microbiology, Washington, D.C., 1983, pp. 509–514.

27. H. Insam, "Microorganisms and Humus in Soils," in A. Piccolo, Ed., *Humic Substances in Terrestrial Ecosystems*, Elsevier, Amsterdam, 1996, pp. 265–292.

28. B. K. G. Theng, K. R. Tate, and P. Sollins, "Constituents of Organic Matter in Temperate and Tropical Soils," in D. C. Coleman, J. Malcolm Oades, and G. Uehara, Eds., *Dynamics of Soil Organic Matter in Tropical Ecosystems*, University of Hawaii Press, Honolulu, 1989, pp. 5–32.

29. W. G. Hale, "Collembola," in A. Burges and F. Raw, Eds., *Soil Biology*, Academic Press, London, 1967, pp. 397–411.

30. A. Tunlid and D. C. White, "Biochemical Analyses of Biomass, Community Structure, Nutritional Status, and Metabolic Activity of Microbial Communities in Soil," in G. Stotzky and J.-M. Bollag, Eds., *Soil Biochemistry*, Vol. 7, Marcel Dekker, New York, 1992, pp. 229–262.

31. W. R. Horwath and E. A. Paul, "Microbial Biomass," in R. W. Weaver, Ed., *Methods of Soil Analysis, Part 2: Microbiological and Biochemical Properties*, Soil Science Society of America, Madison, Wisconsin, 1994, pp. 753–773.

32. J. P. Martin and D. D. Focht, "Biological Properties of Soils," in L. F. Elliott and F. J. Stevenson, Eds., *Soils for Management of Organic Wastes and Waste Waters*, American Society of Agronomy, Madison, Wisconsin, pp. 114–169.

33. J. L. Smith and E. A. Paul, "The Significance of Soil Microbial Biomass Estimations," in J.-M. Bollag and G. Stotzky, Eds., *Soil Biochemistry*, Vol. 6, Marcel Dekker, New York, 1990, pp. 357–396.

34. W. N. Miller and L. E. Casida, *Can. J. Microbiol.*, **16**, 299 (1970).

35. J. Skujins and L. E. Casida, *Can. J. Microbiol.*, **16**, 299 (1970).

36. C. C. Lee, R. F. Harris, J. D. H. Williams, D. E. Armstrong, and J. K. Syers, *Soil Sci. Soc. Amer. Proc.*, **35**, 82 (1971).

37. E. A. Paul and R. L. Johnson, *Appl. Environ. Microbiol.*, **34**, 263 (1977)

38. S. Kiss, M. Dragan-Bularda, and D. Radulescu, *Adv. Agron.*, **27**, 25 (1975)

39. V. F. Kuprevich and T. A. Shcherbakova, "Comparative Enzymatic Activity in Diverse Types of Soil," in A. D. McLaren and J. Skujins, Eds., *Soil Biochemistry*, Vol. 2, Marcel Dekker, New York, 1971, pp. 167–201.

40. M. A. Tabatabai, "Soil Enzymes," in R. W. Weaver, Ed., *Methods of Soil Analysis, Part 2: Microbiological and Biochemical Properties*, Soil Science Society of America, Madison, Wisconsin, 1994, pp. 775–833.

41. J. P. E. Anderson and K. H. Domsch, *Soil Biol. Biochem.*, **10**, 207, 215 (1978).

42. J. P. E. Anderson and K. H. Domsch, *Soil Sci.*, **130**, 211 (1980).

43. D. S. Jenkinson and J. M. Oades, *Soil Biol. Biochem.*, **11**, 193, 201 (1979).

44. D. S. Jenkinson and D. S. Powlson, *Soil Biol. Biochem.*, **8**, 167, 179, 209 (1976).

45. P. Nannipiere, R. L. Johnson, and E. A. Paul, *Soil Biol. Biochem.,* **10,** 223 (1978).

46. D. Parkinson and E. A. Paul, "Microbial Biomass," in A. L. Page, R. H. Miller, and D. R. Keeney., Eds., *Methods of Soil Analysis, Part 2. Chemical and Biological Properties,* 2nd ed., American Society of Agronomy, Madison, Wisconsin, 1982, pp. 821–830.

47. R. Martens, *Soil Biol. Biochem.,* **17,** 57 (1985).

48. J. A. Ocio and P. C. Brookes, *Soil Biol. Biochem.,* **22,** 685 (1990).

49. G. P. Sparling and A. W. West, *Soil Biol. Biochem.,* **20,** 337 (1988).

50. K. R. Tate, D. J. Ross, and C. W. Feltham. *Soil Biol. Biochem.,* **20,** 319 (1988).

51. J. M. Duxbury, M. S. Smith, and J. W. Doran, "Soil Organic Matter as a Source and Sink of Plant Nutrients," in D. C. Coleman, J. Malcolm Oades, and G. Uehara, Eds., *Dynamics of Soil Organic Matter in Tropical Ecosystems,* University of Hawaii Press, Honolulu, 1989, pp. 33–67.

52. A. Kretzschmar, Ed., *Fourth International Symposium on Earthworm Ecology. Soil Biology and Biochemistry,* **24**(12), 1992.

53. C. A. Edwards and J. R. Lofty, *Biology of Earthworms,* Chapman & Hall, London, 1977.

54. M. A. Cole, unpublished observations.

55. N. R. French, R. K. Steinhorst, and D. M. Swift, "Grassland Biomass Trophic Pyramids," in N. R. French, Ed., *Perspectives in Grassland Ecology,* Springer-Verlag, New York, 1979, pp. 59–87.

56. J. C. Moore and D. E. Walter, *Ann. Rev. Entomol.,* **33,** 419 (1988).

57. R. E. Ingham, "Nematodes," in R. W. Weaver, Ed., *Methods of Soil Analysis, Part 2: Microbiological and Biochemical Properties,* Soil Science Society of America, Madison, Wisconsin, 1994, pp. 459–490.

58. A. R. Moldenke, "Arthropods," in R. W. Weaver, Ed., *Methods of Soil Analysis, Part 2: Microbiological and Biochemical Properties,* Soil Science Society of America, Madison, Wisconsin, 1994, pp. 517–542.

59. D. G. Milchunas, W. K. Lauenroth, J. S. Singh, C. V. Cole, and H. W. Hunt, *Plant Soil,* **88,** 353 (1985).

60. R. H. Whittaker and P. L. Marks, "Methods of Assessing Terrestrial Productivity," in H. Lieth and R. H. Whittaker, Eds., *Primary Productivity of the Biosphere,* Springer-Verlag, New York, 1975, pp. 55–118.

61. J. M. Whipps and J. M. Lynch, *New Phytologist,* **95,** 605 (1983).

62. T. Haller and H. Stolp, *Plant Soil,* **86,** 207 (1985).

63. A. Zsolnay, "Dissolved Humus in Soil Waters," in A. Piccolo, Ed., *Humic Substances in Terrestrial Ecosystems,* Elsevier, Amsterdam, 1996, pp. 171–223.

64. E. A. Elkhatib, O. L. Bennett, V. C. Baligar, and R. J. Wright, *Soil Sci. Soc. Amer. J.,* **50,** 297 (1986).

65. J. Wolt and J. G. Graveel, *Soil Sci. Soc. Amer. J.,* **50,** 602 (1986).

66. R. G. Burns, "Enzyme Activity in Soil: Some Theoretical and Practical Considerations," in R. G. Burns, Ed., *Soil Enzymes,* Academic Press, New York, 1978, pp. 295–340.

67. A. D. McLaren, *Chem. Sci.,* **8,** 97 (1975).

68. D. J. Ross, *Soil Biol. Biochem.,* **3,** 97 (1971).

69. J. Skujins, *Bull. Ecol. Res. Comm. NFR,* **17,** 235 (1973).

70. D. S. Jenkinson and J. H. Rayner, *Soil Sci.,* **122,** 298 (1977).

71. J. A. E. Molina et al., *Soil Sci. Soc. Amer J.,* **47,** 85 (1983)

72. J. A. van Veen, J. N. Ladd, and M. Frissel, *Plant Soil,* **76,** 257 (1984)

73. E. A. Paul and N. G. Juma, "Mineralization and Immobilization of Soil Nitrogen by Microorganisms," in F. E. Clark and T. Rosswall, Eds., *Terrestrial Nitrogen Cycles, Ecological Bulletin (Stockholm),* **33,** 179 (1981).

74. W. J. Parton, J. W. B. Stewart, and C. V. Cole, *Biogeochem.,* **5,** 109 (1988).

75. W. J. Parton et. al., "Modelling Soil Organic Dynamics in Tropical Soils," in R. W. Weaver, Ed., *Methods of Soil Analysis, Part 2: Microbiological and Biochemical Properties,* Soil Science Society of America, Madison, Wisconsin, 1989, pp. 153–171.

76. W. J. Parton, D. S. Schimel, C. V. Cole, and D. S. Ojima, *Soil Sci. Soc. Amer. J.,* **51,** 1173 (1973).

77. R. F. Grant, N. G. Juma, and W. B. McGill, *Soil Biol. Biochem.,* **25,** 1331 (1993).

78. J. T. Gilmour, M. D. Clark, and G. C. Sigua, *Soil Sci. Soc. Amer. J.,* **49,** 1398 (1985).

79. G. L. Swartzman and S. P. Kaluzny, *Ecological Simulation Primer,* Macmillan, New York, 1987.

80. B. T. Christensen, *Adv. Soil Sci.,* **20,** 1 (1992).

81. F. J. Stevenson and E. T. Elliott, "Methodologies for Assessing the Quantity and Quality of Soil Organic Matter," in D. C. Coleman, J. Malcolm Oades, and G. Uehara, Eds., *Dynamics of Soil Organic Matter in Tropical Ecosystems,* University of Hawaii Press, Honolulu, 1989, pp. 173–199.

82. D. W. Anderson, S. Saggar, J. R. Bettany, and J. W. B. Stewart, *Soil Sci. Soc. Amer. J.,* **45,** 767 (1981).

83. B. T. Christensen and L. H. Sørensen, *J. Soil Sci.,* **36,** 219 (1985).

84. J. M. Oades, A. M. Vassallo, A. G. Waters, and M. A. Wilson, *Aust. J. Soil Res.,* **25,** 71 (1987).

85. M. Ahmed and J. M. Oades, *Soil Biol. Biochem.,* **16,** 465 (1984).

86. H. Tiessen and J. W. B. Stewart, *Soil Sci. Soc. Amer. J.,* **47,** 509 (1983).

87. B. T. Christensen, *Soil Biol. Biochem.,* **19,** 429 (1987).

88. L. W. Turchenek and J. M. Oades, *Geoderma,* **21,** 311 (1979).

89. M. A. Arshad, J. A. Ripmeester, and M. Schnitzer, *Can. J. Soil Sci.* **68,** 593 (1988).

90. J. A. Baldock, G. J. Currie, and J. M. Oades, "Organic Matter as Seen by Solid State ^{13}C NMR and Pyrolysis Tandem Mass Spectrometry," in M. S. Wilson, Ed., *Advances in Soil Organic Matter Research: The Impact on Agriculture and the Environment,* Redwood Press, Wiltshire, England, 1991, pp. 45–60.

91. J. M. Oades, A. M. Vassallo, A. G. Waters, and M. A. Wilson, *Aust. J. Soil Res.,* **25,** 71 (1987).

92. J. O. Skjemstad, R. C. Dalal, and P. F. Barron, *Soil Sci. Soc. Amer. J.,* **50,** 354 (1986).

93. M. A. Wilson, P. F. Barron, and K. M. Goh, *J. Soil Sci.,* **32,** 419 (1981).

94. W. Zech, L. Haumaier, and R. Hempfling, "Ecological Aspects of Soil Organic Matter in Tropical Land Use," in P. MacCarthy, C. E. Clapp, R. L. Malcolm, and P. R. Bloom, Eds., *Humic Substances in Soil and Crop Sciences: Selected Readings,* American Society of Agronomy, Madison, Wisconsin, 1990, pp. 187–202.

95. J. O. Skjemstad, R. C. Dalal, and P. F. Baron, *Soil Sci. Soc. Amer. J.* **50,** 354 (1986).

96. I. Kögel, R. Hempfling, W. Zech, P. G. Hatcher, and H.-R. Schulten, *Soil Sci.,* **146,** 124 (1988).

97. G. Ogner, *Geoderma,* **35,** 343 (1985).

98. M. A. Wilson, S. Heng, K. M. Goh, R. J. Pugmire, and D. M. Grant, *J. Soil Sci.,* **34,** 83 (1983).

99. I. Kögel-Knabner, W. Zech, P. G. Hatcher, and J. W. de Leeuw, "Fate of Plant Components During Biodegradation and Humification in Forest Soils: Evidence from Structural Characterization of Individual Biomacromolecules," in M. S. Wilson, Ed., *Advances in Soil Organic Matter Research: The Impact on Agriculture and the Environment,* Redwood Press, Wiltshire, England, 1991, pp. 61–70.

100. I. Kögel-Knabner, *Soil Biol. Biochem.,* **8,** 101 (1993).

101. C. M. Preston et al., *Plant Soil,* **158,** 69 (1994).

102. I. Kögel-Knabner, W. Zech, and P. G. Hatcher, *Z. Pflanzenern. Bodenk.,* **151,** 331 (1988).

103. I. Kögel-Knabner, J. W. de Leewu, and P. G. Hatcher, *Sci. Total Environ.,* **117/118,** 175 (1992).

104. L. Beyer, H.-R. Schulten, R. Fruend, and U. Irmler, *Soil Biol Biochem.,* **25,** 589 (1993).

105. C. A. Cambardella and E. T. Elliott, *Soil Sci. Soc. Amer. J.,* **56,** 777 (1992).

106. A. Golchin, J. M. Oades, J. O. Skjemstad, and P. Clarke, *Aust. J. Soil Res.,* **32,** 285 (1994).

107. M. M. Wander, S. J. Traina, B. R. Stinner, and S. E. Peters, *Soil Sci. Soc. Amer. J.* **58,** 1130 (1994).

108. Y. Inbar, Y. Chen, and Y. Hadar, *Soil Sci. Soc. Amer. J.,* **53,** 1695 (1989).

109. Y. Inbar, Y. Chen, and Y. Hadar, *Soil Sci.,* **152,** 272 (1991).

110. M. Krosshavn, M. Wartel, and L. Gengembre, *J. Soil Sci.,* **43,** 485 (1992)

111. H.-R. Schulten et al., *Z. Pflanzenern. Bodenk.,* **153,** 97 (1990).

112. M. A. Wilson, "Application of Nuclear Magnetic Resonance Spectroscopy to Organic Matter in Whole Soils," in P. MacCarthy, C. E. Clapp, R. L. Malcolm, and P. R. Bloom, Eds., *Humic Substances in Soil and Crop Sciences: Selected Readings,* American Society of Agronomy, Madison, Wisconsin, 1990, pp. 221–260.

113. F. J. Stevenson, *Humus Chemistry: Genesis, Composition, Reactions,* 2nd ed., Wiley, New York, 1994.

114. R. M. Silverstein, G. C., Bassler, and T. C. Morrill, *Spectrometric Identification of Organic Compounds,* Wiley, New York, 1981.

115. C. M. Preston, M. Schnitzer, and J. A. Ripmeester, *Soil Sci. Soc. Amer. J.,* **53,** 1442 (1989).

116. G. I. Ågren, E. Bosatta, and J. Balesdent, *Soil Sci. Soc. Amer. J.,* **60,** 1121 (1996).

117. J. M. Bracewell, K. Haider, S. R. Larter, and H.-R. Schulten, "Thermal Degradation Relevant to Structural Studies of Humic Substances," in H. M. B. Hayes, P. MacCarthy, R. L. Malcolm, and R. S. Swift, Eds., *Humic Substances II: In Search of Structure,* Wiley, New York, 1989, pp. 181–222.

118. H.-R. Schulten, *J. Anal. Appl. Pyrol.,* **12,** 149 (1987).

119. J. M. Bracewell and G. W. Robertson, *J. Soil Sci.,* **35,** 549 (1984).

120. J. M. Bracewell and G. W. Robertson, *Geoderma,* **40,** 333 (1987).

121. G. Halma, M. A. Posthumus, R. Miedema, R. van de Westeringh, and H. L. C. Meuzelaar, *Agrochimica,* **22,** 372 (1978).

122. J. M. Bracewell and G. W. Robertson, *J. Soil Sci.,* **38,** 191 (1987).

123. J. M. Bracewell, G. W. Robertson, and B. R. Williams, *J. Anal. Appl. Pyrol.,* **2,** 53 (1980).

124. R. J. Hempfling, F. Ziegler, W. Zech, and H.-R. Schulten, *Z. Pflanzenernähr. Bodenk.,* **150,** 179 (1987).

4

ENVIRONMENTAL ASPECTS OF THE SOIL CARBON CYCLE

Soil has often been used as the receptacle for various organic waste products, some of which can pose a threat to human and animal health and environmental quality. Prior to enactment of laws and creation of regulations beginning in the 1960s in the United States and other countries, disposal of many organic wastes was unregulated and haphazard. The financial heritage of these activities has been estimated at $400 billion to clean up sites in Europe, exclusive of the Soviet Union,[1] for example. Interest in organic substances in the soil environment is centered on five main concerns:

1. Use of agricultural land for managed disposal of organic wastes and waste products
2. Fate of synthetic organic chemicals applied to soil to control pests and weeds
3. Accidental or willful contamination of soil with toxic or hazardous industrial chemicals
4. Contribution of soil organic matter degradation as a cause of increased atmospheric CO_2
5. Possible synthesis of carcinogenic compounds by soil microorganisms

These topics are the subject of a voluminous literature as well as extensive efforts to identify appropriate disposal methods[2] and correct problems that were created by inappropriate activities in the past.[3] Emphasis will be given to the problem of land disposal of organic wastes, with lesser attention being given to the fate of synthetic organic chemicals, organic matter decay as a contributor to atmospheric CO_2, and synthesis of carcinogens.

DISPOSAL OF ORGANIC WASTE PRODUCTS IN SOIL

Coping with the vast quantities of organic wastes that are produced each year from domestic and municipal sewage, animal and poultry production, food-processing industries (canneries, etc.), pulp and paper mills, municipal solid waste, and miscellaneous industrial organics is a major problem of global proportions.[4-8] A common practice in the past in developed countries, and at present in many developing countries, has been to discharge these unwanted wastes into streams, lakes, and the open sea, but this method is no longer acceptable because of environmental deterioration of aquatic systems and proliferation of disease-causing microorganisms and parasites from untreated wastes.

Problems of disposing of municipal sewage and solid waste—publicized in the popular press—are not restricted to the urban community. A similar situation exists regarding disposal of farm wastes, such as those generated from feedlot operations. For example, farm animals in the United States produce 10 times more organic wastes than people, the amount being equivalent to what would be produced by a population of 2.6 billion people. A cow will generate over 16 times more wastes than a human, while annual waste production from seven chickens equals that of one person.

The total amount of organic wastes produced in the United States from livestock and poultry production is over 1.5 billion tons per year, about one-half of which is generated in concentrated production zones in a few states. Solid wastes generated by major farm animals and by treatment plants for human wastes (sewage) are recorded in Table 4.1.

As a result of increased consumption of processed foods, the food industry also produces enormous quantities of organic wastes. The waste from the cottage cheese whey industry alone is equivalent to the domestic waste from 83 million people.

Many organic wastes are potentially valuable as sources of major plant nutrients (N, P, and K), and they can contribute positively to soil quality in

TABLE 4.1   Estimated Solid Waste Generated by Major Farm Animals in the United States[a]

Animal	Number of Animals	Manure Produced
	millions	millions of tons/year
Cattle	109	1,087
Hogs	47	379
Sheep	21	64
Poultry		
Broilers	2,568	12
Turkeys	116	3
Layers	340	16

[a]From McCalla et al.[9] For comparative purposes, amount of sewage sludge produced from human wastes is on the order of 7.3×10^6 metric tons/year.[4]

other ways, such as improvement in soil structure. Their utilization as soil amendments offers a number of advantages to society in terms of modest disposal costs, enhanced soil fertility and crop productivity, and improved water quality as a result of decreased disposal into surface waters.

Specific Problems

The application of organic wastes to soil, such as crop residues, composts, and manures, is an historical and accepted practice and one that is being used extensively today for maintaining soil fertility. Manures, when properly utilized, have a favorable effect on soil productivity because they are good sources of nutrients for plant growth, including N, P, S, and certain trace elements. In the case of soils with poor physical structure or water-holding capacity, organic additions have favorable effects; other benefits include maintenance of soil organic matter and enhancement in biological life.

The monetary value attributed to organic wastes in earlier days can be illustrated by the following statement from the *1938 Yearbook of Agriculture* on soils.[10]

> One billion tons of manure, the annual product of livestock on American farms, is capable of producing $3,000,000,000 worth of increase in crops. The potential value of this agricultural resource is three times that of the Nation's wheat crop and equivalent to $440 for each of the country's 6,800,000 farm operators. The crop nutrients it contains would cost more than six times as much as was expended for commercial fertilizers in 1936. Its organic matter content is double the amount of soil humus annually destroyed in growing the nation's grain and cotton crops.

Because of technological and economic considerations, farmers now are less interested in the monetary value and soil-building value of manures and other wastes, with the result that they are now regarded as a liability rather than an asset. Changes in perspective regarding the value of farmyard manure have occurred because:

1. More widespread use of PO_4^{3-}-, NH_3-, and NO_3^--containing fertilizers that are produced at modest costs in developed countries. Concurrently, costs associated with handling animal wastes have increased so that they are no longer competitive in price with chemical fertilizers as sources of plant nutrients.
2. Farming has become highly specialized and livestock and poultry production has become concentrated in large-scale confinement-type enterprises, which means that large volumes of wastes are generated and must be disposed of in a relatively small area and in geographically discrete regions. In the past, many farms were mixed grain–livestock operations in which the animal wastes were utilized as fertilizer, thereby

creating an internal (on-farm) cycle for return of the nutrients removed by cropping. Now many farmers grow either a cash crop or specialize in livestock production, with the result that the potential for on-farm use of manures has declined.

Limitations to the use of farmland for disposal of animal manures and sewage sludges can be summarized as follows:[4–7]

1. Animal manures and sewage sludges, as sources of plant nutrients, are bulky, low-grade fertilizers of variable composition and frequently have a high water content. Accordingly, they cannot be transported very far before transportation costs exceed the fertilizer value. Many animal confinement facilities and many large cities lack agricultural land within economically practical transportation distances from the production site.

2. Concentrations of nutrients, soluble salts, trace elements, and water vary tremendously from location to location and are seldom analyzed prior to land application. Optimum rates of application are thus difficult to predict.

3. Both animal manures and sewage sludges contain soluble salts, which can cause problems in their use as fertilizers, particularly in irrigated soils of arid regions where soluble salts are already present in irrigation waters. In many soils, leaching of nutrients, especially NO_3^-, to ground water may limit the application rate.

4. Sewage sludges may contain toxic metals and/or synthetic organic chemicals that are retained in soils and may accumulate to levels that are toxic to some plants and thus restrict the type of crop that can be grown. As a result of uptake by plants, at concentrations not necessarily deleterious to the plant, the chemicals may be harmful to animals and humans, and the crop may be unsafe for consumption.

5. Sewage sludges and manures of poultry and nonruminant animals (e.g., hogs) contain pathogenic bacteria, viruses, and parasites that may be a public health risk to farm workers and the public via the food chain. The degree of risk depends on the method for processing the sludge and manures. Experiences with sludge application in a number of locations suggest that the risk is low.

6. On-farm management problems arise due to the physical properties of many organic wastes. Application techniques are inefficient and time-consuming. When liquid sludges are applied to the soil surface, a drying interval is required, and this loss of time can result in delays in seedbed preparation, etc. Production of animal manures and sewage sludges is continuous, while the need for fertilizers is seasonal.

7. Odors and associated nuisances, both real and imagined, create conflicts between urban residents and farmers who could advantageously use the wastes. Once resistance and fear have been generated in a community,

the people there have difficulty in accepting the idea that risks can be low in a properly managed system of utilizing the sludges or manures on land.

8. An obstacle to the use of sewage sludges for crop production is the requirement by environmental protection agencies that both crop and water quality be monitored. This monitoring can be expensive, and farmers who would otherwise be willing to use the material cannot or are unwilling to pay the additional monitoring cost.

The following were suggested by CAST[4] as ways of promoting the beneficial uses of animal manures and sewage sludges on farmland.

1. An increase in quality of organic wastes (e.g., animal manures and sewage sludges) and adequate quality control would make them more competitive with chemical fertilizers. Conservation of the N that usually volatilizes as NH_3 would greatly increase the value of many products. Decreases in the salt and trace element contents of feed rations and improvements in systems of collecting and treating waste waters or control of trace elements at the source (before industrial waste waters are put into sewage systems) would also be helpful. Improvements in the physical condition of many products would allow better control of placement and application rates.

2. Development of new management systems that would not delay other farm operations would create a better image for these materials in the minds of farmers and facilitate use of the materials.

3. Appropriate guidelines for applying sewage sludges and animal manures to croplands would promote or enhance their beneficial use. The guidelines should be based on facts and acceptable risks rather than on fears or unsubstantiated claims of environmental damage. This recommendation was the subject of an exhaustive study, with the result that a comprehensive set of guidelines for use of sewage sludge exists in the Unitetd States.[11]

There are many unanswered questions regarding the application of animal wastes to farmland at high rates, such as effects on crop quality, safety, and yield, on pollution of surface and ground waters, and on long-term productivity of the soil.

Contamination of Lakes and Streams with Organics

An environmental problem common to most organic wastes is that when they are introduced into streams or waterways, either directly through surface run-off or dumping or indirectly through leaching, water quality is adversely affected. The consequences include malodorous conditions, nutrient enrichment (which promotes growth of undesirable aquatic plants and microorganisms), and introduction of pathogens and parasites.

The pollution potential of organic wastes is usually characterized in terms of BOD or COD values.[12]

> *BOD* (*Biological Oxygen Demand*): The amount of O_2 consumed by microbial oxidation during a five-day incubation period. This serves as a measure of readily degraded organic compounds and reduced forms of N and S (NH_3 and HS^-).
>
> *COD* (*Chemical Oxygen Demand*): A measure of total oxidizable organic and inorganic materials is estimated by chemical oxidation with sulfuric acid–potassium dichromate. COD is used less frequently than BOD.

If the waste has a high BOD and is introduced into water in large quantities, the O_2 content of the water may be reduced to such a low value that fish and other aquatic organisms will die (most fish require about 5 mg/L of dissolved O_2 for survival).

A BOD value for water of 1 mg/L (i.e., 1 mg/L of O_2 consumed in a five-day incubation period) is regarded as high-quality water, whereas a BOD value of 5 indicates water of doubtful purity. Runoff entering a stream is considered objectionable if the BOD exceeds 20 mg/L.

Animal wastes have relatively high BOD values; values for digested sewage sludge are relatively low because the organic fraction has been stabilized during waste water treatment. In contrast, wastes and runoff from barnyards and feedlots can have BOD values as high as 10,000, depending on dilution and degree of decomposition. These materials have a devastating effect on aquatic biota and water quality. Similarly high BOD values occur in the effluent from food processing industries.

Build-up of Toxic Metals

Long-time application of organic wastes, particularly sewage sludges, leads to the accumulation of various metals in the soil, some of which are toxic to animals and man.[5,7,13–16] Manures from animals where trace elements (e.g., Cu) have been used as feed additives will also be enriched with metals[17] and thereby create similar problems.

Repeated applications of sewage sludges (and other wastes) to soil might build up the concentration of heavy metals to levels toxic to crops or to man and animals that consume the crops. The elements of most concern are As, Cd, Cr, Cr, Hg, Mo, Ni, Pb, Se, and Zn. Cadmium and mercury are especially toxic to man and animals, and their entry into the food chain and the environment must be kept within acceptable limits. Many of the metals in sludge are bound to the organic matter and may be released to available forms as the sludge decays over a period of years. In many cases, bioavailability does

**TABLE 4.2   Range in Amounts of Miscellaneous
Trace Elements in Sewage Sludges and Municipal
Solid Waste Composts**[a]

Constituent	Sludge	Compost
	mg/kg	mg/kg
As	6–230	1–4.8
Cd	4–846	1–13.2
Cr	17–99,000	8.2–130
Hg	1–10,600	0.5–3.7
Ni	10–3,515	7–101
Pb	13–19,730	22–913

[a]From McCalla et al.[9] and Epstein et al.[19]

not increase with time after application, and the risk to food production remains low.[18]

The metals found in sewage sludges, municipal wastes, and composts derived from these materials vary from city to city, depending on the industrial contribution and source-separation of metal-rich products like batteries. Sewage sludge from Chicago, for example, generally contains rather high amounts of Cr, Zn, Cu, Pb, Ni, and Cd. Typical values for the amounts of As, Ba, Ni, Cr, Cd, Pb, and Mg in sewage sludges and municipal solid waste composts[9,19] are given in Table 4.2. Ranges of five trace elements in sludges from industrial and nonindustrial cities are shown in Table 4.3 to illustrate the large impact of industrial activities on the quality of municipal sludges.

Substantial differences exist in opinion and governmental policy between the United States and some European countries regarding acceptable levels of toxic metals in land-applied organic wastes.[22,23] The United States policy, exemplied by USEPA 503 regulations, is that addition of metals is acceptable if there is little risk of food-chain transfer to other organisms; the approach is risk-based with respect to humans.[24] With this model, on-site and off-site ecological effects on biota other than food-producing organisms are not con-

**TABLE 4.3   Concentration Ranges of Four Metals in Sewage Sludges
as Influenced by Source**[a]

Toxic Metal	Sludges from Industrial and Nonindustrial Cities	Sludges from Nonindustrial Cities
	mg/kg	mg/kg
Cd	5–2,000	5–10
Cu	250–17,000	250–1,000
Ni	25–8,000	25–200
Zn	500–50,000	500–2,000

[a]From CAST[4]

sidered, and an increase in the metal content of the soil above preapplication levels is acceptable, within limits. In contrast, the policy in the Netherlands is that organic wastes cannot be applied to agricultural land if the application leads to an increase in toxic metals above naturally occurring levels. The logic of this policy is conservative and protective of the food supply and environment. The policy is highly restrictive in that many wastes contain elevated levels of toxic metals and thus cannot be land-applied.

The data below give an example of the implications of the two policies. An annual application of 20 metric tons/ha of a typical digested sewage sludge (an amount that supplies sufficient N for a high-yielding crop like corn) for 20 years would increase the trace element content of the soil (to plow depth) to the amounts shown below. By United States standards, these increases may be acceptable; by standards of the Netherlands, the increases in Cu, Zn, and Cr would be unacceptable because levels in the soil would be higher than observed naturally.

Metal	Soil Content (mg/kg)		Metal	Soil Content (mg/kg)	
	Before[a]	After		Before	After
Co	<3–30	18	Mn	70–1500	90
Pb	<10–70	270	Zn	25–170	890
Cu	3–100	180	Cr	10–150	540

[a]Range of 95% of surficial U.S. samples.[20]

Vegetable crops are the least tolerant to excesses of toxic metals; grasses are the most tolerant. Most grain crops are relatively tolerant, but there is a rather broad range between plant species and varieties. Long-term use of soil for disposal of sewage sludge may place a limit on the type of crop that can be grown.

Water-soluble organic constituents in sludge, or formed during decomposition, may move metals through the soil and into ground waters as soluble metal–chelate complexes (see Chapter 11).

Biological Pollutants

Biological pollutants include many pathogens and parasites of man and animals. When stored in a lagoon or biologically oxidized, most pathogens die rapidly. Thus, little public health hazard will normally result from application of digested sewage or lagooned animal wastes to soil.[25] The bacterium *Escherichia coli* is a normal constituent of the intestinal flora of vertebrates, but it does not occur in large numbers in uncontaminated soil or water. For this reason, *E. coli* is a common indicator organism for fecal contamination of soil and water. Most *E. coli* biotypes, fecal coliform, and fecal streptococci bacteria are not particularly pathogenic, but, as noted above, they signify that contamination has occurred and that infectious organisms may be present also.

TABLE 4.4   Average Contents of Fecal Coliforms and Fecal Streptococci in Fecal Wastes of Several Farm Animals[a]

Animal	Fecal Coliform	Fecal Streptococci
	Millions/g	Millions/g
Cow	0.23	1.3
Pig	3.3	84.0
Sheep	16.0	38.0
Poultry	1.3	3.4
Turkey	0.3	2.8

[a] From Wadleigh[8]

Average values for fecal coliform and fecal streptococci populations in wastes of various farm animals are given in Table 4.4; similar populations are found in untreated human fecal wastes. Populations of *E. coli* in treated sewage sludge must be less than 1000 organisms/gram for the material to be considered safe for land application.[27]

Infectious agents of animals that may pollute streams include microorganisms and parasites that can cause anthrax, brucellosis, encephalitis, foot-rot, histoplasmosis, mastitis, Newcastle disease, and salmonellosis, among others.[5] All of these diseases are serious or fatal, and the presence of these organisms in water constitutes a serious threat to human and animal health.

Enrichment of Surface and Ground Water with Nutrients

The practice of spreading large quantities of manures and other wastes, such as sewage sludge, on soil may add to the nutrient enrichment of natural waters through leaching and runoff.

Data giving the nutrient content of two typical types of organic wastes (feedlot manure and municipal sewage sludge) are shown in Tables 4.5 and 4.6, respectively. From the feedlot data, it can be calculated that annual manure application of 40 metric tons per hectare would add about 540 kg N/ha. A corn crop (grain yield of 4600 kg/ha) would remove about 180 kg of N per hectare, or about one-third of the applied amount. The remaining N has the potential to be a water pollutant. In this case, an application rate that supplied N for the crop with little excess would be about 8 metric tons/ha, which means that the amount of land required for disposal will be nearly five times greater than required for the 40 metric tons/hectare application rate. This example illustrates very well the incongruity between environmental protection and the practicalities of using land for disposal of organic wastes.

TABLE 4.5   Nutrient Content of Feedlot Manure[a]

Nutrient	Range %	Average %	Amount in 10 Metric Tons kg
N	1.16–1.96	1.34	134
P	0.320–.85	0.53	53
K	0.75–2.35	1.50	150
Na	0.29–1.43	0.74	74
Ca	0.81–1.75	1.30	130
Mg	0.32–0.66	0.50	50
Fe	0.09–0.55	0.21	21
Zn	0.005–0.012	0.009	0.9

[a]From McCalla et al.[9] and Mathers et al.[21] Moisture contents ranged from 20.9 to 54.5%. Data are from 23 feedlots.

TABLE 4.6   Nutrient Composition of Fresh, Heated, Anaerobically Digested Sewage Sludge[a]

	Concentration Range, %	Typical Sludge (dry-weight basis)[b]	
		%	kg/Mt
Organic N	2–5	3	27
NH_4^+-N	1–3	2	18
Total N	1–6	5	45
P	6–8	3	27
K	0.1–0.7	0.4	4
Ca	1–8	3	27
Mg	2–5	1	9
S	0.3–1.5	0.9	8
Fe	0.1–5	4	36
	mg/kg	mg/kg	
Na	800– 4,000	2,000	2
Zn	50–50,000	5,000	5
Cu	200–17,000	1,000	1
Mn	100– 800	500	0.5
B	15– 1,000	100	0.1

[a]Adapted from Thorne et al.[26]
[b]MT = metric ton. For conversion of kg/MT to lbs/MT, multiply by 2.205.

A 2-inch application of anaerobically digested sewage sludge (< 10% solids) will supply approximately the following amounts of major plant nutrients:

NH_4^+-N	252–280 kg/ha, subject to rapid nitrification
Org. N	336 kg/ha, slowly released during degradation
Total P	200–336 kg/ha, 80% in the sludge organic matter
Total K	45–90 kg/ha

Values vary according to source, treatment, and other factors, but it is evident that the amount of N that was applied in the above example would be far in excess of plant requirements. Additional data for Zn, Cu, and miscellaneous nonessential trace elements were given in Tables 4.2 and 4.3. Thus, for two reasons (trace element content and excess nutrient supply), careful control and management of land for sewage sludge applications is imperative.

When N is applied in excess of plant requirements, any NO_3^- left in the soil after the growing season is subject to leaching and movement into water supplies. The presence of NO_3^- in surface and ground waters is considered undesirable because of the possible influence of NO_3^- in promoting the unwanted growth of aquatic plants (eutrophication) and health hazards associated with the presence of NO_3^- in drinking waters (methemoglobinemia), as discussed in more detail in Chapter 8.

It is appropriate to mention that the NO_3^- in natural waters is normally derived from multiple sources, including municipal and rural sewage, feedlots or barnyards, food processing wastes, septic tank effluent, agricultural land (runoff and leaching), natural NO_3^- (caliche of semiarid regions), sanitary facilities of recreational areas, landfills, and miscellaneous industrial wastes.[7] The relative contribution of each will vary widely, depending on conditions existing at the particular site under consideration. The contribution from municipal sewage can be appreciable when raw or digested sewage is discharged directly into lakes or streams.

Soluble Salts

Animal wastes and sewage sludges contain inorganic salts of K, Na, Ca, and Mg. At high application rates, more salts may be added to the soil than are leached out by irrigation water or natural precipitation.[28] The buildup of soluble salts to unacceptable levels (inhibitory to plant growth) is most likely to occur in areas of low rainfall, such as the western United States (west of a line extending from central North Dakota through central Texas). For soils naturally high in salts, even a small application of wastes can increase salt concentration sufficiently to reduce yields.

The salt content of organic wastes is highly variable and is dependent in part on salt added to the system and management of the wastes. The salinity hazard of animal wastes can be reduced to some extent by lowering the salt content of food rations for the animals.

General Chemical Composition of Organic Wastes

All organic wastes have some common properties. All of them are made up of C, H, and O, and most contain N, P, and S as well. Residues from agriculturally oriented industries are made up of many of the constituents found in crop residues, such as lignin and cellulose, although not necessarily in the same proportion. The wastes from paper and sugar mills contain high amounts of carbohydrates; those of meat packing and dairy processing industries contain relatively large amounts of fatty acids and protein. Wood residues are relatively rich in lignin. Because of differences in degradation rate and environmental impact of the components, each of these wastes has somewhat different requirements for safe disposal by land application.

Animal Wastes   The feces of farm animals consist mostly of undigested food or partially digested components that escaped bacterial action during intestinal passage. In monogastric animals like poultry and hogs (and humans), this undigested food is mostly cellulose or lignin fibers, although some modification of lignin to humic substances has occurred.[29] The feces also contain the cells of microorganisms.

Nitrogen in manure solids occurs largely in organic forms (undigested proteins and microbial cells); liquid manure may also contain significant amounts of NH_3, the latter having been formed by hydrolyis of urea.

Animal wastes are more concentrated than the original feed in lignins and minerals. Lipids are present, along with humic-like substances. Manures also contain a variety of trace organics, such as antibiotics and hormones.

The manure applied to cropland varies greatly in nutrient content, depending on animal type, ration fed, amount and type of bedding material, and storage conditions. Both N content and availability of the N to plants decrease with losses of NH_3 through volatilization and NO_3^- through leaching. The two other major plant nutrients, P and K, are as bioavailable as fertilizer sources of these nutrients. Manures aged by cycles of wetting and drying and subjected to leaching with rainwater may have lost so much N that very little will be available to the crop in the year of application.

The gross chemical composition of animal manures is approximately as shown in Table 4.7. The trace element contents of most manures are somewhat higher than those of crop residues.[30,31] Swine and chicken manures will con-

TABLE 4.7   Gross Chemical Composition of Animal Manures[a]

Constituent	Typical Range, %
Ether-soluble compounds	1.8–2.8
Cold water-soluble compounds	3.2–19.2
Hot water-soluble compounds	2.4–5.7
"Hemicellulose"	18.5–23.5
Cellulose	18.7–27.5
"Lignin"	14.2–20.7
Total N	1.1– 4.1
Ash	9.1–17.2

[a]From McCalla et al.[9]

tain high amounts of Cu and Zn when these elements are used as feed additives.[17]

Sewage Sludge   Historically, the term *sewage sludge* has referred to heated, anaerobically digested sludge of the type produced in a typical sewage disposal facility; the term *biosolids* is more commonly used now. A flow diagram showing a typical method for treating domestic sewage is given in Fig. 4.1. The following steps are commonly used:

Primary Treatment: The simple settling and screening of solids. The effluent (liquid) contains solids that do not settle, plus organic and inorganic nutrients dissolved in the water. Pathogens and other organisms are present in the effluent and solid phase.

Secondary Treatment: This process reduces the amount of solid matter and removes a high percentage of BOD and COD. Some pathogens and parasites are killed and new microbial cells are formed. The effluent still contains high amounts of nutrients. Most sewage sludges used for application to farmland are the residual solids from this treatment.

Tertiary or Advanced Treatment: This is an additional treatment sometimes used for further purification of the effluent, including removal of inorganics, especially P. Among the treatments used singly or in combination are lime precipitation of P (as calcium phosphate), air stripping of NH_3, filtration to remove microbial cells and suspended organics, and activated carbon adsorption of organics.

Because of the large volume of material produced in a sewage treatment plant, serious storage and disposal problems are being encountered by many municipalities. Utilization on agricultural land offers an attractive alternative to other disposal methods, many of which are expensive and/or damaging to the environment.

At the outset, it should be noted that little if any environmental damage is expected when sewage sludge is applied to agricultural land at rates which

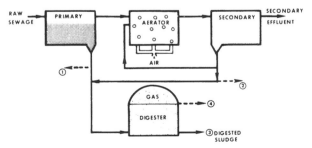

Fig. 4.1   Flow diagram of biological sewage treatment. From Dean and Smith.[13]

supply adequate but not excessive amounts of any given major nutrient or trace element. The main environmental concern is that there may be harmful effects from repeated applications over a prolonged period, particularly at high rates.

The suitability of residuals from domestic sewage as soil amendments is greatly increased by previous biological treatment, which not only stabilizes the material but also reduces pathogens and obnoxious odors. Municipal sewage is treated to reduce the amounts of suspended solids, kill pathogenic bacteria, and reduce BOD to acceptable levels for discharge into streams or lakes. With more stringent water quality standards, greater emphasis in the future will be given to disposal on land.

The product obtained from biological treatment of domestic sewage, which is the material available for land disposal from most municipal treatment plants, is a partially stabilized material which does not contain raw, undigested solids, but which frequently has an unpleasant odor. Liquid sewage sludge is blackish in color and contains colloidal materials and suspended solids. Most sludges, as produced in a sewage treatment plant, contain from 2 to 5% solids.

The solid portion of sewage sludge consists of approximately equal parts of organic and inorganic material. The latter include numerous elements, such as N, P, K, S, Cl, Zn, Cu, Pb, Cd, Hg, Cr, Ni, Mn, B and others. The organic component is a complex mixture consisting of (1) undigested constituents that are resistant to anaerobic decomposition, (2) dead and live microbial cells, and (3) compounds synthesized by microbes during the digestion process. The organic material is rather rich in N, P, and S. The C/N ratio of digested sludge ranges from 7 to 12, but is usually about 10.

Table 4.8 gives the range of chemical composition for digested sewage sludge collected at two sampling dates from the Sanitary District of Greater Chicago.[32] The composition of individual sludges can vary appreciably from the values shown.

TABLE 4.8   Chemical Composition of Two Samples of Digested Sewage Sludge Collected in 1971 and 1974[a]

Constituent	% of Organic Matter	
	1973	1974
Fats, waxes and oils	19.8	19.1
Resins	3.8	8.2
Water-soluble polysaccharides	3.2	14.4
Hemicellulose	4.0	6.0
Cellulose	3.5	3.2
Lignin-humus	16.8	14.5
Protein (N × 6.25)	24.1	39.6
Total recovered	75.1	105.0

[a]From Varanka et al.[32]

As with animal manures, the bioavailability of N in sludges decreases as the content of NH_4^+ and NO_3^- decreases and as the organic N becomes more stable as a result of digestion during biological waste treatment.

Sewage sludges contain relatively large quantities of minor and trace elements, as indicated above in Tables 4.2 and 4.3. Some of the elements are essential for plant growth, but nearly all can be toxic at some concentration. Zinc, Cu, Ni, Cd, Hg, and Pb may occur in quantities sufficient to adversely affect plants and soils. The availability of any given metal in soil will be influenced by such soil properties as pH, organic matter content, type and amount of clay, content of other metals, cation exchange capacity, and variety of crop grown.[22]

Degradation of Organic Wastes in Soil

Most organic wastes will follow the same pathways of degradation discussed in Chapter 1. However, due to previous biological transformations and also to the higher content of inorganic material, digested sewage sludge and composts (and to some extent animal manures) are more resistant to decay than crop residues in soils. This behavior is illustrated by the case of sewage sludge,[33] where an average retention of 80% of the C was obtained after a six-month incubation period (Fig. 4.2), a value much higher than is expected for raw organic materials added to soil. The percentage of the sludge C evolved as CO_2 was highly correlated with "degree days," but in no case was over 20% of the C recovered as CO_2 during the test period.

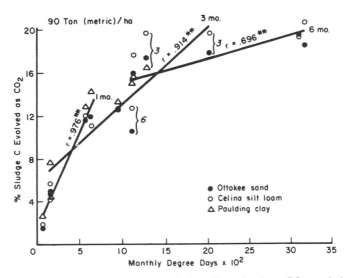

Fig. 4.2 Relationship between percent sludge C evolved as CO_2 and degree days after 1, 3, and 6 months of incubation. From Miller,[33] reproduced by permission of the American Society of Agronomy.

These data provide some insight into the quantities of C that would be expected to be retained in soil at moderate loading of organic wastes. Extent of decomposition (and subsequently C retention) would be expected to vary when applied as massive rates vs. repeated smaller additions, and to be affected by soil texture and by natural cycles of wetting and drying and freezing and thawing. Less C would be expected to be retained in warm humid climates compared to cold humid climates.

The effect of mean annual temperature on the relative rate of C (organic matter) loss at select locations for a north to south transect of the eastern United States[34] is shown in Fig. 4.3. Because of temperature effects on microbial activity, greater decay will occur as the mean annual temperature increases. Another important factor is that soils of the northern sections of the United States are frozen for part of the year and the mean annual temperature is relatively low, as seen from Fig. 4.4. Accordingly, decay rates will be higher and C retention will be lower in soils of the more southern regions, where temperatures are seldom low enough to inhibit microbial activity completely.

Management of Organic Wastes for N Conservation

In many places, permissible loading rates for the application of manures (and other organic wastes) to agricultural soils are based on the amount of a given nutrient (e.g., N, P, K) that will become available to the crop during the year of application. In other cases, rates are restricted on the basis of toxic element

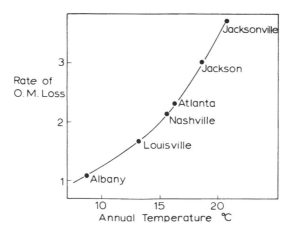

Fig. 4.3 Effect of mean annual temperature on the relative rate of organic matter loss at select locations for a north to south transect of Eastern United States. See Fig. 4.4 for monthly temperatures at the various locations. From Thomas,[34] reproduced by permission of the American Society of Agronomy.

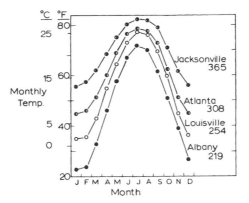

Fig. 4.4 Mean monthly temperatures for several locations along a north to south transect of Eastern United States. For mean annual temperatures see Fig. 4.3. From Thomas,[34] reproduced by permission of the American Society of Agronomy.

content (e.g., Cd). A discussion of alternative decision making factors for waste loading rates is given elsewhere.[4–6,28]

The following example is a case where NO_3^- leaching is a problem and application rates must be restricted on the basis of fertilizer N equivalents for the waste under consideration. Emphasis will be given to farmyard manure, but the principles will apply equally well to other nitrogenous organic wastes. The problem is extremely complex, for a variety of reasons. The N will exist in various forms during microbial transformations, and there is much difficulty in predicting what these forms will be at any one time or how rapidly the N will become available as a result of biodegradation of the manure. Various aspects of the recycling of N through land application of organic wastes are discussed by Smith and Peterson[35] and Hansen and Henriksen.[36]

In general, the net amount of NO_3^- produced, and subsequently available to plants or available for microbial immobilization, leaching, and denitrification, will represent the difference between the total amount of N applied and the amount that is immobilized and tied up in organic forms (i.e., amount of residual N in the soil after the growing season). The quantity of N retained in the soil in organic forms will be directly related to C retention, which for soils of the temperate region of the earth will be about one-third of the C applied (see Chapter 1). There is evidence, however, that a higher percentage ($\sim 50\%$) will remain when manures are applied at high rates, possibly because manures contain humic-like substances that are relatively resistant to decomposition. That is, some of the C in the waste is in a biologically active (available) form, but some fraction has the characteristics of more persistent soil organic matter.

The significance of the above-mentioned observations is that N is preserved along with C in the ratio of 10 parts C to one part N, which is near the average C/N ratio of the microbial biomass. Thus, for each 10 metric tons (dry weight) of organic wastes added to the soil, at least one-third (3300 kg) will remain behind in a modified form after the first year. Assuming that 90% of the dry matter in manure is organic material with a C content of 50%, approximately 1500 kg of C will be preserved, and this C will retain 150 kg of N. If one-half of the C is preserved, about 225 kg of N will be retained.

Tables 4.9 and 4.10 give the amounts of potential inorganic N released by application of 10 and 20 metric tons of manure (dry-weight basis) containing variable amounts of N (retention of one-third and one-half of the C, respectively). Feedlot manure will typically contain from 2.5 to 3.5% N, whereas fresh manure will contain 4.0% N or more. The moisture content of feedlot manure will be about 50%, while that of fresh manure can be 75% or more.

Table 4.11 gives the approximate metric tons of manure (dry-weight basis) that would be required to provide 90 and 180 kg (200 and 400 lbs) of available mineral N. It can be seen that the quantity required does not follow a direct 1:1 relationship to N content. With increasing N content, a lower percentage of the N will remain in the soil at the end of the first season.

It should be emphasized that the data given in Tables 4.9 to 4.11 apply to a single application of manure. For repeated applications over a period of years, these relationships do not hold, because the residual organic matter remaining in the soil after the first year—representing one-third to one-half of the manure initially applied—undergoes further decay during subsequent years with the release of bound N. As a rough approximation, one-third to one-half of the N remaining after each year will be mineralized the succeeding year until near-complete "humification" has occurred after about five years.

TABLE 4.9   Nitrogen Balance for Application of 10 and 20 Metric Tons of Animal Manure Containing Variable Amounts of N: Retention of One-Third of C (in kg)[a]

	10 Metric Tons			20 Metric Tons		
% N[b]	Total N	Retained in Residues	Inorganic N	Total N	Retained in Residues	Inorganic N
1.5	150	150	b	300	300	b
2.0	200	150	50	400	300	100
2.5	250	150	100	500	300	200
3.0	300	150	150	600	300	300
3.5	350	150	200	700	300	400
4.0	400	150	250	800	300	500
5.0	500	150	350	1,000	300	700

[a]Dry-weight basis. For conversion to pounds, multiply by 2.205.
[b]For N contents of 1.5% or less, there may be a net loss of mineral N from the soil through immobilization.

TABLE 4.10   Nitrogen Balance for Application of 10 and 20 Metric Tons of Animal Manure Containing Variable Amounts of N. Retention of One-Half of C (in kg)[a]

	10 Metric Tons			20 Metric Tons		
		N Retained in	Inorganic		N Retained in	Inorganic
% N[b]	Total N	Residues	N	Total N	Residues	N
2.0	200	225	b	400	450	b
2.5	250	225	25	500	450	50
3.0	300	225	75	600	450	150
3.5	350	225	125	700	450	250
4.0	400	225	175	800	450	350
5.0	500	225	275	1,000	450	500

[a]Dry-weight basis. For conversion to pounds multiply by 2.205.
[b]For N contents of 2% or less, there may be a net loss of mineral N from the soil through immobilization.

In practical terms, this means that application rates of new waste to provide an amount of inorganic N equivalent to that released from the original application will need to be reduced each succeeding year for five years, after which the rate would be constant and equivalent to the addition of an equal quantity of inorganic N. In other words, after five years, essentially all of the N added in the manure must be considered to be "available" because the amount immobilized will be compensated for by mineralization of residual N from earlier applications.

Successive applications for five or more years commonly occur when a small land area is used for disposal. For best utilization of nutrients, the manure should be spread over a broad area at a low loading rate.

TABLE 4.11   Estimated Metric Tons of Manure Required per Hectare (Dry-Weight Basis) to Provide 90 and 180 kg (200 and 400 lbs) of Available Mineral N at Two Levels of C Retention

	One-Third Retention of C		One-Half Retention of C	
N Content of Manure, %	90 kg N	100 kg N	90 kg N	180 kg N
2.0	20	40	a	a
2.5	10	20	40	50
3.0	7	14	13	26
3.5	5	10	8	16
4.5	4	8	6	12

[a]Very high rates required because of net immobilization.

SYNTHESIS OF CARCINOGENIC COMPOUNDS

Several specific types of organic compounds that are carcinogenic, mutagenic, or acutely toxic in very low amounts can be produced by microorganisms. They include methylmercury, dimethylarsine, dimethylselenide, and nitrosamines. These substances are not normal constituents of water or agricultural soils, but can be produced under certain circumstances in aqueous sediments, and possibly in polluted soils. The role of microorganisms in synthesizing hazardous substances from innocuous precursors is discussed by Alexander.[37]

Methylmercury

Discharge of mercury (Hg) in industrial effluents results in the formation of methylmercury (CH_3Hg^+) in waterways and streams through microbial action.[38] This deadly poisonous compound is fat-soluble and accumulates in fish and, when consumed by humans, can lead to serious illness or death. Human fatalities due to methylmercury exposure have been documented in the literature.

The bottom sediments of many lakes and waterways are badly contaminated with Hg, and slow conversion to the methylated form is possible for many years to come. Sources of Hg in agricultural soils include organic pesticides and organic wastes such as municipal sewage sludges.

Many schemes have been proposed for the formation of methylmercury by microorganisms.[39,40] The probable pathway suggested by Wood[40] is transfer of the methyl group from methyl cobalamin (CH_3—B12), which is a common coenzyme in both anaerobic and aerobic bacteria, as shown in Fig. 4.5.

Dimethylarsine

Arsenic is of environmental concern because of its previous widespread use in pesticides and defoliants, its presence in many different ecosystems, and

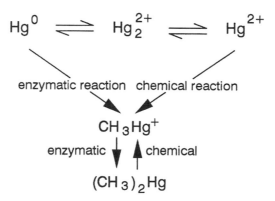

Fig. 4.5 Proposed mechanism for the formation of methylmercury by microorganisms. From Wood.[40]

its toxicity to man.[41] A problem similar to that mentioned for methylmercury exists when arsenic and its derivatives are introduced into environments containing an active flora of anaerobic organisms. Like methylmercury, dimethylarsine is deadly poisonous and can enter the food chain through accumulation in fish.[42,43]

The mechanisms of dimethylarsine formation are similar to that involving Hg in that methyl transfer from methyl B12 can occur,[40,42] as shown in Fig. 4.6.

Dimethylselenide

Anionic forms of selenium are also subject to microbial alkylation, the product being demethylselenide.[40,41,44,45] Selenium occurs in many soils as selenate and selenite; in some regions, such high concentrations are present that plants accumulate the element to levels that are toxic to grazing animals. Like NO_3, selenite is relatively mobile in soil and can be transported from irrigated soils to adjacent surface water. Francis et al.[44] suggest that the microbial methylation of selenium is potentially widespread and that losses of selenium can occur from soils as the volatile dimethylselenide.

Nitrosamines

Nitrosamines are formed by the chemical reaction of amines ($R–NH_2$) with NO_2^-; both chemicals must be present concurrently (and under relatively acidic conditions) in order for these carcinogenic and toxic substances to be formed. Nitrite required for the reaction is produced as an intermediate during biochemical N transformations, but the compound seldom persists in soil. However, temporary accumulations are possible in microsites where NO_2^--oxidizing autotrophs (*Nitrobacter* sp.) are inhibited by free NH_3. Studies on the formation of dimethylamine and diethylamine in soils treated with pesticides that yield amines upon partial degradation indicate that the potential

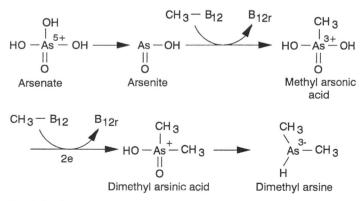

Fig. 4.6 Proposed mechanism for the formation of dimethylarsine.

exists for nitrosamine synthesis. However, it has yet to be established that these compounds are produced in this way in natural soils (see Chapter 8).

PERSISTENT SYNTHETIC ORGANIC CHEMICALS

A new dimension to the C cycle in soil has been introduced as a consequence of the widespread use of synthetic organic chemicals to control pests (insecticides, fungicides) and weeds (herbicides), as well as dumping of hazardous organics on soil. Generically, these chemicals are referred to as *xenobiotics,* i.e., these are synthetic, man-made chemicals with substantially different structures and environmental behavior when compared to most natural organic compounds. They include such synthetic organic compounds as plasticizers (e.g., phthalates), dyes of various types that are used for clothing and foods, flame retardants, impurities in pesticide formulations (e.g., chlorinated dioxins), and halogenated solvents. Chlorinated dioxins are widely recognized as dangerous substances in the environment, the most notorious being 2,3,7,8-tetrachlorodibenzo-*p*-dioxin (TCDD), whose structure is shown below.

TCDD   Typical Phthalate Ester

Two classes of organic chemicals that are of particular interest from the standpoint of pollution are the phthalic acid esters (polyesters of the benzene dicarboxylic acids) and a wide variety of halogenated organic compounds. Phthalates have been under close scrutiny because large quantities are produced each year for use in construction and housing and the medical sciences, as carriers in pesticide formulations, and as additives to plastics.[46] It is now realized that these phthalates are general environmental contaminants because they do not degrade readily and some have health adverse effects.[46] Many scientists believe that they do not pose an eminent threat to human health, although it is recognized that certain individuals may be more affected than others. Questions have also arisen concerning possible subtle effects of repeated exposure to very low concentrations of phthalates or derivatives obtained therefrom by partial degradation.

The quantity of the top 15 synthetic organic chemicals produced in the United States alone during the period 1986–1996 was about 1.4 trillion pounds, and pesticide production was about 12.7 billion pounds.[47] Over the past 50 years, over 600,000 metric tons of industrial organic chemicals have been produced, about 15% of which are estimated to have entered the mobile environmental reserve.[48]

Synthetic organic chemicals can be divided into two general types as far as their persistence in soil is concerned, namely, those that are biodegradable and those that are recalcitrant (i.e., degrade slowly if at all). The division between the two is somewhat arbitrary in that certain organic chemicals are biodegradable under some soil conditions but not others. For example, groups of synthetic organics derived from alkanes or simple aromatic compounds are destroyed at a much slower rate in flooded sediments and under anaerobic soil conditions than in well-aerated soils, while the reverse may be true for other classes of organic compounds such as halogenated compounds. The persistence of some synthetic organic compounds in poorly drained soils and wet sediments can be attributed to several causes, the most important being the need for O_2 in key enzymatic reactions during degradation of alkanes and aromatic rings. Wet sediments contain a less varied micro- and macroflora than well-drained soils; furthermore, there is less modification of the environment through various processes associated with crop production (mixing of the soil by plowing, incorporation of crop residues, etc.).

For the most part, biodegradable organic chemicals undergo the same type of cleavage reactions as biogenic molecules (e.g., decarboxylation, deamination, hydroxylation, β-oxidation, and hydrolysis of ester linkages).[37] Typical type reactions are given in Table 4.12; examples of cleavage reactions are listed in Table 4.13. The fate and biodegradability of a synthetic organic chemical can be predicted to some extent from its chemical structure, and this concept is an integral part of the development of new pesticides, for which persistence in a biologically active form is undesirable.

A popular concept in the past (prior to the 1960s) was that essentially every organic chemical introduced into the biosphere would be metabolized within a reasonably short time by microorganisms. Later this was found to be incorrect; in fact, some synthetic organic chemicals are recalcitrant and persist in soil for long periods. They include chlorinated aromatic compounds of various types including polychlorinated biphenyls, (PCBs), certain pesticides (particularly chlorinated insecticides), plastics and related synthetic polymers, and an array of industrial organic chemicals.[49,50] More recently, many of the presumably persistent organic compounds that are halogenated have been found to be biodegradable under some conditions.[51,52] It is still reasonable to consider many of these compounds as "recalcitrant" because, under many natural conditions, their degradation is very slow and their environmental residence times can be years.

Comparison of the persistence of individual pesticides in Fig. 4.7 demonstrates the long-time persistence of organochlorine insecticides, notably DDT. As a result of poor degradation, use of many of these compounds in the United States is highly restricted or prohibited, although production and shipment to many developing countries continue. Chemical formulas for some of the more persistent pesticides are given in Fig. 4.8.

In addition to chemical structure characteristics that affect microbial degradation, a major factor affecting the persistence of synthetic organics in soils is adsorption to clay and organic matter surfaces.[54,55] The herbicides Diquat and Paraquat persist in soils for long periods (data not shown), primarily because they adsorb strongly and rapidly to soil. Adsorption depends upon

TABLE 4.12   Type Reactions for Cleavage of Chemicals of Environmental Importance[a]

Category	Reaction[b]	Example
Dehalogenation	$RCH_2Cl \rightarrow RCH_2OH$	Propachlor
	$RCl_nRCl_n \rightarrow RCl_nRCl_{n-1} + Cl^-$	Chlorinated solvents
	$ArCl \rightarrow ArOH$	Nitrogen
	$ArF \rightarrow ArOH$	Flamprop-methyl
	$ArCl_5OH \rightarrow ArCl_4OH + Cl^-$	Pentachlorophenol
	$Ar_2CHCH_2Cl \rightarrow Ar_2C{=}CH_2$	DDT
	$Ar_2CHCHCl_2 \rightarrow Ar_2C{=}CHCl$	DDT
	$Ar_2CHCCl_2 \rightarrow Ar_2CHCHCl_2$	DDT
	$Ar_2CHCCl_2 \rightarrow Ar_2C{=}CCl_2$	DDT
	$RCCl_2 \rightarrow RCOOH$	DDT, N-Serve
Deamination	$ARNH_2 \rightarrow ArOH$	Fluchloralin
Decarboxylation	$ArCOOH \rightarrow ArH$	Bifenox
	$RCH(CH_3)COOH \rightarrow RCH_2CH_3$	Dichlorfop-methyl
	$ArN(R)COOH \rightarrow ArN(R)H$	DDOD
Methyl oxidation	$RCH_3 \rightarrow RCH_2OH$ and/or	Bromacil, diiso-
	$\rightarrow RCHO$ and/or	propylnaphthalene
	$\rightarrow RCOOH$	pentachlorobenzyl alcohol
Hydroxylation	$ArH \rightarrow ArOH$	Benthiocarb, dicamba
β-oxidation	$ArO(CH_2)_nCH_2CH_2COOH \rightarrow$	1-(2,4-Dichloro-
	$ArO(CH_2)_nCOOH +$	phenoxy)-alkanoic
	CH_3COOH	acids
Triple bond reduction	$RC{\equiv}CH \rightarrow RCH{=}CH_2$	Buturon
Double bond reduction	$Ar_2C{=}CH_2 \rightarrow Ar_2CHCH_3$	DDT
	$Ar_2CHCl \rightarrow Ar_2CH_2Cl$	DDT
Double bond hydration	$Ar_2C{=}CH_2 \rightarrow Ar_2CHCH_2OH$	DDT

[a] From Alexander[37]
[b] R = organic moiety; Ar = aromatic.

the chemical structure of the organic compounds and properties of the soil system (kind and amount of clay, organic matter content, pH, and type of exchange cations).

Adsorption by soil organic matter has been shown to be a key factor in the behavior of synthetic organics in soil.[56-58] It has been well established, for example, that the application rate for an adsorbable pesticide that will achieve adequate pest control can vary as much as 20-fold, depending upon the nature of the soil and the amount of organic matter it contains. Soils that are black in color (e.g., most Mollisols) have higher organic matter contents than those that are light in color (e.g., Alfisols), and application rates must often be adjusted upward on the darker soils in order to achieve the desired degree of pest control.

TABLE 4.13 Type Reactions for Cleavage of Chemicals of Environmental Importance[a]

Substrate	Reaction[b]	Example
Ester	RC(O)OR' → RC(O)OH + HOR'	Malathion, phthalates
Ether	ArOR → ArOH + R	Chlomethoxynil, 2,4-D
	ROCH$_2$R' → ROH + HOCH$_2$R'	Dichlorfop-methyl
C—N bond	R(R')NR'' → R(R')NH + R''H	Alachlor,
	and/or → R(R')H + R''NH$_2$	trimethylamine
	RN(Alk)$_2$ → RNHAlk + Alk	Chlorotoluron, Trifluralin
	RNH$_2$CH$_2$R' → RNH$_2$ + Ch$_2$R'	Glyphosate
Peptide,	RNHC(O)R' → RNH$_2$ + HOOCR'	Benlate, dimetholate
carbamate	R(R')NC(O)R'' → R(R')NH +	Benzoylprop-ethyl
	HOOCR''	
R=NOC(O)R	RCH=NOC(O)R → RCH=NOH	Aldicarb
	+ C(O)R	
C—S bond	RSR' → ROH and/or HSR'	Benthiocarb, Kitazin P
C—Hg bond	RHgR' → RH + HgR'	Ethylmercury, phenyl-
		mercuric acetate
C—Sn bond	R$_3$SnOH → R$_2$SnO + R or	Tricyclohexyltin
	RSnO$_2$H	hydroxide
C—O—P[c]	(AlkO)$_2$P(S)R → AlkO(HO)P(S)R	Gardona, malathion
	and/or → (HO)$_2$P(S)R	
P-S	RSP(O)(R')OAlk →	Hinosan, Kitazin
	HOP(O)(R')OAlk	
Sulfate ester	RCH$_2$OS(O$_2$)OH → RCH$_2$OH	Sesone
	and/or HOS(O$_2$)OH	
S—N	ArS(O$_2$)NH$_2$ → ArS(O$_2$)OH	Oryzalin
S—S	RSSR → 2 RSH	Thiram

[a] From Alexander.[37]
[b] R = organic moiety; Alk = alkyl.
[c] For the reaction, S = sulfur or oxygen.

Adsorption by soil colloids (humus and clay) poses severe problems for ascertaining the long-time fate of synthetic organics in the environment. For example, adsorbed organics can be transported during soil erosion to lakes and reservoirs where conditions may be unfavorable for microbial detoxification. The adsorption phenomena must also be taken into account in assessing the impact of industrial processes that lead to widespread dispersal of organics into the ecosystem, such as conversion of coal to natural gas.

In some cases, compounds are partially degraded and become part of the organic matter of soil.[59] In this form, they are frequently unextractable by water or organic solvents and are referred to generically as "bound residues," whose exact structure is difficult or impossible to determine.

A full accounting of the fate of synthetic organics in soil is beyond the scope of the present chapter. Numerous books and reviews have been written on degradation of pesticides and industrial chemicals (primarily solvents and

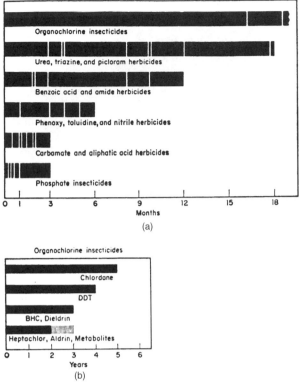

Fig. 4.7 Persistence of several groups of pesticides (*a*) and specific organochlorine insecticides (*b*). Adapted from Kearney et al.[53] and Sethunathan et al.[56]

oil-derived products) and pollution of soils by these compounds. Some key reviews and books are listed in References 53 to 63.

CONTRIBUTION OF SOIL TO CO$_2$ OF THE ATMOSPHERE

It is well known that the CO$_2$ content of the atmosphere has been increasing at a steady rate over the past century.[64-69] This is a matter of grave concern because of the possible warming of the Earth's surface by the so-called greenhouse effect. Global temperature increases of 2 to 3°C are predicted for early in the next century, when the content of CO$_2$ in the atmosphere is expected to have doubled. The fear exists that the increased climatic warming will have a calamitous effect and should be suppressed.[64-68] An opposing view has been expressed by others (e.g., Idso[69]), who have concluded that an increase in CO$_2$ will have a beneficial effect on agricultural productivity.

Current increases in the CO$_2$ content of the atmosphere cannot be attributed to losses of soil C, for the reason that the decline in humus content for the vast majority of the world's cultivated soils occurred prior to the mid-

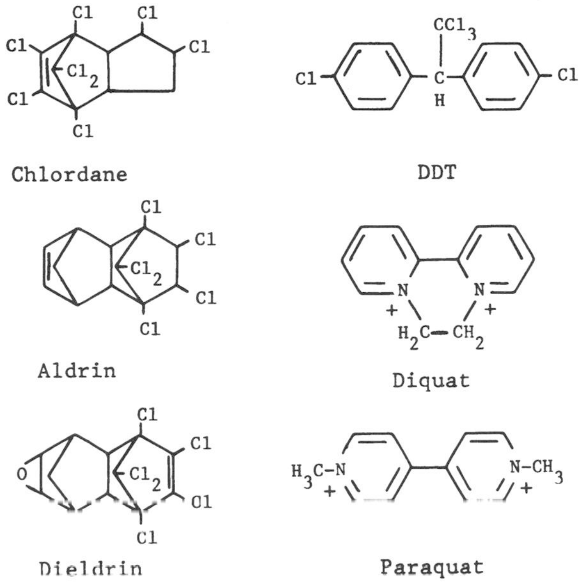

Fig. 4.8 Chemical structures of some of the more persistent pesticides in soil.

nineteenth century (see Chapter 2); i.e., before the rapid increase in CO_2 content of the atmosphere. Under steady-state conditions, the net contribution of soil organic matter to CO_2 in the atmosphere will be negligible. Bolin[64] estimates that from 10 to 40×10^9 tons of C have been added to the atmosphere since the early nineteenth century through changes in soil C reserves. Conrad[70] and Bouwman[71] provide information about the role of soils and microorganisms in atmospheric chemistry.

Recent increases in atmospheric CO_2 content appear to be due mainly to the burning of fossil fuels (e.g., coal and oil-derived fuels), although agricultural practices of various types also play a role. These include the harvest of forests and subsequent oxidation of humus on the forest floor, slash-and-burn agriculture in the tropics, drainage of wetlands with accelerated decomposition of organic matter, and utilization of peat for agricultural purposes. Modifications in vegetation types as a result of clearing of land for agriculture and changes in the duration of the active photosynthetic (growth) season can have an effect on the atmosphere because annual crops, with a short growing season and no accumulation of persistent biomass (as seen with woody plants and herbaceous perennial plants), do not remove as much CO_2 from the atmosphere as the native vegetation would have removed.

SUMMARY

Large quantities of organic wastes are generated each year from domestic and municipal sewage, animal and poultry enterprises, food processing industries, pulp and paper mills, municipal garbage, and others. It is imperative to have safe and economic practices for the utilization and disposal of these wastes. With increased fertilizer and energy costs, some organic wastes (e.g., animal manures and selected municipal organic wastes) are again being considered as beneficial soil amendments. Use of some products (e.g., municipal and domestic organic wastes) incurs special problems, such as addition of toxic metals and pathogens, nutrient imbalances, soluble salts, odors, and can have difficulty in gaining social acceptance.

Organic wastes must be managed so that they do not pollute air or soil–water resources, nor should they be used in any way that might introduce toxins or pathogens into the food supply. There is a challenge to find more effective ways to manage organic wastes in soil so as to conserve soil resources and protect the environment.

Microorganisms in soil and water can form metabolic products such as dimethylmercury that are substantially more toxic than the parent compounds.

An array of synthetic organic chemicals are deliberately or inadvertently introduced into soil, many of which are recalcitrant and can persist for long periods. Persistence of a given organic chemical is also affected by adsorption to clay and organic colloids.

The organic matter content of most agricultural soils is at steady state. Accordingly, current increases in the CO_2 content of the atmosphere cannot be attributed to losses of soil C. The main causes are the burning of fossil fuels and deforestation.

REFERENCES

1. D. J. Glass, T. Raphael, R. Valo, and J. Van Eyk, "International Activities in Bioremediation: Growing Markets and Opportunities," in R. E. Hinchee, J. A. Kittel, and H. J. Reisinger, Eds., *Applied Bioremediation of Petroleum Hydrocarbons,* Battelle Press, Columbus, Ohio, 1995, pp. 11–33.

2. D. W. Tedder and F. G. Pohland, Eds., *Emerging Technologies in Hazardous Waste Management III,* American Chemical Society, Washington, D.C., 1993.

3. Anonymous, *Ranking Hazardous Waste Sites for Remedial Action,* National Academy Press, Washington, D.C., 1994.

4. CAST, *Utilization of Animal Manures and Sewage Sludges in Food and Fiber Production, Report 41,* Council for Agricultural Science and Technology, Ames, Iowa, 1975.

5. CAST, *Application of Sewage Sludge to Cropland: Appraisal of Potential Hazards of the Heavy Metals to Plants and Animals, Report 64,* Council for Agricultural Science and Technology, Ames, Iowa, 1976.

6. L. F. Elliott and F. J. Stevenson, Eds., *Soils for Management of Organic Wastes and Waste Waters,* American Society of Agronomy, Madison, Wisconsin, 1977.

7. T. L. Willrich and G. E. Smith, Eds., *Agricultural Practices and Water Quality,* Iowa State University Press, Ames, Iowa, 1970.

8. C. H. Wadleigh, *Wastes in Relation to Agriculture and Forestry,* USDA Misc. Publ. No. 1065, U.S. Government Printing Office, Washington, D.C., 1968, pp. 1–112.

9. T. M. McCalla, J. R. Peterson, and C. Lue-Hing, "Properties of Agricultural and Municipal Wastes," in L. F. Elliott and F. J. Stevenson, Eds., *Soils for Management of Organic Wastes and Waste Waters,* American Society of Agronomy, Madison, Wisconsin, 1977, pp. 9–43.

10. R. M. Salter and C. J. Schollenberger, "Farm Manure," in *Soils and Men: Yearbook of Agriculture,* U.S. Government Printing Office, Washington, D.C., 1938, pp. 445–461.

11. Anonymous, *Use of Reclaimed Water and Sludge in Food Crop Production,* National Academy of Sciences Press, Washington, D.C., 1966.

12. Anonymous, "Aggregate Organic Constituents," in *Standard Methods for the Examination of Water and Wastewater,* American Public Health Association, Washington, D.C., 1992, pp. 5-1–5-10.

13. R. B. Dean and J. E. Smith, Jr., "The Properties of Sludges," in *Recycling Municipal Sludges and Effluents on Land,* National Association of State Universities and Land-Grant Colleges, Washington, D.C., 1973, pp. 39–47.

14. T. M. McCalla, L. R. Frederick, and G. L. Palmer, "Manure Decomposition and Fate of Breakdown Products in Soil," in T. L. Willrich and G. E. Smith, Eds., *Agricultural Practices and Water Quality,* Iowa State University Press, Ames, Iowa 1970, pp. 241–255.

15. A. L. Page, *Fate and Effects of Trace Elements in Sewage Sludge when Applied to Agricultural Lands,* Environmental Protection Agency Technology Series, Program Element 1B2043, EPA, Cincinnati, Ohio, 1974.

16. G. W. Leeper, *Managing the Heavy Metals on the Land,* Sheaffer & Roland, Chicago, Illinois, 1978.

17. A. L. Sutton, D. W. Nelson, V. B. Mayrose, and D. T. Kelly, *J. Environ. Qual.,* **12,** 198 (1983).

18. R. L. Chaney and J. A. Ryan, "Heavy Metals and Toxic Organic Pollutants in MSW-Composts: Research Results on Phytoavailability, Bioavailability, Fate, etc.," in H. A. J. Hoitink and H. M. Keener, Eds., *Science and Engineering of Composting,* Renaissance Publications, Worthington, Ohio, 1993, pp. 451–506.

19. R. Epstein, R. L. Chaney, C. Henry, and T. J. Logan, *Biomass and Bioenergy,* **3,** 227 (1992).

20. H. T. Shacklette, J. C. Hamilton, J. G. Boerngen, and J. M. Bowles, *Elemental Composition of Surficial Materials in the Conterminous United States,* Geological Survey Professional Paper 574-D, U.S. Government Printing Office, Washington, D.C., 1971.

21. A. C. Mathers, B. A. Stewart, J. D. Thomas, and B. J. Blair, "Effect of Cattle Feedlot Manure on Crop Yields and Soil Conditions," *Proc. Symp. Animal Waste Management, Texas Tech.,* Rep. **11,** 1973, pp. 1–13.

22. J. A. Ryan and R. L. Chaney, "Regulation of Municipal Sewage Sludge under the Clean Water Act Section 503: A Model for Exposure and Risk Assessment for MSW-Compost," in H. A. J. Hoitink and H. M. Keener, Eds., *Science and Engineering of Composting,* Renaissance Worthington, Ohio, 1993, pp. 422–450.

23. F. A. M. de Haan and S. E. A. T. M. van de Zee, "Compost Regulations in View of Sustainable Soil Use," in H. A. J. Hoitink and H. M. Keener, Eds., *Science*

and Engineering of Composting, Renaissance Worthington, Ohio, 1993, pp. 507–522.

24. R. L. Chaney and J. A. Ryan, *Risk Based Standards for Arsenic, Lead and Cadmium in Urban Soils,* DECHEMA, Frankfurt am Main, Germany, 1994.

25. R. H. Miller, "Soil Microbiological Aspects of Recycling Sewage Sludges and Waste Effluents on Land," in *Recycling Municipal Sludges and Effluents on Land,* National Association of State Universities and Land-Grant Colleges, Washington, D.C., 1973, pp. 79–90.

26. M. D. Thorne, T. D. Hinesly, and R. L. Jones, "Utilization of Sewage Sludge on Agricultural Land," *Illinois Agricultural Experiment Station Agronomy Fact Sheet* **SM-29,** 1975, pp. 1–8.

27. Anonymous, "Standards for the Use or Disposal of Sewage Sludge," *Federal Register* **58**(32), 9248-9404, U.S. Government Publishing Office, Washington, D.C., 1993.

28. B. A. Stewart and B. D. Meek, "Soluble Salt Considerations with Waste Application," in L. F. Elliott and F. J. Stevenson, Eds., *Soils for Management of Organic Wastes and Waste Waters,* American Society of Agronomy, Madison, Wisconsin, 1977, pp. 218-232.

29. S. L. Jansson, *Lantbruks. Hogskol Ann., 36,* 51, 135 (1960).

30. H. J. Atkinson, G. R. Giles, and J. G. Desjardins, *Can. J. Agr. Sci., 34,* 76 (1954).

31. D. C. Whitehead, "Nutrient Minerals in Grassland Herbage," *Commonwealth Bureau of Pastures and Field Crops Mimeo. Publ. No. 1, 1966,* pp. 1–83.

32. M. W. Varanka, Z. M. Zablocki, and T. D. Hinesly, *J. Water Pollut. Control Fed.,* **48,** 1728 (1976).

33. R. H. Miller, *J. Environ. Qual., 3,* 376 (1974).

34. G. W. Thomas, "Land Utilization and Disposal of Organic Wastes in Hot, Humid Regions," in L. F. Elliott and F. J. Stevenson, Eds., *Soils for Management of Organic Wastes and Waste Waters,* American Society of Agronomy, Madison, Wisconsin, 1977, pp. 492–507.

35. J. H. Smith and J. R. Peterson, "Recycling of Nitrogen Through Land Application of Agricultural, Food Processing, and Municipal Wastes," in F. J. Stevenson, Ed., *Nitrogen in Agricultural Soils,* American Society of Agronomy, Madison, Wisconsin, 1982, pp. 791–831.

36. J. A. Hansen and K. Henriksen, Eds., *Nitrogen in Organic Wastes Applied to Soils,.* Academic Press, New York, 1989.

37. M. Alexander, *Science,* **211,** 132 (1981).

38. P. J. Craig, "Organomercury Compounds in the Environment," in P. J. Craig, Ed., *Organometallic Compounds in the Environment,* Longman, Essex, England, 1986, pp. 65–110.

39. A. Jernelov and A. L. Martin, *Ann. Rev. Microbiol., 29,* 61 (1975).

40. J. M. Wood, *Science,* **183,** 1049 (1974).

41. D. P. Cox and M. Alexander, *J. Microb. Ecol., 1,* 136 (1974).

42. M. O. Andreae, "Organoarsenic Compounds in the Environment," in P. J. Craig, Ed., *Organometallic Compounds in the Environment,* Longman, Essex, England, 1986, pp. 198–228.

43. W. R. Cullen and K. J. Reimer, *Chem. Rev.,* **89,** 713 (1989).

44. A. J. Francis, J. M. Duxbury, and M. Alexander, *Appl. Microb.,* **28,** 248 (1974).

45. Y. K. Chau, "Organic Group VI Elements in the Environment," in P. J. Craig, Ed., *Organometallic Compounds in the Environment,* Longman, Essex, England, 1986, pp. 254–278.

46. Anonymous, "Additives in the Environment," *Plastics Technol.,* July 1990, pp. 49–59.

47. Anonymous, *Chemical & Engineering News,* June 23, 1997, pp. 40–45.

48. National Research Council (NCR). *Polychlorinated Biphenyls,* National Academy of Sciences, Washington, D.C., 1979.

49. M. Alexander, *Biodegradation and Bioremediation,* Academic Press, San Diego, l994, pp. 159–176, 272–286.

50. R. Bartha and A.V. Yabannavar, "Biodegradation Testing of Polymers in Soil," in J. D. Hamilton and R. Sutcliffe, Eds., *Ecological Assessment of Polymers,* Van Nostrand Reinhold, New York, 1997, pp. 53–66.

51. V. Tandoi et al., *Environ. Sci. Technol.,* **28,** 973 (1994).

52. W. W. Mohn and J .M. Tiedje, *Microbiol. Reviews,* **56,** 482 (1992)..

53. P. C. Kearney, E. A. Woolson, J. R. Plimmer, and A. R. Isensee, *Residue Reviews,* **28,** 137 (1969).

54. B. L. Sawhney and K. Brown, Eds., *Reactions and Movement of Organic Chemicals in Soils.* Soil Science Society of America, Madison, Wisconsin, 1989.

55. D. M. Linn et al., Eds., *Sorption and Degradation of Pesticides and Organic Chemicals in Soil,* Soil Science Society of America, Madison, Wisconsin, 1993.

56. N. Sethunathan et al., *Residue Reviews,* **68,** 91 (1977).

57. W. D. Guenzi, Ed., *Pesticides in Soil and Water,* American Society of Agronomy, Madison, Wisconsin, 1974.

58. L. G. Morrill, B. Mahilium, and S. H. Modiuddin, *Organic Compounds in Soils: Sorption, Degradation, and Persistence,* Ann Arbor Science, Ann Arbor, Michigan, 1982.

59. J. Dec and J.-M. Bollag, "Dehalogenation of Chlorinated Phenols During Binding to Humus," in T. A. Anderson and J. R. Coats, Eds., *Bioremediation Through Rhizosphere Technology,* American Chemical Society, Washington, D.C., 1994, pp. 102–111.

60. M. R. Overcash, Ed., *Decomposition of Toxic and Non-toxic Organic Compounds in Soils,* Ann Arbor Science, Ann Arbor, Michigan, 1981.

61. E. Riser-Roberts, *Bioremediation of Petroleum-Contaminated Sites,* CRC Press, Boca Raton, Florida, 1992.

62. R. E. Hinchee, R. E. Hoeppel, and D. B. Anderson, Eds., *Bioremediation of Recalcitrant Organics,* Battelle Press, Columbus, Ohio, 1995.

63. R. E. Hinchee, A. Leeson, and L. Semprini, *Bioremediation of Chlorinated Solvents,* Battelle Press, Columbus, Ohio, 1995.

64. B. Bolin, *Science,* **196,** 613 (1977).

65. W. S. Broecker, T. Takahashi, H. J. Simpson, and T.-H. Peng, *Science,* **206,** 409 (1979).

66. F. Lyman, *The Greenhouse Trap,* Beacon Press, Boston, 1990.

67. S. Joffre, "Climate Change," in D. Stanners and P. Bourdeau, Eds., *Europe's Environment: The Dobřiš Assessment,* European Environmental Agency, Copenhagen, 1995, pp. 512–522.

68. G. M. Woodwell et al., *Science,* **199,** 141 (1978).

69. S. B. Idso, *Carbon Dioxide: Friend or Foe?,* IBR Press, Institute for Biospheric Research, Tempe, Arizona, 1982.

70. R. Conrad, *Microbiol. Reviews,* **60,** 609 (1996).

71. A. F. Bouwman, Ed., *Soils and the Greenhouse Effect,* Wiley, Chichester, England, 1990.

5

THE NITROGEN CYCLE IN SOIL: GLOBAL AND ECOLOGICAL ASPECTS

The N cycle in soil (Fig. 5.1) is an integral part of the global N cycle. The ultimate source of soil N is the atmosphere, where the very stable molecule (N_2) is the predominant gas (79.1% by volume). Nitrogen is significant because, after C, H, and O, no other element is so intimately associated with reactions carried out by living organisms. The cycling of other nutrients, notably P and S, is closely associated with biochemical N transformations.

Although considered as a sequence of reactions of an individual N atom, an "N cycle" for a single N atom does not exist in nature. Rather, any given N atom moves from one form to another in an irregular or random fashion. Also, the soil contains an internal cycle that is distinct from the overall cycle of N but that interfaces with it (see Chapter 6). A key feature of the internal cycle is the turnover of N through mineralization performed by microorganisms and soil fauna and immobilization carried out by microorganisms.

Gains in soil N occur by microbial fixation of molecular N_2 and by addition of ammonia (NH_3), nitrate (NO_3^-), and nitrite (NO_2^-) in rainwater; losses occur through crop removal, leaching, and volatilization. Biological N_2 fixation converts N_2 to combined forms (NH_3 and organic N); this process is performed by bacteria and cyanobacteria or by plant–bacterial associations with these organisms. Organic forms of N, in turn, are converted to NH_3 and NO_3^- by mineralization. The conversion to organic N to NH_3 is termed *ammonification;* the oxidation of NH_3 to NO_3^- is termed *nitrification.* Utilization of NH_3 and NO_3^- by plants and soil organisms constitutes assimilation and immobilization, respectively. Nitrogen (as NO_3^- or NO_2^-) is ultimately returned to the atmosphere as molecular N_2 by biological denitrification, thereby completing the cycle.

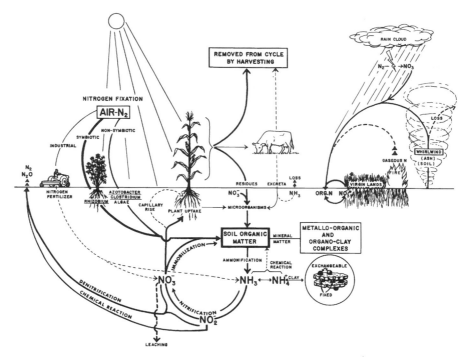

Fig. 5.1   The N cycle in soil. From Stevenson.[1]

Microorganisms perform a vital role in soil N transformations, but not all transformations are mediated by microorganisms. Ammonia and NO_2^-, produced by microbial transformations of nitrogenous materials, can undergo chemical reactions with organic substances, in some cases leading to the evolution of N gases. Through the association of humic materials with mineral matter, organo–clay complexes are formed whereby the N compounds are protected against attack by microorganisms. The positively charged ammonium ion (NH_4^+) undergoes substitution reactions with other cations of the exchange complex, and it can be fixed by clay minerals.

The basic feature of biological N transformations centers on oxidation and reduction reactions. In the oxidized compounds of N (N_2O, NO_2^-, and NO_3^-), the outer electrons of N serve to complete the electron shells of other atoms, particularly oxygen; in reduced compounds of N (NH_3 and most organic-N compounds), the three electrons required to fill the outer shell are supplied most commonly by H or C.

Topics covered in this chapter include global aspects of the soil N cycle, gains and losses of N from the soil–plant system, and the flow of soil N to other ecosystems. The next two chapters deal specifically with the "internal" cycle (Chapter 6) and environmental aspects (Chapter 7). Universal aspects of the N cycle are covered in numerous articles and reviews.[1–12]

GEOCHEMICAL DISTRIBUTION OF N

Nitrogen is an important constituent of the four recognized spheres of the earth, namely, the lithosphere, atmosphere, hydrosphere, and biosphere. The inventory of N in the four spheres is given in Table 5.1.

A striking characteristic of the distribution pattern is the enormous size of the inert reservoir (e.g., N contained in primary rocks) and the large organic-N pool of soil and water. The amount of N held by igneous rocks of the crust and mantle is about 50 times that present in the atmosphere. Relatively small amounts of N are found in the hydrosphere and biosphere.

Origin of Soil N

The original source of combined N in soils (and sediments) was atmospheric N (mainly N_2, with smaller contributions from N oxides). This N, in turn, is believed to have originated from fundamental rocks of the earth's crust and mantle.[11] One popular theory is that the earth was formed by the accretion of solid particles called "planetismals" and that the atmosphere arose through gradual off-gassing from the interior as the newly formed earth was warmed by the heat generated by compression, by decay of radioactive elements, and

TABLE 5.1   Inventory of N in the Four Spheres of the Earth[a]

Sphere	N Content, $\times 10^{16}$ kg
Lithosphere	16,360
Igneous rocks	
a. of the crust	100
b. of the mantle	16,200
Core of the earth	13
Sediments (fossil N)	35–55
Coal	0.007
Sea-bottom organic compounds	0.054
Terrestrial soils	
a. Organic matter	0.022
b. Clay-fixed NH_4^+	0.002
Atmosphere	386
Hydrosphere	
a. Dissolved N_2	2.19
b. Combined N	0.11
Biosphere	0.028–0.065

[a]Most estimates are from Burns and Hardy[3] and Soderlund and Svensson.[9] The values for terrestrial soils are from Stevenson.[1]

possibly by other exothermic processes. Vapors and gases were driven from the interior through evaporation as the temperature rose. Later, as the earth cooled, the vapors condensed to form the oceans. The N, which probably consisted mostly of NH_3, was ejected largely during the early stages of the earth's existence; small quantities have been liberated during the course of geological times, and the process is continuing today. Molecular O_2 of the atmosphere was probably formed through photosynthesis, as well as by photochemical dissociation of water vapor in the atmosphere. As the atmosphere became enriched with O_2, reduced N (NH_3) became oxidized to molecular N_2. Small additions of N have been made to the atmosphere over geologic times by volatilization of N compounds from meteorites during entry into the earth's atmosphere.[11]

Nitrogen in the Hydrosphere

Nitrogen in aquatic systems occurs as inorganic N_2, NH_4^+, NO_2^-, and NO_3^-, as well as dissolved and particulate organic matter.[8] Most of this N (95%) is dissolved N_2 (see Table 5.1), which is inert and is unaffected by chemical or most biological activity in the water. The remaining N occurs in various organic and inorganic forms, but mostly the former. Only the N present as NH_3, NO_3^-, NO_2^-, and organic matter belongs to what might be called the active N fraction.

For all practical purposes, the N reserve of the ocean can be considered to be in a state of quasi-equilibrium. Variations in abundance of the different forms of N occur with depth, season, biological activity, and other factors. However, in the long run, the amount of each form remains relatively constant.

New sources of N to the ocean are the atmosphere and land, from which combined N is carried by rivers and rain. The total amount of N added each year is about 78×10^9 kg, of which 19×10^9 kg, or about one-fourth of the total, is transported by rivers.

The loss of N by deposition of organically bound N into sediments is about 8.6×10^9 kg per year. The unaccounted-for N, about 70×10^9 kg, is the estimate of losses by bacterial denitrification.

Nitrogen in the Biosphere

Many difficulties are encountered in determining the distribution of N in living matter (the biosphere). Unlike other spheres, the biosphere is in a constant and often rapid state of change. Also, living organisms are not uniformly distributed, and the N contents of different organisms vary widely. The estimate in Table 5.1 for total N in the biosphere (0.028–0.065×10^{16} kg) is at best an approximation.

RESERVOIRS OF N FOR THE PLANT-SOIL SYSTEM

A summary of N in various pools of the soil–plant–animal system is given in Table 5.2. The amount of N contained in soils in organic forms or as clay-fixed NH_4^+ far exceeds that which is present in plant-available inorganic forms (NO_3^- and exchangeable NH_4^+). Somewhat more N resides in plant biomass than in animal biomass. In forest systems, plant biomass is a variable-sized pool, depending upon forest type and stand age, but in natural forests, the soil reservoir is usually greater than the plant pool.

The values in Table 5.2 for the amounts of N in the organic matter of terrestrial soils (3.0–5.5×10^{14} kg) are of the same order as those given in Table 5.1 (2.0×10^{14} kg), even though they were derived from entirely different data. The former are from the work of Burns and Hardy[3] and Soderlund and Svensson;[9] those of Table 5.1 were estimated from data for total organic C in soil associations of the world[1] by using the assumption that the average C/N ratio was 10 for mineral soils and 30 for organic soils (Histosols). An average of 10% of the N in the mineral soils was assumed to occur as clay-fixed NH_4^+ (see Chapter 6 for details).

It should be pointed out that the absolute amount of organic N in soils varies greatly and is influenced by the soil-forming factors of climate, topography, vegetation, parent material, and age (see Chapter 2). Organic matter and N content usually decline when soils are first placed under cultivation, with establishment of new steady-state levels that are often substantially lower than those of the virgin soil. The specific cropping system is an important factor in the new levels that are found. The effectiveness of soil management practices in maintaining soil N reserves depends on such factors as rate and frequency of organic residue application, the chemical composition and stability of the materials, N content (C/N ratio), manner of incorporation of the residues into the soil, soil characteristics, and seasonal variations in temperature and moisture.

The distribution of total, organic, and clay-fixed NH_4^+ in soil associations

TABLE 5.2 Total Amounts of N (kg) in Various Pools of the Soil–Plant–Animal System

Compartment	Burns and Hardy[3]	Soderlund and Svensson[9]
Plant biomass	1.0×10^{14}	1.1–1.4×10^{13}
Animal biomass	1.0×10^{12}	2.0×10^{11}
Litter	—	1.9–3.3×10^{12}
Soil organic matter	5.5×10^{14}	3.0×10^{14}
Soil biomass	—	5.0×10^{11}
Fixed NH_4^+	—	1.6×10^{13}
Soluble inorganic	1.0×10^{12}	?

of the world is given in Table 5.3. Because of their high organic matter and N contents, Histosols are major contributors to the total soil N (28.7×10^{12} kg, or more than 10% of the total). It should be noted that Histosols are often considerably deeper than the one-meter depth upon which the traditional calculations were made. Hence, their contribution to the global N pool is probably larger that the estimates indicate. The relatively low amounts of N in the mineral soils of South America are due to the fact that most of them are tropical soils low in organic matter. When calculated on N content per hectare,

TABLE 5.3   Nitrogen in Some Soil Associations of North America, South America, and Other Areas (Values Represent Amounts to a Depth of 1 Meter)[a]

Association	Area, 10^5 km^2	N, 10^{12} kg		
		Organic	Fixed NH$_4^+$	Total
North America				
Histosols	13.3	8.9	—	8.9
Podzols	31.5	5.9	0.7	6.6
Cambisols	9.4	5.0	0.6	5.6
Haplic Hastanozems	32.0	4.6	0.5	5.1
Eutric Gleysols	10.9	3.0	0.3	3.3
Gelic Regosols	15.9	2.9	0.3	3.2
Dystric Cambisols	3.6	1.9	0.2	2.1
Phaeozems	10.4	1.9	0.2	2.1
All others	76.0	10.7	1.2	11.9
Totals	**203.0**	**44.8**	**4.0**	**48.8**
South America				
Ferrasols	89.8	9.7	1.1	10.8
Dystric Histosols	3.8	2.5	—	2.5
Cambisol–Andisols	6.2	3.1	0.4	3.5
Cambisols	4.2	2.3	0.3	2.6
Acrisol–Xerosol–				
Kastanozems	16.6	1.2	0.1	1.3
All others	61.4	3.9	0.4	4.3
Totals	**182.0**	**22.7**	**2.3**	**25.0**
Asia, Africa, Europe, Oceania				
Histosols	26.0	17.3	—	17.3
Cambisols	47.0	25.2	2.8	28.0
Podzols	130.0	24.3	27.0	27.0
Kastanozems	131.0	18.9	2.1	21.0
Chernozems	51.0	18.0	2.0	20.0
Ferrasols	141.0	15.3	1.7	17.0
Cambisol–Vertisols	28.0	9.9	1.1	11.0
All others	282.0	19.8	2.2	22.0
Totals	**836**	**148.7**	**14.6**	**163**
Combined Totals		**216.2**	**20.9**	**237.1**

[a]From Stevenson.[1]

TABLE 5.4   Estimated Amounts of N to Depths of 15 cm and 1 m for the Major Soil Associations of the United States[a]

Soil Association	Approximate Area	Average Amount of N per ha		Total N in Association	
		to 15 cm	to 1 m	to 15 cm	to 1 m
	($\times 10^6$ ha)	($\times 10^3$ kg)	($\times 10^3$ kg)	($\times 10^9$ kg)	($\times 10^9$ kg)
Brown forest	73	3	8	204	546
Red and yellow	61	2	5	134	273
Prairie	46	4	18	178	891
Chermozem and Chermozem-like	50	5	18	249	891
Chestnut	41	3	12	136	496
Brown	21	3	1	59	19
Totals				**960**	**3,213**

[a]From Stevenson.[1]

some tropical soils contain substantially less plant-available N than temperate-zone soils, which is a serious problem in trying to grow crops in the tropics. The so-called "Green Revolution" is based largely on increased fertilizer N input and development of crop varieties that respond to heavy fertilizer additions.

Estimated amounts of N to depths of 15 and 100 cm for the major soil associations of the United States are recorded in Table 5.4. Most of the N resides in Mollisols (listed as Prairie, Chernozem, and Chestnut soils).

According to Rosswall,[7] about 95% of the N that cycles annually within the pedosphere interacts solely within the soil–microbial–plant system. On this basis, only 5% of the total global flow is concerned with exchanges to and from the atmosphere and hydrosphere. However, losses of fertilizer N through leaching and denitrification in some agricultural systems are much higher than the 5% global value might suggest.

The average mean residence time (MRT) for N in soils has been estimated to be about 175 years.[7] Some components will have much greater ages, perhaps 1000 years or more.

NITROGEN AS A NUTRIENT

Nitrogen occupies a unique position among the soil-derived elements essential for plant and microbial growth because of the rather large amounts required by most agricultural crops and microorganisms in comparison to other elements. A deficiency of N in plant tissue is shown by yellowing of the leaves and by slow and stunted growth. For many microorganisms, N is a key limiting factor for vegetative growth and population increase. Other factors being favorable, an adequate supply of N in the soil promotes rapid plant growth and the development of dark-green color in the leaves. Major roles of N in plant nutrition include, being (1) a component of chlorophyll, DNA, and RNA, (2) essential for carbohydrate utilization, (3) a constituent of enzymes and other proteins, vitamins, and hormones, (4) stimulative of root development and activity, and (5) supportive to the uptake of other nutrients.[13]

Assimilative Nitrate Reduction

This process is catalyzed by a wide range of organisms, including most plants and many bacteria, algae, and fungi, which have the ability to reduce NO_3^- to NH_3, which is subsequently used for synthesis of protein and other cellular components.[14-15a] Nitrogen gas (N_2) is not an intermediate in this pathway. The enzymes that carry out this process are different than the enzymes used for dissimilatory NO_3^- reduction (denitrification, discussed below), and the end result is immobilization of N as organic N. In contrast, denitrifying organisms produce N_2O, N_2, and/or NH_4^+ as end products. Nitrate is the main form of N taken up by most crop plants, the most notable exception being lowland rice, which uses primarily NH_4^+–N. Assimilation of NO_3^-–N is a three-step process:[15a]

$$NO_3^- + 2e^- \xrightarrow[\text{reductase}]{\text{nitrate}} NO_2^-$$

$$NO_2^- + 6e^- \xrightarrow[\text{reductase}]{\text{nitrite}} NH_3$$

$$NH_4^+ + \text{organic-C acceptor} \xrightarrow[\text{enzymes}]{\text{several}} \text{amino acids}$$

Reduction of NO_3^- to NO_2^-—the rate-limiting step in the transformation—is catalyzed by nitrate reductase, a metalloprotein containing Fe and Mo that is found in both root and aerial tissues.[14,15]

The second step is catalyzed by nitrite reductase in the leaf tissue. Although the reaction involves the transfer of six electrons, no other free intermediates (e.g., hydroxylamine or hyponitrite) are known. In some systems, the electrons required for reduction are provided by ferredoxin, an Fe–S protein. Nitrite may be protonated to HNO_2 before transport across the chloroplast membrane. The overall sequence is:

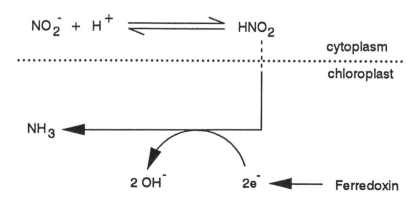

Since NO_2^- formation is the step that limits the rate of synthesis of reduced $N(NH_4^+$, amino acids), the activity of nitrate reductase is sometimes regarded as a good indicator of plant growth rate.[17] The level of nitrate reductase in plants varies during the day, over the course of the growing season, and during periods of moisture or heat stress. Considerable energy is required for these reactions; thus, reduction is closely linked to photosynthesis.

The NH_4^+ produced by nitrite reductase seldom accumulates in cells but is rapidly metabolized and incorporated first into the amino acids glutamine, glutamic acid and aspartic acid, and then into other amino acids and N-containing biochemicals. Viets and Hageman[17] and Lea[15] discuss pathways of N metabolism in plants.

Nitrogen Requirements of Plants

The amount of N consumed by plants varies greatly from one species to another, and for any given species, the amount varies with genotype and the environment.[17] Also, considerable variation exists in the relative amount of the N contained in the different plant parts (grain, stems, leaves, roots, etc.). Examples of N removals in the harvested portion and residues of some important crops under conditions of good yields are given in Fig. 5.2. Substantial variation from the reported values can occur depending on soil N status, fertilization practice, and climate. In general, more N is contained in the harvested portion than in the stover, vines, straw, or roots because most grain crops transfer most of the N into the seeds. Nitrogen accumulation by plants

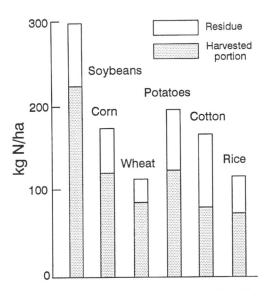

Fig. 5.2 Nitrogen contained in the harvested portion and residue of good yields of some major agricultural soils. Adapted from Olson and Kurtz.[16]

is very rapid during the period of rapid vegetative growth, as illustrated for three crops in Fig. 5.3.

GAINS IN SOIL N THROUGH BIOLOGICAL N_2 FIXATION

Except for additions from irrigation water and flooding, practically all of the N that enters the soil by natural processes on a global basis is derived from biological N_2 fixation and atmospheric deposition of NH_3, NH_4^+, and NO_3^-. Nitrogen is also added in crop residues and animal manures, but these represent recycling within the soil–plant system.

Although a vast supply of N occurs in the earth's atmosphere (386×10^{16} kg), it is present as an inert gas and cannot be used by either higher plants or animals. The covalent triple bond of the N_2 molecule ($N{\equiv}N$) is highly stable and can be broken chemically only at elevated temperatures and pressures. Nitrogen-fixing microorganisms, on the other hand, perform this difficult task at ordinary temperatures and pressures. Nitrogen fixation in prehistoric times created the combined N that is currently present in many commercially important natural deposits, such as coal, petroleum, and the caliche of the Chilean desert.

Because of the extensive research on biological N_2 fixation, only a brief résumé can be given here. For detailed information, the reader is referred to several books and reviews on the subject.[3,18–33]

Data given later on global N fluxes will show that the total amount of N added to the earth each year through biological N_2 fixation is about 139×10^9 kg, of which about 60% (89×10^9 kg) is contributed by nodulated legumes grown for grain, pasture, hay, and other agricultural purposes.[3,4,8] Esti-

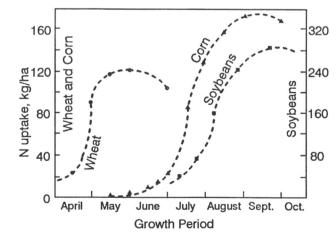

Fig. 5.3   Average rates of N accumulation in the aboveground crop of nonirrigated wheat and irrigated corn and soybeans in Nebraska. Adapted from Olson and Kurtz.[16]

mated average rates of biological N_2 fixation for some specific organisms and associations are given in Table 5.5.

The ability of a few bacteria, actinomycetes, and cyanobacteria to fix molecular N_2 can be regarded as second in importance only to photosynthesis for the maintenance of life on Earth. This ability is confined to procaryotic organisms, with the exception of one N_2-fixing fungus. The two basic biochemical processes in nature are often considered to be photosynthesis and respiration; to this list should be added biological N_2 fixation and possibly denitrification because of their pivotal role in the fate of N in the environment.

Numerous approaches have been used to estimate the extent of biological N_2 fixation in natural ecosystems, including:[20,34]

1. The increase in total combined N in the system under study. The Kjeldahl method for total N is usually used for this purpose. The method is imprecise, since only net N accumulation is measured; i.e., the value is:

Apparent N fixed = Actual N fixed + other N-sources − N losses

TABLE 5.5   Estimated Rates of Biological N_2 Fixation for Specific Organisms and Associations[a]

Organism or System	N_2 Fixed, kg/ha/yr
Blue-green algae	25
Free-living microorganisms	
Azotobacter	0.3
Clostridium pasteurianum	0.1–0.5
Plant–algal associations	
Gunnera	12–21
Azolla	313
Lichens	39–84
Legumes	
Soybeans (*Glycine max L. Merr.*)	57–94
Cowpeas (*Vigna, Lespedeza, Phaseolus,*	
and others)	84
Clover (*Trifolium hybridum L.*)	104–160
Alfalfa (*Medicago sativa L.*)	128–600
Lupines (*Lupinus sp.*)	150–169
Nodulated nonlegumes	
Alnus	40–300
Hippophae	2–179
Ceanothus	60
Coriaria	150

[a]From Evans and Barber.[19] Results reported for some legumes (e.g, clover, alfalfa, lupines) are unusually high (see results given in Table 5.9).

This method has other limitations, the most serious of which is that N from sources other than N_2 gas will contribute. These sources include atmospheric deposition and movement of N from the subsurface, both of which result in artificially high values for N_2 fixation.

2. Accumulation of N using nodulating and nonnodulating soybean isolines. In this case, the amount fixed is equal to the N content of the nodulating strain (N derived from N_2 and from soil and/or atmospheric deposition of NH_3–N and NO_3^-–N) minus that of the nonnodulating strain (N from soil and atmospheric deposition only).

3. The amount of ^{15}N–N_2 taken up by the test plant when grown in a closed chamber containing ^{15}N–N_2. This method is the definitive test for N_2-fixing activity.

4. Growth of legumes in soil where the organic matter has been enriched with ^{15}N. A lower ^{15}N content of the plant tissue suggests an alternative N source (i.e., ^{14}N–N_2 from the atmosphere).

5. Estimates of available soil N (A-values) as measured from the amounts of applied ^{15}N taken up by a nonleguminous plant and the legume. Conceptually, this approach is similar to item 2, but it requires the assumption that the nonlegume and legume have equal abilities to accumulate soil N, which is not always true.

6. Delta ^{15}N values (see Chapter 6) for nonleguminous and leguminous plants grown on the same soil. This approach is based on the observation that the ^{15}N content of the soil N is higher than for N_2 of the atmosphere.

7. The acetylene reduction method. In the presence of the enzyme nitrogenase, acetylene (CH≡CH) is reduced to ethylene (CH_2=CH_2). Determination of ethylene by gas–liquid chromatography provides a simple, relatively inexpensive, and sensitive estimate for nitrogenase activity, although the values cannot be used directly to assess quantitatively the fixation of N_2.[25,26]

Each method has its advantages and disadvantages, as discussed elsewhere.[20,34] The most direct approach is through use of ^{15}N–N_2, but a closed system is required, so this method is not suitable for large plants, extended experiments, or during high sunlight conditions where heating inside the chamber can modify plant physiology. A popular technique used at the present time for demonstrating N_2 fixation under field conditions is acetylene reduction.[34] An open cylinder is buried in the soil, plants are grown, the soil surface is sealed, and, at intervals, an acetylene–air mixture is allowed to flow through the nodulated root zone. The gas leaving the exit port is analyzed for ethylene, from which N_2 fixation rates are estimated.[20] Since ethylene is a phytohormone, this method can give aberrant results because of modified plant morphology or physiology.

The biochemical process whereby molecular N$_2$ is converted to a reduced form (NH$_3$) has been worked out in great detail for a variety of free-living organisms and symbiotic systems.[26] The overall equation is:

$$N_2 + 8e^- + 8H^+ + \sim 25\ ATP \xrightarrow{\text{nitrogenase}} 2NH_3 + H_2 + \sim 25\ ADP + \sim 25P_i$$

The nitrogenase system consists of a four-protein complex, with two molecules of each of two different proteins per active enzyme. There are substantial differences in the molecular weight of the enzyme from different organisms, but in all cases, nitrogenase is a large and very complex enzyme. The first subunit has a molecular weight of about 180,000 and contains both Mo and Fe; the second has a molecular weight of about 51,000 and contains nonheme Fe. The larger Mo–Fe protein is the N$_2$ reductase enzyme, while the smaller Fe protein provides the electrons for reduction. Other electron transporting agents include ferredoxin, an Fe–S protein, or flavodoxin.

A postulated reaction sequence in biological N$_2$ fixation for free-living anaerobic bacteria (eg., *Clostridium*) is illustrated in Fig. 5.4.[41] In this organism, the electrons (reductant) are produced during sugar fermentation and metabolism of pyruvate; ATP furnishes energy for various steps of the reduction process. The NH$_3$ thus formed is used for the synthesis of amino acids and other N-containing biochemicals, as noted for assimilatory nitrate reductase.

Other compounds with triple bonds and atoms of a similar size as N are also reduced:

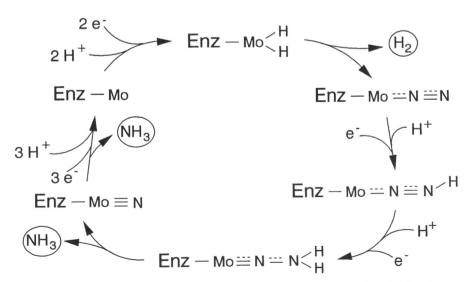

Fig. 5.4   Scheme for reduction of N$_2$ by nitrogenase. Adapted from Richards.[41]

$$\text{HC}\equiv\text{CH} + 2e^- + 2\text{H}^+ \rightarrow \text{H}_2\text{C}=\text{CH}_2$$
acetylene   ethylene

$$\text{C}\equiv\text{N}^- + 8e^- + 8\text{H}^+ \rightarrow \text{CH}_4 + \text{NH}_4^+$$
cyanide

The organisms that fix N_2 are conveniently placed into three groups:

1. Nonsymbiotic (free-living) fixers, or those that fix molecular N_2 apart from a specific host plant
2. Associative fixers, which form relatively casual and poorly structured relationships with roots or aerial portions of plants
3. Symbiotic fixers, which fix N_2 in organized associations with higher plants, including some nonlegumes.

The nonsymbiotic fixers include photoautotrophs like the cyanobacteria of the family Nostocaceae and various photosynthetic bacteria (e.g., *Rhodospirillum*), as well as heterotrophic aerobic bacteria (e.g., *Azotobacter, Beijerinckia, Derxia*) and anaerobic bacteria, including some *Clostridium* species and the sulfate-reducing bacteria. Some free-living actinomycetes and fungi have been reported to fix N_2, but in many cases, these claims were not verifiable when the $^{15}N–N_2$ method was used, and therefore the apparent N_2 fixation may have resulted from the acquisition of combined N from air, water, or culture media components.

Cyanobacteria

Many cyanobacterial species are N_2 fixers, including species of the genera *Anabaena, Nostoc,* and *Aulosira. Cylindrospermum,* and *Calothrix,* among others.[35,36] Nitrogen-fixing cyanobacteria function as microsymbionts in various genera of the gymnosperm family *Cycadacea*, in the angiosperm genus *Gunnera* (family Haloragaceae), and in the aquatic fern, *Azolla*.

The cyanobacteria are a group of archaic organisms that have persisted during long epochs of the earth's history. They can be found in practically every environment where sunlight is available for photosynthesis, such as surface waters, uninhabited wastelands, and barren rock surfaces. They can colonize virgin landscapes because they are completely autotrophic and thereby able to synthesize all of their cellular material from CO_2, N_2, water, and mineral salts. In addition, they form symbiotic relationships with a variety of other organisms, such as the lichen fungi.

Geographically, lichens are widely distributed over land masses of the earth. They make up a considerable portion of the vegetation in northern arctic and high-altitude regions (tundra). Besides being the pioneering plants on virgin landscapes, they bring about the disintegration of rocks to which they are attached, thereby forming soil in which higher plants can get a start. In desert areas of the southwestern United States, they form surface crusts of varying density that stabilize the sand surface, and they cling to surface stones. Crusts of cyanobacteria have been found in semiarid soils of eastern Australia and the Great Plains of the United States, where their favorite habitat is the undersurface of translucent pebbles.

The importance of cyanobacteria in supplying fixed N in most agricultural soils is probably limited to the initial stages of soil formation. These organisms fix N$_2$ only in the presence of sunlight; consequently, their activity is confined almost exclusively to superficial layers of the earth's crust where there is little or no competition for light from higher plants or where the material is too compacted to permit penetration of plant roots. There is abundant evidence to indicate that cyanobacteria are important agents in fixing N$_2$ in rice paddy fields, as well as eutrophic lakes and other aquatic systems.

Azolla, a freshwater fern, is used extensively in parts of Southeast Asia as a green manure crop and substitute for N fertilizers in rice paddy culture.[36–39] This plant forms a symbiotic relationship with the cyanobacterium *Anabaena azollae.* The microbe inhabits a cavity on the ventral surface of *Azolla* fronds and lives symbiotically with the fern while fixing N$_2$.[40,41] The practical importance of the relationship to rice production in many parts of Asia cannot be overemphasized. The fern grows readily over the paddy water, often covering the surface completely as an *Azolla* "bloom." *Azolla* is the principal source of N for over one million hectares of rice in China.[42] *Anabaena* is capable of fixing N$_2$ in a free-living state (unlike many of the bacterial symbionts described below), but it is through association with *Azolla* that most if not all of the N$_2$ is fixed (encapsulation within *Azolla* fronds restricts predation by invertebrates and fish, in contrast to free-living cells, which are easy prey).

Free-Living Bacteria

Classical examples of N$_2$-fixing, free-living bacteria are the photosynthetic *Rhodospirillum,* the anaerobic heterotroph *Clostridium,* and the aerobic heterotroph *Azotobacter,* as well as *Beijerinckia* and *Derxia.*[43]

Photosynthetic N$_2$-fixing bacteria require light and anaerobic conditions, both of which restrict their activities to shallow, eutrophic waters, the surfaces of shallow sediments, or estuarine muds. They generally are found as a layer overlying the mud and covered by a layer of algae; fixation of N$_2$ is possible because pigments of the photosynthetic bacteria absorb light in the region of the spectra not absorbed by pigments of the overlying algae. In comparison to the cyanobacteria, their contribution to the N-economy of aquatic systems is small.

The anaerobic fixer *Clostridium* is universally present in soils, including those too acidic for *Azotobacter.* The normal condition of *Clostridium* is the spore form, vegetative growth occurring only during brief anaerobic periods following rains, and therefore its contribution to the soil N-economy is very limited.

Azotobacter is also widely distributed in soils. The most common species, *A. chroococcum*, is found world-wide but mostly in near-neutral or slightly alkaline soils. In contrast, *Beijerinckia* and *Derxia* are typical inhabitants of acidic soils. *Beijerinckia* has been found in soils of India, Southeast Asia, tropical Africa, South America, and northern Australia, but is essentially absent from temperate-zone soils.

Under natural soil conditions, N_2 fixation by free living heterotrophic bacteria is greatly limited by the low amount of available energy. For these organisms, the energy required to fix N_2 is similar to that required for the synthesis of all other cell components. Hence, the activity of these organisms so far as N_2 fixation is concerned is low, even in environments with relatively high organic matter contents. In an early discussion of the subject, Jensen[45] suggested that many of the estimates for N_2 fixation by nonsymbiotic N_2-fixing bacteria—frequently as high as 20 to 50 kg/ha/yr—were much too high. He concluded that the level of available organic matter in most soils was too low to support fixation of this magnitude. The consensus of many soil scientists is that 6 kg N/ha/yr or less are added to soils of the United States by the combined activities of nonsymbiotic N_2-fixing microorganisms; in semiarid soils, no more than 3 kg N/ha/yr is fixed. On this basis, the amount of N_2 fixed by nonsymbiotic N_2 fixers in soils under intensive cultivation would appear to be too low to have much practical impact. However, for noncultivated soils of semiarid regions, 3 kg N/ha/yr represents about 50% of the new N added each year.

Conditions for optimum N_2 fixation by free-living microorganisms include the presence of adequate energy sources (light or readily degraded organic materials), low levels of available soil N, adequate mineral nutrients, near-neutral pH, and suitable moisture. In view of the large number of microorganisms that have been reported to fix molecular N_2, it would appear that N gains under field conditions result from the cumulative action of numerous organisms fixing rather small amounts of N rather than through fixation by only one or two organisms. Gains in soil N through the activities of free-living N_2-fixing microorganisms are reviewed elsewhere.[46,47]

A variety of arguments support the view that free-living bacteria do not provide large amounts of combined N for plants in most cultivated soils, including:

1. Free-living N_2 fixers are heterotrophic and inefficient users of carbohydrates, with only 1 to 10 mg of N being fixed per gram of carbohydrate used.

2. Heterotrophic N_2 fixers must compete with other bacteria, actinomycetes, and fungi for available organic matter, which is normally in short supply in most soils.

3. Fixation of N$_2$ is greatly reduced in the presence of readily available combined N because synthesis of the nitrogenase system is repressed (prevented). In productive agricultural soils, levels of available N are often sufficiently high to seriously inhibit fixation of N$_2$.

Substantial gains in N, frequently on the order of 66 to 112 kg/ha/year, have been reported in many soils in the apparent absence of legumes or other plants known to form a symbiotic relationship with N$_2$-fixing microorganisms. Some investigators believe that these increases cannot be attributed entirely to errors inherent in measuring N gains under field conditions, but that, under certain circumstances, significant amounts of N can be added to the soil through the combined activities of free-living, N$_2$-fixing organisms. For example, the incorporation of organic residues in soil may enhance fixation when used as an energy source by nonsymbiotic N$_2$ fixers. Results in this case have been mixed, with some studies indicating that additional C-sources enhance N$_2$ fixation, with other studies showing no benefit.

Considerable difficulty is encountered in evaluating reports for gains in soil N under field conditions. In addition to faulty experimental techniques, lack of statistical control, and errors inherent in measuring small increases in soil N by the Kjeldahl method against a background of large amounts of total N in soil, the possibility exists that N had been added by other means, such as upward movement of NO$_3^-$ in solution, recycling via plant roots, and accretion from the atmosphere or laboratory air.

There are indications that nonsymbiotic N$_2$ fixers can make significant contributions to the soil N because appreciable gains in soil N have been observed for legume-free grass sods, which suggests that extensive (albeit, erratic) fixation can occur in the rhizosphere of crop plants. In addition, the ability to grow crops continuously on the same land for years without N fertilizers, and without growing legumes, is well known and may be due in part to nonsymbiotic N$_2$ fixation.

At one time, Russian scientists claimed that inoculation of soils and seeds with *Azotobacter* and *Clostridium* resulted in improved growth of wheat and cotton. Extensive programs of inoculation were carried out in Russia and elsewhere from 1930 to the 1970s. Attempts were made at various times to confirm the early Russian claims, but, in general, the findings failed to show increases in growth that could be ascribed positively to N$_2$ fixation.[48]

Data summarizing N gains in soil as reported by Moore[47] are given in Table 5.6. In some, but not all cases, nonsymbiotic organisms may have been responsible for fixation of N$_2$. For example, some research was conducted on grass plots kept legume-free in order to assess the exact contribution of nonsymbiotic fixers.

The values in Table 5.6 represent the typical or usual range reported by numerous investigators. Both lower and higher N gains were reported, with some of the higher values exceeding 220 kg/ha/yr. The actual amount of N$_2$ fixed under most agricultural systems is probably much less than suggested by values in Table 5.6.

TABLE 5.6   Gains in Soil N Through Biological N_2 Fixation[a]

Number of References Cited	Conditions	N Gains, kg/ha/yr
5	Soil amended with crop residues	15–78
	Field plots under sod-like crops	14–56
4	Lysimeter studies	25–67
4	Stands of *Pinus* sp. or other monoculture, nonnodulated trees	36–67

[a]As recorded by Moore.[47]

Many of the unexplained increases in soil N have been associated with the growth of grasses, indicating that a specialized association may exist between the grasses and certain N_2-fixing microorganisms. An associative symbiosis has been observed between certain rhizosphere bacteria and the root surfaces of corn, wheat, and tropical grasses.[20,43,44] An additional complication in interpretation of older results with nonleguminous plants is the discovery of *Azospirillum,* a bacterium that forms relatively casual associations with the roots of some grasses and that can fix substantial amounts of N_2 under some conditions.[27,31,50,51]

The rhizosphere, or that part of the soil influenced by plant roots,[52] would appear to be a particularly favorable site for N_2 fixation because of organic material excreted or sloughed off by roots. In natural plant communities, a relatively low rate of fixation (<10 kg/ha/year) may be adequate, whereas under intensive farming conditions, fixation of rather large amounts (often >200 kg/ha/yr) would be required for optimum yields. Losses of substantial amounts of organic compounds from the roots may be detrimental to the plant, however, since those compounds could also be used to produce more plant material. In this case, any benefits derived from N_2 may be offset by loss of energy-rich compounds for plant biomass production or root growth.

Symbiotic N_2 Fixation by Leguminous Plants

The symbiotic partnership between bacteria of the genera *Rhizobium* and *Bradyrhizobium* and leguminous plants has a long history in agriculture and the basic sciences. The importance of this relationship is emphasized by the fact that even with the tremendous increase in fertilizer N usage over the past four decades, legumes are a major source of fixed N for a large portion of the world's soils. On the conservative estimate of an average fixation of 55 kg of N for the 84 million hectares of legumes planted each year in the United States, a total of over 2×10^9 kg of N are fixed annually. Legumes of agricultural significance can be broadly divided into grain and forage legumes, the distinction being that seeds of the former are harvested for food (e.g., peas and beans), whereas the foliage of the latter is used for animal feed.

LaRue and Patterson[53] and Peoples et al.[49] suggested that legumes as sources of N are certain to increase in importance. Factors leading to more extensive use of legumes or other N-fixing plants include the need to (1) make marginal lands more productive, (2) control erosion and desertification, (3) reduce fertilizer costs, especially as fertilizer N becomes more expensive, and (4) reclaim drastically disturbed land, such as from strip mining.

The Leguminosae family contains from 10 to 12 thousand species, most of which are indigenous to the tropics. Thus far, only about 1200 species have been examined for nodulation, of which about 90% have been found to bear nodules. Less than 100 species are used in commercial food production. Earlier work on this association is summarized in Quispel.[27] Whereas greatest attention has been given to the cultivated legumes, wild species are of considerable importance for fixation of N$_2$ in natural ecosystems and may become important new genetic resources for molecular biologists.

Norris[54] summarized available information regarding the global distribution of the Leguminosae, from which he compiled the tribal and species distribution. A summary is given in Table 5.7. A smaller number of genera and species are indigenous to the temperate regions of the earth than to the tropics and subtropics. Some generalizations regarding the distribution of the Leguminosae are possible. With the exception of one genus, the subfamily Mimosoideae (141 species) is entirely tropical and subtropical, while 89 of the 95 genera in the subfamily Caesalpinioideae (over 95% of the species) are confined to the tropics and subtropics. In the subfamily Papilionate, 141 genera of plants (3084 species) occur in temperate regions, while 176 genera (2430 species) occur in the tropics and subtropics.

The bacterial symbionts, primarily members of the genus *Rhizobium* or *Bradyrhizobium,* are Gram-negative, nonspore-forming rods. The species listed in Table 5.8[55,56] are generally recognized. A group of leguminous species that exhibit specificity for a common *Rhizobium* or *Bradyrhizobium* spe-

TABLE 5.7   Distribution of the Leguminosae^a

Subfamily	Number of Genera and Species	Genera and Species in Tropics and Subtropics	Genera and Species in Temperate Regions	Genera Occurring in Both Tropic and Temperate Zones
Mimosoideae	31–1341	31–1200	1–141	1
Caesalpinioideae	95–1032	89–988	7–44	1
Papilionateae	305–6514	176–2430	141–3084	12
Total	431–8887	296–4618	149–3269	14

^{a}From Norris.[54]

TABLE 5.8 Classification Scheme of Common *Rhizobium*–Legume Associations

Current Species Name	Cross-Inoculation Group (historic)	Host Genera	Historic Species Name
Rhizobium meliloti	Alfalfa	*Medicago*, Alfalfa *Melilotus*, sweet clover *Trigonella*, fenugreek	*R. meliloti*
R. leguminosarum bv. viciae	Pea	*Pisum*, pea; *Vicia*, vetch *Lathyrus*, sweetpea	*R. leguminosarum*
bv. trifolii	Clover	*Trifolium*, clovers	*R. trifolii*
bv. phaseoli	Bean	*Phaseolus*, beans	*R. phaseoli*
Bradyrhizobium japonicum	Soybean	*Glycine*, soybean *Vigna*, cowpea	*R. japonicum*
B. elkanii		*Lespedeza*, lespedeza	
R. fredii		*Crotalaria*, crotalaria	
R. xinjiangensis (broad host range among these species)	*Pueraria*, kudzu	*Arachis*, peanut *Phaseolus*, lima bean	
R. tropici	Bean	*Phaseolus vulgaris*, *Leucaenae*	*R. phaseoli*
R. etli	Bean	*Phaseolus vulgaris*, common bean	*R. phaseoli*
R. loti		*Lotus*	

cies is referred to as a "cross-inoculation group." The validity of these groupings has often been challenged because the boundaries between the groups overlap, some bacterial isolates that are genetically distant from each other nodulate the same host plants, and some strains of rhizobia form nodules on plants occurring in several different groups.

Estimates for N fixation by various legumes are tabulated in Table 5.9.[49] Much of the work has been carried out using lysimeters or controlled experimental plots, and extrapolation of the data to agricultural soils in general is speculative. LaRue and Patterson[53] and Peoples et al.[49] concluded that:

1. Few legume crops satisfy all N requirements by fixation, with the balance of the required N being obtained from soil. The highest percentages (80%) of N$_2$-derived nitrogen are typical of low-fertility soils or soils made artificially low in plant-available N by amendment with carbonaceous organic residues.

2. Some legumes (e.g., soybeans) may actually deplete soil N because N removal in the grain exceeds the amount of N derived from N$_2$ fixation.

Most rhizobia are capable of prolonged independent existence in the soil without the host plant; however, N$_2$ fixation takes place only when symbiosis is established with the plant. Maintenance of a satisfactory population of any given *Rhizobium* species in the soil depends largely upon the previous occurrence of the appropriate leguminous plant. High acidity, lack of necessary nutrients, poor physical condition of the soil, elevated temperatures of sandy soils, and attack by bacteriophages contribute to their disappearance. The

TABLE 5.9   Estimates of N$_2$ Fixation by Some Typical Legumes[a]

Forage Crops		Pulses	
Species	kg N/ha	Species	kg N/ha
Alfalfa (*Medicago sativa*)	148–290	*Phaseolus vulgaris*	10
White clover (*Trifolium repens*)	128–268	*Pisum sativum*	17– 69
Ladino clover (*Trifolium repens*)	165–189	*Vicia faba*	121–171
Red clover (*Medicago pratense*)	17–154	Lupine	121–157
Subclover (*Trifolium subterraneum*)	21–207	Chick pea	67–141
Egyptian clover (*Trifolium alexandrimum*)	62–235	Lentil	62–103
Vetch (*Vicia villosa*)	184	*Arachis hypogea*	87–122

[a]Adapted from LaRue and Patterson,[53] from which specific references can be obtained. A wide range was reported for the amounts fixed by soybeans (15 to over 200 kg N/ha).

desirability of legume inoculation to ensure nodulation with host specific effective rhizobial strains is not well established.[57,58] The subject of legume seed inoculants is covered by Roughley.[59]

The major factors that affect symbiotic N_2 fixation are those parameters that define optimal conditions for plant growth, such as light, temperature, water relations, and soil pH, among others.[32] Maximum fixation is obtained only when the supply of available mineral N in the soil is low because high levels of combined-N depress nodulation formation and inhibit N_2 fixation by existing nodules. However, during the early stages of plant growth, small amounts of fertilizer N may improve nodulation and N_2 fixation, especially on N-poor soils. Presumably, the added N alleviates the N starvation period that can occur between the exhaustion of seed N and the onset of N_2 fixation. The optimum amount of N required varies with the leguminous species.

In many places in the world, an abundance of fertilizer N at reasonable cost has prompted a reevaluation of the role of legumes in crop production. It now seems certain that, under certain circumstances, legumes in a rotation can be replaced effectively by nonlegumes, provided that chemically fixed N is applied. If crop residues are returned to the soil following harvest, increased plant production brought about by adequate fertilization may allow nonlegumes to assume some of the functions historically assigned to legumes, namely, to improve soil tilth, prevent erosion, increase the storehouse of soil N, and enhance the activities of desirable microorganisms in such a way that they can enhance the soil as a dynamic medium for the growth of plants.

The extent to which fertilizer N will replace legumes in crop rotations will depend upon the availability of inexpensive N fertilizers, the need for increased acreage of nonleguminous crops, and the ability of legume-free cropping systems to maintain soil fertility and prevent erosion.

Microorganisms Living in Symbiosis with Nonleguminous Plants

Nitrogen fixation in a symbiotic relationship analogous to the leguminous association has been demonstrated for many angiosperms, including plants belonging to the families Betulaceae, Casuarinaceae, Coriariaceae, Elaegnaceae, Myricaceae, Rhamnaceae, and Rosaceae. Many species are shrubs or trees that occur naturally on poor soils, where they function as "pioneering plants" and become established due to their ability to obtain N through biological N_2 fixation. Few of these plants are of agricultural importance, although they have been used to provide N to non-N_2-fixing tree species during colonization.[60] The geographical distribution of the nonleguminous families for which N_2 fixation has been confirmed is outlined in Table 5.10. Nodulated nonlegumes occur in 14 genera or 7 families of dicotyledon plants. The genera of nodulated nonlegumes include about 300 plant species, of which about one-third have been reported to bear nodules. The endophyte is an actinomycete (*Frankia*) that can be isolated from nodules of only a few genera of plants bearing actinorrhizal nodules.[61] Documentary evidence for N_2 fixation

TABLE 5.10   Distribution of Nodulated Nonlegumes

Family and Genera	Incidence of Nodulating Species[a]	Geographical Distribution
Betulaceae		
Alnus	25/35	Cool regions of the northern hemisphere
Casuarinaceae		
Casuarina	14/45	Topics and subtropics, extending from East Africa to the Indian Archipelago, Pacific Islands and Australia
Coriariaceae		
Coriaria	12/15	Widely separated regions, chiefly Japan, New Zealand, Central and South America, the Mediterranean region
Elaeagnaceae		
Elaeagnus	9/45	Asia, Europe, North America
Hippophae	1/1	Asia, and Europe, from the Himalayas to the Arctic Circle
Shepherdia	2/3	Confined to North America
Myricaceae		
Myrica	12/35	Temperate regions of both hemispheres
Rhamnaceae		
Ceanothus	30/55	Confined to North America
Discaria	1/10	Temperate, subtropical and tropical regions
Rosaceae		
Cerocarpus	1/20	Cool regions of temperate zone
Dryas	3/4	Cool regions of temperate zone
Purshia	2/2	Cool regions of temperate zone

[a]Incidence refers to ratio of species bearing nodules to total number of species as reported by Silver.[63] See also Becking.[62]

by these associations is given in several reviews.[20,62,63] The family Betulaceae, the birches and alders, is found almost entirely in the cool temperate and arctic zones of the northern arctic hemisphere. Thus far, only the alder (*Alnus*) has been found to bear nodules. Crocker and Major[64] estimated an annual gain of 62 kg/ha by *A. crispa* during colonization of the recessional moraines of Alaskan glaciers.

Fourteen of the 45 species of Casuarinaceae, the main nonleguminous angiosperm family of nodulating plants occurring in tropical and subtropical areas, are N₂-fixing. Plants of this family are of great ecological significance in the Australian environment.[65]

Twelve of the 15 species of the family Coriariaceae (genus *Coriaria*) have been found to bear nodules. Bond and Monsterrat[66] suggested that the discontinuous distribution of this family indicates that in ancient times it made a far greater contribution to the supply of fixed-N than at present.

The family Elaeagnaceae, consisting of the genera *Elaeagnus, Hippophae,* and *Shepherdia,* is distributed widely in the temperate regions of both hemispheres. The genus *Elaeagnus,* with 45 species (9 of which bear nodules), occurs in Asia, Europe, and North America. *Shepherdia,* a plant confined to North America, consists of 3 species, 2 of which nodulate. Crocker and Major[64] reported that *Shepherdia,* in company with *Alnus,* colonized the moraines of receding glaciers in Alaska.

The family Myricaceae (the galeworts) is distributed widely in the temperate regions of both hemispheres. It consists of about 35 *Myrica* species, of which 12 species have nodules. A few species occur in the tropics. Bog myrtle (*M. gale*) may be involved in N_2 fixation in acid peats.

The family Rhamnaceae has 40 genera of trees and shrubs that are spread over most of the globe. However, 30 of the 31 species in this family that have been reported to nodulate occur in the genus *Ceanothus,* consisting of about 55 species confined to North America. More than half of the species are found in the southwestern United States. The accumulation of N during soil development on the Mount Shasta mudflows in California has been attributed to N_2-fixing *Ceanothus* species.[67]

The family Rosaceae occurs typically in cool regions of the temperate zone. Species occurring in three genera have been reported to nodulate.

Future Trends in Biological N_2 Fixation

Considerable attention has been given to ways of maximizing biological N_2 fixation as a source of combined N for plants. Interest in this subject has developed from the urgency to solve practical problems related to energy, the environment, and world food requirements. Some ideas for modification of naturally occurring biological N_2 fixing systems, as developed during the late 1970s and early 1980s, are as follows:[19,68,69]

1. Transfer N_2-fixing genes from bacteria to higher plant cells, thus endowing the plant with the capability of using molecular N_2 without forming a symbiotic association with the bacteria.
2. Transfer N_2-fixing genes into a parasitic bacterium capable of invading plant cells and establishing an effective N_2-fixing system analogous to the leguminous nodule.
3. Use protoplast fusion methods to create new symbiotic associations between microorganisms and higher plants.
4. Select or develop, by genetic means, N_2-fixing bacteria capable of living on the roots of such cereal crops as corn and wheat and providing adequate fixed N for optimum plant growth.
5. Develop, by genetic manipulation, rhizobial strains and/or legume varieties that are insensitive to soil NH_4^+ and NO_3^- concentrations that normally inhibit nodulation and N_2 fixation.

6. Develop, by use of plant-breeding methods, legumes that have increased photosynthetic capabilities and, therefore, greater capacities for providing energy to N_2-fixing bacteria in the nodules.

As further basic information about N_2-fixing systems became available, it became apparent that many of these ideas would not bear fruit for an extended time because the systems were far more complicated than originally envisioned.[70,71] For example, energy supply for N_2 fixation is a critical limitation to the amount of N_2 that is fixed. Hence, transferring genes into a new plant species would not necessarily result in high rates of N_2 fixation and good crop yields unless photosynthetic capacity was also increased. Because of the high energy requirement of N_2 fixation, a corn plant must sacrifice energy to support N_2 fixation, energy that would otherwise be used for grain production, with the result that yield would decline. In many cases, fertilizer N may be less expensive than the so-called "free N" from N_2 fixation. The number of collateral genes necessary for N_2 fixation (those in addition to the nitrogenase enzyme) are greater than initially envisioned, with the result that the difficulty of successfully transferring all genes of the N_2-fixation complex between organisms is dramatically higher. Although the early dreams have not been fulfilled, the motivation for research of this type still exists, since most of the world's population uses cereals (which are not N_2 fixers) as a primary food source. Eventually, research may provide a solution to these problems, but for the present, man is obliged to rely upon naturally existing systems to obtain the benefits of N_2 fixation.

ATMOSPHERIC DEPOSITION OF COMBINED N

Combined N, consisting of NH_3, NO_2^-, NO_3^-, and organically bound N, is a common constituent of atmospheric precipitation.[72–74] Nitrite occurs in trace amounts and is usually ignored or included with the NO_3^- determination. The organically bound N is probably associated with cosmic dust and does not represent a new addition to land masses of the world, nor is the quantity deposited of ecological significance.

The amount of N added to the soil each year in atmospheric precipitation (per hectare basis) is normally too small to be of significance in crop production. However, this N may be of considerable importance to the N economy of mature ecosystems, such as unfertilized forests, deserts, and native grasslands. Natural plant communities, unlike domesticated crops, are not subject to continued large losses of N through cropping and grazing, and the N in precipitation serves to restore the small quantities that are lost by leaching and denitrification.

Eriksson[72] summarized earlier measurements (to 1952) for combined N in atmospheric precipitation. For the United States and Europe, the estimates for NH_4^+ plus NO_3^-–N ranged from 0.78 to 22.0 kg N/ha/yr. Many high values recorded in the literature may represent analytical or sampling errors, although higher than normal amounts would be expected near highly industrialized areas due to N_2O release from burning of fossil fuels.

The concentration of N in precipitation decreases with increasing latitude. Tropical air contains 10 to 30% more mineral N than polar air and nearly twice as much as arctic air.[73] In temperate regions of the earth, the mineral N in precipitation is highest during the warmer periods, and, for a single rainfall event, concentrations decrease progressively with the duration of precipitation. Rain contains higher quantities of NH_3 and NO_3^- than snow, a result that may be due to their greater adsorption in the liquid phase.

Hutchinson[74] gives the following sources of combined N in atmospheric precipitation:

1. From soil and the ocean
2. From fixation of atmospheric N:
    electrically
    photochemically
    in the trail of meteorites
3. From industrial contamination

Important sources of NH_3 in the atmosphere include volatilization from land surfaces, combustion of fossil fuel, and natural fires. The quantity of N fixed in the trail of meteorites is negligible. According to Hutchinson,[74] only 10 to 20% of the NO_3^- in precipitation can be accounted for by electrical discharge during thunderstorm activity.

Soils have the ability to absorb atmospheric NH_3, and earlier investigators (notably Leibig[75]) placed considerable emphasis on this process as a means of providing N to plants. The opinion of most soil scientists is that the process is of little practical significance except under special circumstances. High values for NH_3 sorption have been observed for soils near industrial areas,[77-79] as well as downwind from cattle feedlots.[76,77] Substantial retention of NH_3 within a crop canopy was reported by Denmead et al.,[78] who observed that NH_3 produced near ground level was almost completely absorbed within the plant cover.

In conclusion, N gains from precipitation plus gaseous adsorption can vary from insignificant amounts to 50 kg N/ha/yr or more, depending on geographical location and proximity to an NH_3 source.

NITROGEN LOSSES FROM SOIL

Of all the nutrients required for plant growth, N is by far the most mobile and subject to greatest loss by physical, chemical, and/or biological processes from the soil–plant system. Even under the best circumstances, no more than two-thirds of the N added as fertilizer can be accounted for by crop removal or recovered in the soil at the end of the growing season; losses of as much as one-half of the applied amount are not uncommon. Numerous attempts have been made to understand the low recoveries, and it is now known that available mineral forms of N, whether added as fertilizer or produced through decay of organic matter or soil-applied wastes, will not remain very long in most soils.

There are two biological processes for reduction of oxidized N forms (e.g., NO_3^- and NO_2^-). One of these, assimilatory NO_3^- reduction, is discussed above. The other process is dissimilatory NO_3^- reduction, more commonly known as denitrification.

Five main processes for N loss are discussed in this section, including bacterial denitrification, chemodenitrification, NH_3 volatilization, leaching, and erosion; minor volatile losses from plants also occur. At the outset, it should be noted that none or all of these processes may operate in any given soil. Both conditions and N-forms change with time, and there is competition for N among different reactions (Fig. 5.5). As a general rule, N loss from any given soil occurs in more than one way, such as by a combination of leaching and denitrification from medium to heavy-textured soils of humid and semi-humid zones. Fixation of NH_4^+ and assimilation of N are competitive processes with those that promote N loss.

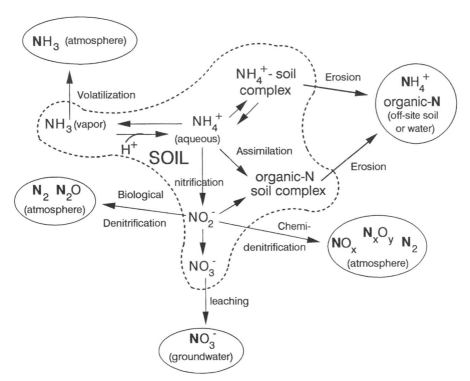

Fig. 5.5 Major processes for N loss from soil. Bold letters indicate N loss from the system.

Bacterial Denitrification

Under suitable conditions, NO_3^- is lost rapidly from soil through denitrification.[79-86] The ability to convert NO_3^- to N_2 and N_2O is limited to organisms that are able to utilize the N in NO_3^-, NO_2^-, and N_2O as a terminal electron acceptor instead of O_2. Like biological N_2 fixation, this process has been studied extensively, and a voluminous literature is available.[87-92]

Denitrification is a geochemically important process because it is the primary mechanism for the return of N_2 (originally derived from biological N_2 fixation) to the atmosphere. Just as C from organic compunds is returned to the atmosphere as CO_2 through metabolism, combined N is returned to the atmosphere as N_2 through denitrification.

Some scientists believe that N_2 of the earth's atmosphere originated from the continued activities of denitrifying microorganisms throughout geological history. Most of the atmospheric N_2 has likely passed at least once through the denitrification cycle. The annual exchange of N between the atmosphere and the biosphere has been reported to range from 170 to 340 mg/m^2/ yr.[74] This corresponds to a cycle length between 44 million and 220 million years, or from one-tenth to one-half of the time span from the Cambrian to the present.

Some specific requirements must be met for denitrification to occur:[81,88,89,93]

1. Presence of bacteria possessing the requisite metabolic ability. This requirement is easily met because denitrifying organisms are abundant in most soils and aquatic systems and their diversity is quite high.
2. Presence of suitable electron donors, such as organic C compounds, reduced S compounds, and molecular H_2. This requirement is a major constraint (along with items 3 and 4) on denitrification in the environment.
3. Existence of small and localized zones in otherwise aerobic soils and in waterlogged sediments where anaerobic conditions prevail. Many denitrifiers are facultative organisms that can grow in the presence or absence of O_2; if O_2 is present, the O_2 represses (prevents) synthesis of the enzymes that reduce the N atom.
4. Presence of N as NO_2^-, NO_3^-, or N_2O to serve as terminal electron acceptors.

About 33 genera of bacteria have the capacity to denitrify; common ones are listed in Table 5.11. The organisms primarily involved are heterotrophic (live off organic matter) and belong to the genera *Alcaligenes, Agrobacterium, Bacillus,* and *Pseudomonas.* Several chemoautotrophs (i.e., species of *Thiobacillus*) are also capable of utilizing NO_3^-, with production of N gases. However, they are of little importance in most agricultural soils.

The probable sequence of bacterial denitrification and reduction of the nitrogen atom is as follows:

$$NO_3^- \rightarrow NO_2^- \rightarrow NO \rightarrow N_2O \rightarrow N_2$$
$$(+5) \quad (+3) \quad (+2) \quad (+1) \quad (0)$$

The initial step in denitrification (reduction of NO_3^- to NO_2^-) is catalyzed by the enzyme dissimilatory nitrate reductase. This enzyme contains labile sulfide groups, Mo, and Fe in both heme and nonheme forms. Two types of enzymes are believed to be involved in the reduction of NO_2^- to NO: cytochrome c and a Cu-containing protein. The enzymes responsible for NO and N_2O reduction have been thoroughly characterized.[92] The highly reactive nature of NO has made the isolation of NO reductase difficult. The enzyme responsible for N_2O reduction is labile and cannot be easily isolated from cells. Physiologically, the N atom is a terminal electron acceptor that replaces O_2. In aerobic metabolism electrons are generated and transferred to carriers and ultimately to O_2:

$$e^- \text{ from substrate oxidation} \rightarrow e^- \text{ carrier} \rightarrow e^- \text{ transport system} \rightarrow O_2 \rightarrow H_2O$$

During denitrification, the same sequence occurs, except that the final electron acceptor is the N atom, not O_2, and the ultimate product is N_2, not H_2O:

$$e^- \text{ from substrate oxidation} \rightarrow e^- \text{ carrier} \rightarrow$$
$$e^- \text{ transport system} \rightarrow 2NO_x \rightarrow N_2$$

Not all denitrifiers have the ability to reduce NO_2^- to N_2O or to reduce N_2O to N_2. As a result, under laboratory conditions, NO_2^- or N_2O may be found as the final products of NO_3^- reduction. However, in natural soil and

TABLE 5.11   Genera of Bacteria Capable of Denitrification[a]

Genus	Comments
Alcaligenes	Commonly found in soils
Agrobacterium	Commonly found in soils
Azospirillum	Capable of N_2 fixation, commonly associated with grasses
Bacillus	Thermophilic denitrifiers reported
Flavobacterium	Denitrifying species recently isolated
Halobacterium	Requires high salt concentrations for growth
Hyphomicrobium	Grows on one-carbon substrates
Paracoccus	Capable of both chemolithotrophic and heterotrophic growth
Propionibacterium	Fermentors capable of denitrification
Pseudomonas	Commonly found in soils
Rhodopseudomonas	Photosynthetic bacteria
Thiobacillus	Generally grow as chemoautotrophs

[a]Adapted from Firestone.[81]

aquatic conditions, N_2 is the principal product. In some cases, N_2O has been reported as an end product of denitrification in soils, but the N_2O formed is probably an artifact that results from use of soils that were dried, stored in the laboratory, and subsequently remoistened. Under these conditions, accumulation of NO_2^- and N_2O is frequently observed; the N_2O is a product of chemodenitrification (see next section), not biological denitrification.[94]

Nitrous oxide is a gas and can escape from the soil before being reduced further to N_2; N_2O emissions are a source of environmental concern because N_2O is a "greenhouse gas."[87] The ratio of N_2 to N_2O in the gases evolved from soil depends upon such factors as pH, moisture content, Eh, temperature, NO_3^- concentration, and content of available organic C.

A variety of methods have been used in attempts to estimate N losses from soils through bacterial denitrification.[34,81,84]

1. By mass–balance calculations. The amount of N lost from the soil–plant system is determined from the difference between N inputs and outputs, including crop removal and leaching. Systematic errors for all analyses are incorporated into the estimate for gaseous loss.[95] This method is susceptible to a multitude of problems because it requires analysis of a wide variety of N compounds.

2. By following the disappearance of NO_3^- from the system under study. This approach is generally unsatisfactory because some of the NO_3^- may be reduced to NH_4^+ and assimilated to form organic N.[96,97]

3. By estimates of N_2 and N_2O production. The experiments are usually carried out in closed chambers under an N_2-free atmosphere (i.e., He or Ar gas).[98] The evolved gases have been detected by gas chromatography, although the ^{15}N tracer technique has also been applied. This method also has a number of problems that diminish confidence in the values obtained.[98–101]

4. Through use of ^{15}N-labeled NO_3^- in field studies. The soil is covered with a chamber that allows for sampling and detection of ^{15}N-labeled N_2 in the evolved gases.[102,103] Nitrate highly enriched with ^{15}N is required and the approach is both time-consuming and expensive.

5. By measurement of residual ^{15}N–NO_3^- in field plots. A complication of this approach is the possibility that some NO_3^- may be lost by leaching below the sampling depth. Also, sampling errors can arise because of nonuniform distribution of the applied ^{15}N. The method described in item 3 may also be used here.

6. By the acetylene inhibition method. This approach is based on the observation that acetylene inhibits the reduction of N_2O to N_2 by denitrifying microorganisms. Hence, N_2O instead of N_2 accumulates as the principal product. Like method 3, this procedure has the disadvantage of requiring a closed system. The procedure has been used in both

laboratory and field studies to estimate denitrification.[104–107] Problems associated with this method are discussed by Bremner and Hauck.[34]

Concerns regarding the potentially adverse effects of nitrogenous fertilizers as contributors to environmental problems (increased N_2O concentration in the atmosphere) have stimulated the development of chamber techniques for measuring the evolution of N_2O from field soils. However, the detection of N_2O does not necessarily prove that denitrification has occurred, since denitrifiers are not the only agents for N_2O formation. For many years, it was assumed that N_2O release from soil was the result of denitrification but it is now known that nitrifying organisms can also produce N_2O during oxidation of NH_4^+ to NO_3^-.[86,108,109]

Finally, it should be stated that most laboratory studies have been carried out under optimum conditions for denitrification. In general, results of such studies cannot be directly related to conditions existing in the field. As noted below, the method of sample preparation, such as air drying, can profoundly affect denitrification rates.

Nitrogen losses through denitrification vary greatly and are highly variable over relatively small areas, depending on NO_3^- levels, microbial distribution, available organic matter, temperature, and moisture status of the soil. Optimum conditions for denitrification are as follows:[93,110–113]

Poor Drainage: Moisture is important because of its effect on aeration. As soil moisture content increases, transport of O_2 is slower, and localized anaerobic conditions may result. Denitrification is negligible at moisture levels below about two-thirds of the water-holding capacity but is appreciable in flooded soils. It may occur in anaerobic microenvironments of well-drained soils, such as water-filled pores, the rhizosphere, and the immediate vicinity of decaying plant and animal residues. Nitrate, introduced into aqueous sediments in surface or ground waters, applied as fertilizer, or formed at the oxidizing surface layer in wetlands or during aerobic periods in soils, is particularly susceptible to gaseous loss through denitrification.[113]

Temperature: Denitrification increases at temperatures of 25°C and above, proceeds at a progressively slower rate at lower temperatures and practically ceases at 2°C.

Soil reaction near neutral: Denitrifying bacteria are sensitive to low pH and, as a result, acidic soils contain such a sparse population of denitrifying bacteria that NO_3^- losses by denitrification may be negligible. In addition, nitrification rates are lower in acidic soils, so less N is converted to NO_3^-.

Abundant supply of readily degraded organic matter: The amount of organic matter available to denitrifying microorganisms is generally appreciable in the surface horizon but negligible in the subsoil. Significant amounts of soluble organic matter may be found under cattle feedlots, as well as in the lower horizons of soils amended with large quantities of organic wastes.

The presence of inorganic N as NO_3^-, NO_2^-, or N_2O is a prerequisite for denitrification. When the soil becomes saturated or partially inundated with water (for example, after a heavy rain), any NO_3^- present in the soil is subject to conversion to N_2 and N_2O through denitrification if other requirements for the process are met. Subsequent drainage and reestablishment of aerobic conditions can lead to further nitrification of NH_4^+, thereby providing additional substrate for the denitrifying bacteria during a subsequent anaerobic phase.[113] Losses of soil- and fertilizer-derived N through denitrification would be expected to be especially severe under alternating aerobic and anaerobic soil conditions brought about by intermittent heavy rains. During the aerobic phase, NO_3^- can be formed and subsequently lost during the anaerobic phase.[113]

A second requirement for denitrification is the presence of easily degraded organic matter. A positive correlation has been noted between total organic C in soil and denitrification rate,[114] but better correlations have been obtained with water-soluble or readily degraded C,[116–118] as seen in Fig. 5.6. An increase in denitrification rate occurs when the soil is dried out or frozen prior to the denitrification measurement,[119–120] which can be explained in part by release of soluble organic compounds, as well as changes in microbial ecology as biotic composition changes after drying or freezing.

Nitrogen losses through denitrification are affected by mineralization–immobilization and will be especially severe in soils that contain high

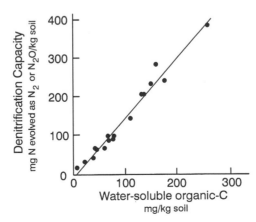

Fig. 5.6   Correlation between water-soluble organic C and the denitrification capacity of the soil. Adapted From Burford and Bremner.[116]

amounts of degradable organic matter with a low C/N ratio. Under such circumstances, net mineralization occurs and NO_3^- can accumulate.[112] Additions of carbonaceous plant residues (i.e., those with a high C/N ratio) will conserve N through net immobilization, thereby reducing denitrification losses. In this case, there is competition for organic C between heterotrophs that do not denitrify and the denitrifiers. A potentially beneficial effect of returning crop residues to the soil following harvest is reduction of N losses through denitrification (and leaching).

There is considerable controversy about the effect that plant roots have on denitrification, since both positive and negative effects have been recorded. Plant roots can influence denitrification in several ways:[81,121]

1. Providing substantial amounts of organic matter, which serves to support activity of denitrifying bacteria
2. Creating anaerobic zones by depletion of O_2 supply
3. Creating a drier soil, thereby increasing the rate of O_2 diffusion to the root zone and promoting NO_3^- formation and depressing denitrification as a result of increased O_2 availability.
4. Depleting NO_3^- supply through plant uptake
5. Increasing O_2 availability in the rhizosphere of aquatic plants in rice soils or anoxic sediments

The net effect of the above-mentioned factors may be positive or negative, thereby accounting for the variable results reported in the literature. Other factors affecting denitrification in the root zone include O_2 diffusion rate and supply of NO_3^- in the rhizosphere.

Denitrification can be considered desirable when it occurs below the rooting zone, because the NO_3^- content of ground water will be decreased. Denitrifying microorganisms are known to be present at considerable depths in soil, and it is possible that some of the NO_3^- leached into the subsoil may be volatilized as N_2 and N_2O before reaching the water table. Meek et al.[122] concluded that much of the NO_3^- leached into the subsoils of some irrigated California soils was lost through denitrification.

In marine and freshwater sediments, the fate of NO_3^-–N may be substantially different than described above.[123,124] In these cases, the end-product of NO_3^- reduction is NH_4^+, not N_2. Many anaerobic sediments contain bacteria that carry out dissimilary NO_3^- reduction to NH_4^+.[125,126] In these organisms, the N atom of NO_3^- serves as the terminal electron acceptor, just as it does in classical denitrification. The advantage to the organisms in this case is that a single N atom can serve as a terminal electron acceptor for 8 electrons, from $N(+5)$ of NO_3^- to $N(-3)$ of NH_4^+, instead of being an acceptor for only 5 electrons from $N(+5)$ to $N(0)$ of N_2. Under conditions where N-containing electron acceptors are scarce, the process would give these organisms a substantial advantage over typical denitrifiers. As a result of dissimilary NO_3^-

reduction of this type, N is not lost from the system as readily, although volatilization of NH_3 may occur.

Chemodenitrification

Chemidenitrification is a process by which N gases are formed in soils by chemical reactions of NO_2^- with organic matter. In most soils, the oxidation of NO_2^- to NO_3^- by *Nitrobacter* proceeds at a faster rate than the conversion of NH_4^+ to NO_2^- by *Nitrosomonas;* consequently, NO_2^- seldom exists in detectable amounts. High levels are sometimes found, however, when NH_3- or NH_4^+-type fertilizers are applied to soil at high application rates.[127–132] Nitrite accumulation has been attributed to inhibition of the second step of the nitrification process, presumably due to NH_3 toxicity to *Nitrobacter*. The buildup of NO_2^- is undesirable because of phytotoxicity[129] and because NO_2^- is relatively unstable and can undergo a series of reactions leading to the formation of nitrogenous gases. The possibility that gaseous loss of N may accompany temporary NO_2^- accumulations has been noted by several investigators.[127,132–135]

Accumulation of NO_2^- has been observed following addition of anhydrous NH_3, NH_4^+-salts, and urea to soils.[128–133] According to Hauck and Stephenson,[130] large fertilizer granules, high application rates, and an alkaline pH in the immediate vicinity of the granule are particularly favorable for NO_2^- accumulation. The importance of soil-fertilizer geometry on NO_2^- accumulations is emphasized by Bezdicek et al.[136]

Several investigators have suggested that blockage of nitrification at the NO_2^- stage occurs under certain field conditions (discussed above) and that nonenzymatic loss of N results from chemodenitrification. The term *sidetracking of nitrification* has sometimes been used to designate N loss by this mechanism.[133]

A factor of some importance in determining the reactivity of NO_2^- is the change in pH accompanying nitrification. Nitrite is particularly reactive at low pH's, a condition that may be attained in localized zones in soil following application of NH_3 or NH_4^+-type fertilizers. Nitrification (which produces H^+), of applied anhydrous NH_3, for example, starts in peripheral zones of moderately high NH_3 concentrations and proceeds inward towards the center of the retention zone. As a result of biological conversion of NH_3 to NO_2^- and NO_3^-, the pH of the soil is lowered; in peripheral zones, values as low as 4.2 to 4.8 have been observed. The pH of the soil immediately outside the retention zone may also be lowered because of migration of NO_2^- and NO_3^-. In these low pH regions, further oxidation of NO_2^- may be hindered by sensitivity of *Nitrobacter* to acidic conditions. A similar sequence may occur at the soil-particle interface of an individual urea or $(NH_4)_2SO_4$ granule.

Reactions of Nitrite with Humic Substances

Some soil organic matter components react chemically with NO_2^- to form N gases, as Fig. 5.7 illustrates. The possibility that losses of N occur through this mechanism is of considerable interest because there is as yet no fully suitable explanation for the losses of fertilizer N from normally aerobic soils.

Evidence for a NO_2^-–organic matter interaction has been obtained when ^{15}N–NO_2^- was applied to soil,[137,138] in which case part of the NO_2^-–N was fixed by the organic matter and part was converted to N gases. The quantity of NO_2^-–N converted to organic forms increases with increasing NO_2^- concentration and decreasing pH and is related to the C content of the soil.[137]

The gases that have been obtained through the reaction of NO_2^- with lignins and soil humus preparations in buffer solutions at pH 6 and 7 are shown in Fig. 5.8; results of a similar study using lignin-building blocks have given a similar array of N gases. The main N gas identified in these studies was NO; other gases included N_2, N_2O, and CO_2.[139] Still another gas (methyl nitrite, CH_3ONO) has been observed in the gases produced from lignins under highly acidic conditions,[140] but there is no evidence that this gas is formed under conditions existing in natural soils.

Mechanisms leading to the evolution of N gases by nitrosation of lignins and humic substances are not fully understood, but the initial reaction may involve the formation of nitroso and oximino derivatives, which subsequently react with excess NO_2^- to give N gases, as seen in this sequence:

An excellent discussion of mechanisms involved in nitrosation is given by Austin.[141]

Classical Nitrite Reactions

Nitric acid (or NO_2^-) reacts with amino acids and other reduced forms of soil N, such as NH_3, to form N gases. The reactions are:

$$NH_3 + HNO_2 \rightarrow NH_4NO_2 \rightarrow 2H_2O + N_2\uparrow \qquad (1)$$

$$RNH_2 + HNO_2 \rightarrow ROH + H_2O + N_2\uparrow \qquad (2)$$

The reaction of HNO_2 with amino acids (reaction (2)) occurs much more readily than the reaction with NH_3 (reaction (1)). Rather low pH values are

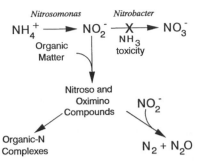

Fig. 5.7 Possible role of organic matter in promoting the decomposition of NO_2^- with production of N_2 and N_2O.

required, and since free amino acids are normally present in soil in only trace quantities, major losses of N by either of the above processes would appear unlikely under most soil conditions. It should be noted, however, that the pH at the surface of clay particles may be somewhat lower than that of the soil proper. Another factor to consider is that the tendency of NO_2^- to convert to NO_3^- and NO by chemical dismutation is considerably greater than its tendency to react with NH_3 or amino compounds. This reaction is:

$$3HNO_2 \rightarrow HNO_3 + N_2O\uparrow + 2NO\uparrow \tag{3}$$

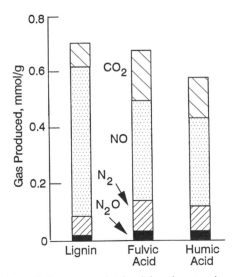

Fig. 5.8 Composition of the gases obtained by the reaction of NO_2^- with lignin, fulvic acid (FA), and humic acid (HA) at pH 6.[139]

The NO formed in reaction (3) will not necessarily escape from the soil, due to of its strong tendency to react with O_2 to form nitrogen dioxide ($2NO + O_2 \rightarrow 2NO_2$), which can react further with water to form NO_2^- and NO_3^-. The cyclic nature of NO_2^- decomposition can thus be depicted as follows (equations not balanced):

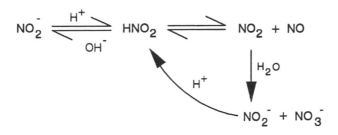

Arguments both for and against appreciable loss of N as NO_2 or NO can be given. Desiccation of the soil following partial nitrification of NH_4^+-type fertilizers would be particularly favorable for conversion of the NO_2^- to gaseous products and for escape of NO and NO_2 into the atmosphere. An imperceptibly slow evolution of N gases that would be difficult to quantify could lead to significant N losses on a per-hectare-year basis.

Ammonia Volatilization

Under suitable conditions, NH_3 can be lost from soils by volatilization.[12,127,142–144] Ammonia entering the atmosphere from world-wide terrestrial sources has been estimated to be:

Source	Amount
Wild animals	2 to 6 $\times$ 10^9 kg/year
Domestic animals	20 to 35 $\times$ 10^9 kg/year
Combustion of fossil fuel	4 to 12 $\times$ 10^9 kg/year
Total	26 to 53 $\times$ 10^9 kg/year

Ammonia losses from field soils have been estimated in several ways. In one approach, a chamber is placed over the soil surface and the ambient NH_3 is trapped in an acid solution or acid-impregnated glass wool or filter paper. Two types of chamber techniques are used. In one, the NH_3 is trapped under static conditions; in the second, NH_3-free air is passed over the soil surface. A criticism of these methods is that artificial conditions are created over and around the area being sampled.[145] This approach has been modified by allowing the soil to remain exposed most of the time to natural conditions but to close the lid of the volatilization chamber only for short intervals while the

NH_3 loss measurements are being made. The second method is suitable only when NH_3 volatilization is substantial; when volatilization losses are low, the amount of NH_3 evolved during a short trapping period may not be sufficient for accurate analysis.

A micrometeorological technique has been used.[12,78,146–148] The approach is similar to one used extensively in meteorological research to measure rates of gas exchange above natural surfaces, and it does not impose unnatural conditions on the study area. In this method, volatilized NH_3 is collected in acid traps placed at various heights above the soil surface and the amount of NH_3 lost on a per-hectare basis is estimated from meteorological data for wind speed and direction.

A summary of results from field studies of NH_3 losses from surface-applied NH_4^+ or urea is given in Table 5.12. Losses have ranged from as little as 3% to as much as 50% of the applied N, depending upon such factors as soil texture, pH, and the amount of fertilizer applied. In general, losses of N through NH_3 volatilization are small when the fertilizer is incorporated into the soil, especially if soils are acidic or neutral. Large amounts of NH_3 may be lost when NH_3-forming fertilizers are applied to the surface of alkaline or calcareous soils. Losses are accentuated when weather conditions favor drying of the soil.

Conditions in rice paddy fields are conducive to NH_3 volatilization.[143] Submergence causes the pH of most soils to converge from initial values to values near neutrality, which, along with turbulence at the water–air interface, leads to NH_3 loss.

TABLE 5.12   Summary of NH_3 Losses as Measured in the Field[a]

Fertilizer Type and System	Added N Evolved as NH_3, %
Urea	
Silt loam soil, pH 6.3	19
Fine sandy loam soil, pH 5.6–5.8	9–40
Loamy sand, pH 7.7	22
Grass sod	20
Forest litter	4–25
Flooded rice soils	6–8
$(NH_4)_2SO_4$	
Silt loam soil, pH 6.3	4
Clay soil, pH 7.6	35
Surface of grass sod	50
Flooded rice soils	3–7
NH_4NO_3	
Loamy sand, pH 7.7	17

[a]Adapted from Nelson,[127] from which specific references can be obtained.

The ability of NH_3 to form electrostatic bonds with clay minerals and organic colloids is an important factor in reducing or eliminating losses of soil and fertilizer N through NH_3 volatilization. The initial reaction is protonation of NH_3 to form NH_4^+, with subsequent exchange reactions with other cations on exchange sites of clay and humus particles:

$$NH_3 + H_2O \rightarrow NH_4^+ + OH^- \qquad (4)$$

$$NH_4^+ + \text{cation-X} \rightarrow NH_4X + \text{cation}^{n+} \qquad (5)$$

where X is an exchange site on either the clay or organic matter.

In acid soils, NH_3 can react directly with H^+ on the exchange complex.

$$NH_3 + HX \rightarrow NH_4X \qquad (6)$$

Ammonium ions in the aqueous phase of soil enter into an equilibrium reaction with NH_3.

$$NH_4^+ \rightleftharpoons NH_{3(aq)} + H^+ \qquad (7)$$

The aqueous NH_3, in turn, is subject to gaseous loss:

$$NH_{3(aq)} \rightarrow NH_{3(air)}\uparrow \qquad (8)$$

The pK_a value for reaction (7) is 9.5. At pH values of 6, 7, 8, and 9, approximately 0.036, 0.36, 3.6, and 36% of the total reduced N in the soil solution will be present as $NH_{3(aq)}$, respectively. Thus, NH_3 loss to the atmosphere will be related to both pH and NH_4^+ concentration of the solution phase. Losses increase as temperature and wind speed over the soil surface increase.

Considerable attention has been given in recent years to loss of NH_3 by application of NH_4^+-containing fertilizers to the surface of agricultural soils.[149–152] As one might expect, losses have been substantially higher from calcareous soils as compared to noncalcareous ones because of this reaction:

$$(NH_4)_2SO_4 + CaCO_3 \rightarrow 2\ NH_3\uparrow + CO_2\uparrow + H_2O + CaSO_4 \qquad (9)$$

Nitrogen is a critical nutrient in most grazing ecosystems, and NH_3 volatilization from urine and dung usually accounts for the major part of the N lost from grazed pastures.[153] Appreciable NH_3 volatilization also occurs when farmyard manure is spread directly on the surface of cultivated soils, even those that are acidic, because of localized high pH resulting from formation of NH_4OH (reaction (4)). Volatilization of NH_3 from cattle feedlots can contribute to the pollution of lakes and streams.[76]

The popularity of anhydrous NH_3 as a fertilizer is well known. In the usual practice, anhydrous NH_3 is injected in bands in the soil to depths of 10 to 15 cm. The liquid volatilizes and some NH_3 invariably escapes to the atmosphere through injection slits and soil cracks.

The facts concerning NH_3 volatilization can be summarized as follows:

1. Losses are of greatest importance on calcareous soils, especially when NH_4^+-containing fertilizers are applied on the soil surface. Only slight losses occur in soils of pH 6 to 7, but losses increase markedly as soil pH increases.
2. Losses increase with temperature, and they can be appreciable when neutral or alkaline soils containing NH_4^+ near the surface dry out.
3. Losses are greatest in soils of low cation-exchange capacities such as sands because these soils have few binding sites for NH_4^+. In contrast, clay and humus absorb NH_4^+ and prevent its volatilization. Even in alkaline soils, little NH_3 will be lost provided adequate moisture is present.
4. Losses can be high when nitrogenous organic wastes, such as farmyard manures, are stored in piles or in lagoons or applied to the soil surface.
5. Appreciable amounts of NH_3 are lost when urea is applied to soils under grass or pasture, a result that has been attributed to hydrolysis of urea by the enzyme urease, with subsequent pH increase and volatilization of NH_3.
6. Losses of soil and fertilizer N through NH_3 volatilization are reduced in the presence of growing plants. Not only are NH_4^+ levels reduced through plant uptake, but some of the evolved NH_3 may be reabsorbed by the plant canopy.[78]

Leaching

Transfer of N from soil to lakes and streams occurs through leaching. The amount of N lost from the total land area of the United States through leaching has been estimated at 2 to 3.7×10^9 kg/yr.

Nitrogen is leached mainly as NO_3^-, although NH_4^+ may be lost from sandy soils and from soils where excessive amounts of NH_4^+ are applied. In intensively cropped soils where fertilizer has not been applied, leaching losses are greatly reduced because the NO_3^- content of the soil is lower and less water passes through the soil.

Leaching losses occur when three prerequisites are met:[95]

1. Soil NO_3^- levels are high.
2. Plant uptake and microbial immobilization cannot remove all of the NO_3^- from solution before it reaches greater depths in the soil profile where there are few roots and decreased microbial activity.

3. Downward movement of water is sufficient to move NO_3^- below the rooting depth.

These conditions are best met in soils of the humid and subhumid zones, to a lesser extent in soils of the semiarid zone, and infrequently if at all in arid zone soils that are not irrigated. Losses from irrigated sandy soils can be substantial. Nitrate seldom accumulates in grassland or forest soils, even those of the humid region, because of rapid plant uptake, which restricts N losses through leaching.

In humid and subhumid regions, any NO_3^- remaining in the soil after the end of the growing season is subject to leaching, denitrification, or both. In some cases, NO_3^- can accumulate in the subsoil and move downward into the ground water, depending on the soil, climate, fertilizer, and management practices.[95] In soils of arid and semiarid regions, residual NO_3^- within the rooting depth represents a potential N-source for the following crop and is used as a basis for assessment of soil N availability.[154]

The leaching of NO_3^- in irrigation agriculture has received considerable attention in recent years because of possible pollution of ground water. McNeal and Pratt[155] found that leaching losses accounted for 13 to 100% of the fertilizer N added to some irrigated California soils and commonly averaged 25 to 50% of the N applied in most cropping situations. Aspects of the management of N for maximum efficiency and minimal pollution are discussed by Keeney[6] and Addiscott et al.[156]

Erosion and Runoff

As well as through leaching, considerable N may be lost from the soil through erosion and surface runoff. Sheet erosion is highly selective in that the eroded fraction of relatively small soil particles contains several times more N than the total soil.

Current estimates indicate that 5×10^{12} kg of soil are lost annually through erosion, with about 80% being lost as waterborne suspended material and the remainder through wind erosion. From one-half to three-fourths of the eroded soil is from agricultural land. Assuming a loss of 3×10^{12} kg of agricultural soil and an average N content of 0.15%, an estimated 4.5×10^9 kg of N would be lost annually.[95] Most of the N lost through soil erosion is in organic forms and will eventually be deposited in streams, lakes, and oceans with little opportunity of being recycled into agricultural systems. Bottom lands are often enriched with nutrients, including N, by periodic flooding.

Loss of N from Plants

Measurements for total N in many crops (e.g., wheat, corn, soybeans) show a rapid accumulation into tissue during vegetative growth, followed by a slight decline after flowering;[16,157] (see Fig. 5.3). Various explanations have been given for the loss, one being volatilization of amines, NH_3, or N oxides from the plant following senescence. Whereas accurate quantitative data for N losses from plants are lacking, the magnitude of any such losses would appear to be small when compared to losses through leaching or denitrification. There may be major exceptions, however.[12,16,157]

FLUX OF SOIL N WITH OTHER ECOSYSTEMS

The soil N cycle (see Fig. 5.1) is connected to the global N cycle through several pathways, the main ones being biological N_2 fixation and denitrification. Other transfer mechanisms include leaching of NO_3^-, volatilization of NH_3 and other gases from soil, soil erosion and accession of organic and mineral forms of N in atmospheric precipitation.

Considerable difficulty is encountered in obtaining accurate values for N transfer between soil and other ecosystems. Nitrogen fluxes are highly dependent not only on the type of ecosystem but also on environmental conditions at any specific location. Accordingly, extrapolations based on N fluxes at the local level are of doubtful validity when expanded on a regional or global basis. For any given ecosystem, inputs through biological N_2 fixation are relatively accurate; those for soil N losses through denitrification are the least reliable. The subjects of N fluxes and N transfers are covered in several reviews.[4-10]

Estimates for N transfers and fluxes for the global terrestrial system are recorded in Table 5.13. Total biological N_2 fixation for the overall terrestrial system amounts to 139×10^9 kg/yr, most of which (89×10^9 kg) is fixed in soils used for agricultural crops. Additional sources of combined N include industrial fixation (36×10^9 kg in 1976) and burning of fossil fuels (20×10^9 kg).

Loss of N from the terrestrial ecosystem occurs through bacterial denitrification (107 to 161×10^9 kg/yr) and river discharge 18 to 33×10^9 kg/yr). Transfers of N to and from the global terrestrial system occur through NH_3 volatilization (113 to 244×10^9 kg/yr) and atmospheric deposition of NH_3, NH_4^+, and NO_x (91 to 186×10^9 kg/yr).

The global terrestrial model can be further subdivided on a regional or subregional basis. Estimated inputs and outputs of N for croplands of the United States (1930–1970 period) are given in Table 5.14. The main trend for this period was an increase in N removed by harvested crops; a slight increase is shown for symbiotic N_2 fixation. No estimates are given for gaseous N loss through denitrification, but the trend would be expected to be significant and upward due to increased fertilizer N use during this period.

Nitrogen fluxes for select agroecosystems in the United States, as recorded for 1978, are given in Table 5.15. As expected, total N inputs varied greatly, depending upon the amount of N added as fertilizer or fixed biologically (soybean ecosystem). Three of the five agroecosystems are shown to have negative N balances. Paul[160] concluded that the flow of N through any given

TABLE 5.15   Nitrogen Fluxes in Selected Agroecosystems in the United States for 1978[a] (kg of N/ha)[a]

	Maize for Grain, Northern Indiana	Soybeans for Grain, Northeast Arkansas	Wheat, Central Kansas	Potatoes, Maine	Cotton, California
N inputs					
Fertilizer	112	—	34	168	179
N_2 fixation	t^b	123	t	t	t
Irrigation water and flooding	10	—	—	—	50
Atmospheric deposition	—	10	6	6	3
Crop residues	41	30	20	65	48
Total inputs	**163**	**163**	**60**	**239**	**280**
N outputs from soil					
Denitrification	15	15	5	15	20
Volatilization	t	t	t	t	t
Leaching	15	10	4	64	83
Runoff (inorganic N)	6	3	1	5	50
Runoff (organic N)	10	13	4	10	t
Wind erosion (dust)	—	t	t	t	t
Total outputs	**172**	**161**	**70**	**239**	**280**
Inputs–outputs	−9	2	−10	0	0

[a] From Thomas and Gilliam.[10]
[b] t = trace amounts.

In addition to symbiotic species, many free-living microorganisms in soil are able to fix atmospheric N_2, but the amounts that they fix generally are too low to be of practical importance.

The main processes of N loss from soils of humid and semihumid regions are leaching and denitrification. Losses through NH_3 volatilization are of greatest importance in grazing ecosystems, when nitrogenous organic wastes are deposited on the soil surface, and when NH_4^+-containing fertilizers are applied to the surface of calcareous soils. Losses of N through chemodenitrification are associated with temporary NO_2^- accumulations and may occur when anhydrous NH_3 or NH_4^+-containing fertilizers are applied to the soil at high rates, particularly in a band.

REFERENCES

1. F. J. Stevenson, "Origin and Distribution of Nitrogen in Soils," in F. J. Stevenson, Ed., *Nitrogen in Agricultural Soils*, American Society of Agronomy, Madison, Wisconsin, 1982, pp. 1–42.

2. W. H. Baur and F. Wlotzka, "Nitrogen," in K. H. Wedepohl, Ed., *Handbook of Geochemistry*, Vol. 2, Springer-Verlag, New York, 1969, pp. 7-A-1–7-0-11.

3. R. C. Burns and R. W. F. Hardy, *Nitrogen Fixation in Bacteria and Higher Plants*, Springer-Verlag, New York, 1975.

4. F. E. Clark and T. Rosswell, Eds., *Terrestrial Nitrogen Cycles: Processes, Ecosystem Strategies, and Management Impacts*, Ecol. Bull. 33, Stockholm, 1981.

5. R. D. Hauck and K. K. Tanji, "Nitrogen Transfers and Mass Balances," in F. J. Stevenson, Ed., *Nitrogen in Agricultural Soils*, American Society of Agronomy, Madison, Wisconsin, 1982, pp. 891–925.

6. D. R. Keeney, "Nitrogen Management for Maximum Efficiency and Minimum Pollution," in F. J. Stevenson, Ed., *Nitrogen in Agricultural Soils*, American Society of Agronomy, Madison, Wisconsin, 1982, pp. 605–649.

7. T. Rosswall, "The Internal Nitrogen Cycle Between Microorganisms, Vegetation, and Soil," R. H. Svensson and R. Soderlund, Eds., *Nitrogen, Phosphorus, and Sulphur-Global Cycles*, SCOPE Report 7, Ecol. Bull. 22 (Stockholm), 1976, pp. 157–167.

8. R. Soderlund and T. Rosswall, "The Nitrogen Cycle," in O. Hutzinger, Ed., *The Handbook of Environmental Chemistry*, Vol. 1, Springer-Verlag, New York, 1982, pp. 61–81.

9. R. Soderlund and B. H. Svensson, "The Global Nitrogen Cycle," in R. H. Svensson and R. Soderlund, Eds., *Nitrogen, Phosphorus, and Sulphur-Global Cycles*, SCOPE Report 7, Ecol. Bull. 22 (Stockholm), 1976, pp. 23–73.

10. G. W. Thomas and J. W. Gilliam, "Agro-ecosystems in the U.S.A.," in M. J. Frissel, Ed., *Cycling of Mineral Nutrients in Agricultural Ecosystems*, Elsevier Scientific, New York, 1978, pp. 182–243.

11. R. W. Fairbridge, Ed., *Encyclopedia of Geochemistry and Environmental Sciences: IVA*, Van Nostrand Reinhold, New York, 1972, pp. 795–801, 836–837, 849.

12. J. R. Freney and J. R. Simpson, Eds., "Gaseous Loss of Nitrogen from Plant-Soil Systems," *Developments in Soil Science*, Vol. 9, Martinus-Nijhoff, The Hague, 1983.

13. M. J. Merrick and R. A. Edwards, *Microbiol. Rev.*, **59**, 604 (1995).

14. D. J. D. Nicholas, "Utilization of Inorganic Nitrogen Compounds and Amino Acids by Fungi," in C. C. Ainsworth and A.S. Sussman, *The Fungi, An Advanced Treatise*, Vol. I, *The Fungal Cell*, Academic Press, New York, 1965, pp. 349–376.

15. P. J. Lea, "Nitrogen Metabolism," in P. J. Lea and R. C. Leegood, Eds., *Plant Biochemistry and Molecular Biology*, Wiley, Chichester, 1993, pp. 155–180.

15a. A. B. Kleinhofs and R. I. Warner, "Advances in Nitrate Assimilation," in B. J. Miflin and P. J. Lea, *The Biochemistry of Plants: A Comprehensive Treatise*, Vol. 1, Academic Press, San Diego, 1990, pp. 89–90. (See also J. S. Pate and D. B. Layzell, "Energetics and Biological Costs of Nitrogen Assimilation," pp. 11–42 in this volume.)

16. R. A. Olson and L. T. Kurtz," Crop Nitrogen Requirements, Utilization and Fertilization," in F. J. Stevenson, Ed., *Nitrogen in Agricultural Soils*, American Society of Agronomy, Madison, Wisconsin, 1982, pp. 567–604.

17. F. G. Viets, Jr., and R. H. Hageman, *Factors Affecting the Accumulation of Nitrate in Soil, Water, and Plants*, Agricultural Handbook No. 413, U.S. Department of Agriculture, Washington, D.C., 1971.

18. W. J. Brill, *Microbiol. Rev.,* **44,** 449 (1980).

19. H. J. Evans and L. E. Barber, *Science,* **197,** 332 (1977).

20. U. D. Havelka, M. G. Boyle, and R. W. F. Hardy, "Biological Nitrogen Fixation," in F. J. Stevenson, Ed., *Nitrogen in Agricultural Soils,* American Society of Agronomy, Madison, Wisconsin, 1982, pp. 365–422.

21. J. M. Vincent, *Nitrogen Fixation in Legumes,* Academic Press Australia, Sydney, 1982.

22. R. W. F. Hardy and A. H. Gibson, Eds., *A Treatise on Dinitrogen Fixation, Sections 3 and 4,* Wiley-Interscience, New York, NY, 1977.

23. J. J. Child, "Biological Nitrogen Fixation," in E. A. Paul and J. N. Ladd, Eds., *Soil Biochemistry,* Vol. 5, Marcel Dekker, New York, 1981.

24. R. H. Burris, "Nitrogen Fixation," in J. Bonner and J. E. Varner, *Plant Biochemistry,* Academic Press, New York, 1965, pp. 961–979.

25. J. I. Sprent and P. Sprent, *Nitrogen Fixing Organisms: Pure and Applied Aspects.* Chapman & Hall, London, 1990.

26. J. R. Postgate, *The Fundamentals of Nitrogen Fixation,* 2nd ed., Cambridge University Press, Cambridge, 1987.

27. A. Quispel, Ed., *The Biology of Nitrogen Fixation,* North-Holland, Amsterdam, 1974.

28. R. J. Smith, "Nitrogen Fixation," in P. J. Lea and R. C. Leegood, Eds., *Plant Biochemistry and Molecular Biology,* Wiley, Chichester, 1993, pp. 129–154.

29. F. A. Skinner, R. M. Boddey, and I. Fendrik, Eds., *Nitrogen Fixation with Non-Legumes,* Kluwer Academic, Dordrecht, 1989.

30. G. H. Elkan and R. G. Upchurch, *Current Issues in Symbiotic Nitrogen Fixation,* Kluwer Academic, Dordrecht, 1997.

31. M. Polsinelli, R. Materassi, and M. Vincenzini, Eds., *Nitrogen Fixation,* Kluwer Academic, Dordrecht, 1991.

32. P. M. Gresshoff, L. E. Roth, G. Stacey, and W. E. Newton, Eds., *Nitrogen Fixation: Achievements and Objectives,* Chapman & Hall, New York, 1990.

33. P. H. Graham, M. J. Sadowsky, and C. P. Vance, Eds., *Symbiotic Nitrogen Fixation,* Kluwer Academic, Dordrecht, 1994.

34. J. M. Bremner and R. D. Hauck, "Advances in Methodology for Research on Nitrogen Transformations in Soils," in F. J. Stevenson, Ed., *Nitrogen in Agricultural Soils,* American Society of Agronomy, Madison, Wisconsin, 1982, pp. 467–502.

35. W. D. P. Stewart, *Plant Soil,* **32,** 555 (1970).

36. W. D. P. Stewart, Ed., *Nitrogen Fixation by Free-Living Microorganisms,* Cambridge University Press, New York, 1975.

37. Y. R. Dommergues and H. G. Diem, Eds., *Microbiology of Tropical Soils and Plant Productivity. Developments in Plant and Soil Sciences,* Vol. 5, Martinus Nijhoff, The Hague, 1982.

38. I. Watanabe et al., "Physiology and Agronomy of *Azolla-Anabaena* Symbiosis," in F. A. Skinner, R. M. Boddey, and I. Fendrik, Eds., *Nitrogen Fixation with Non-Legumes,* Kluwer Academic, Dordrecht, 1989.

39. K. E. Giller and K. J. Wilson, *Nitrogen Fixation in Tropical Cropping Systems,* CAB International, Wallingford, England, 1991.

40. F. Carrapico and R. Tavares, "New Data on the *Azolla-Anabaena* Symbiosis: 1 and 2," in F. A. Skinner, R. M. Boddey, and I. Fendrik, Eds., *Nitrogen Fixation with Non-legumes,* Kluwer Academic, Dordrecht, 1989, pp. 89–94, 95–102.

41. R. L. Richards, "The Chemistry of Nitrogen Fixation," in M. J. Dilworth and A. R. Glenn, Eds., *Biology and Biochemistry of Nitrogen Fixation,* Elsevier, Amsterdam, 1991, pp. 58–75.

42. L. C. Chu, "Use of Azolla in Rice Production in China," in W. G. Rockwood and C. Mendoza, *Nitrogen and Rice,* International Rice Research Institute, Laguna, Philippines, 1979, pp. 375–394.

43. J. G. Torrey, *BioScience,* **281,** 586 (1978).

44. P. B. Vose and A. P. Ruschel, Eds., *Associative N_2-Fixation,* CRC Press, Boca Raton, Florida, 1981.

45. H. L. Jensen, *Trans. 4th. Intern. Congr. Soil Sci.,* **1,** 165 (1950).

46. M. F. Jurgensen and C. B. Davey, *Soils Fert.,* **33,** 435 (1970).

47. A. W. Moore, *Soils Fert.,* **29,** 113 (1966).

48. E. N. Mishustin, *Plant Soil,* **32,** 545 (1970).

49. M. B. Peoples, D. F. Herridge, and J. K. Ladha, *Plant Soil,* **174,** 3 (1995).

50. Y. Okon, *Azospirillum/Plant Associations,* CRC Press, Boca Raton, Florida, 1994.

51. W. Klingmüller, Ed., *Azospirillum IV: Genetics, Physiology, Ecology,* Springer-Verlag, Berlin, 1988.

52. E. A. Curl and B. Truelove, *The Rhizosphere,* Springer-Verlag, Berlin, 1986.

53. T. A. LaRue and T. G. Patterson, *Adv. Agron.,* **34,** 1 (1981).

54. D. O. Norris, *Emp. J. Exp. Agric.,* **24,** 247 (1956).

55. E. Martínez-Romero, *Plant Soil,* **161,** 11, (1994).

56. P. van Rhijn and J. Vanderleyden, *Microbiol. Rev.,* **59,** 124 (1995).

57. R. C. Dawson, *Plant Soil,* **32,** 655 (1970).

58. M. A. Cole, "Legume Seed Inoculation," in R. D. Hauck, Ed., *Nitrogen in Crop Production,* American Society for Agronomy, Madison, Wisconsin, 1984, pp. 379–388.

59. R. J. Roughley, *Plant Soil,* **32,** 675 (1970).

60. J. M. Friedrich and J. O. Dawson, *Can. J. For. Res.,* **14,** 864 (1984).

61. D. R. Benson and W. B. Silvester, *Microbiol. Rev.,* **57,** 293 (1993).

62. J. H. Becking, *Plant Soil,* **32,** 611 (1970).

63. W. S. Silver, "Physiological Chemistry of Non-Leguminous Symbiosis," in J. R. Postgate, Ed., *The Chemistry and Biochemistry of Nitrogen Fixation,* Plenum Press, New York, 1971.

64. R. L. Crocker and J. Major, *J. Ecol.,* **43,** 427 (1955).

65. D. O. Norris, "The Biology of Nitrogen Fixation," in *A Review of Nitrogen in the Tropics with Particular Reference to Pastures,* Bull. 46, Commonwealth Agricultural Bureaux, Harpenden, England, 1962, pp. 113–129.

66. G. Bond and P. Montserrat, *Nature,* **182,** 474 (1958).

67. B. A. Dixon and R. L. Crocker, *J. Soil Sci.,* **4,** 142 (1953).

68. A. Hollaender, *Genetic Engineering for Nitrogen Fixation,* Plenum Press, New York, 1977.

69. J. M. Lyons et al., Eds., *Genetic Engineering of Symbiotic Nitrogen Fixation and Conservation of Fixed Nitrogen,* Plenum Press, New York, 1981.
70. E. W. Triplett, *Plant Soil,* **186,** 29 (1996).
71. F. J. de Bruijn, Y. Jing, and F. B. Dazzo, *Plant Soil,* **174,** 225 (1995).
72. E. Eriksson, *Tellus,* **4,** 215 (1952).
73. A. A. Ångstrom and L. Hogberg, *Tellus,* **4,** 31 (1952).
74. G. E. Hutchinson, *Amer. Scient.,* **32,** 178 (1944).
75. J. von Liebig, *Organic Chemistry and Its Application to Agriculture and Physiology,* 4th ed., Taylor, London, 1847.
76. G. L. Hutchinson and F. G. Viets, Jr., *Science,* **166,** 514 (1969).
77. L. F. Elliott, G. E. Schuman, and F. G. Viets, Jr., *Soil Sci. Soc. Amer. Proc.,* **35,** 752 (1971).
78. O. T. Denmead, J. R. Freney, and J. R. Simpson, *Soil Biol. Biochem.,* **8,** 161 (1976).
79. B. A. Bryan, "Physiology and Biochemistry of Denitrification," in C. C. Delwiche, Ed., *Denitrification, Nitrification, and Atmospheric Nitrous Oxide,* Wiley, New York, 1981, pp. 67–84.
80. C. C. Delwiche and B. A. Bryan, *Ann. Rev. Microbiol.,* **30,** 241 (1976).
81. M. K. Firestone, "Biological Denitrification," in F. J. Stevenson, Ed., *Nitrogen in Agricultural Soils*, American Society of Agronomy, Madison, Wisconsin, 1982, pp. 289–326.
82. D. D. Focht and W. Verstraete, "Biochemical Ecology of Nitrification and Denitrification," in M. Alexander, Ed., *Advances in Microbial Ecology,* Vol. 1, Plenum Press, New York, 1977, pp. 135–214.
83. B. A. Haddock and C. W. Jones, *Bacteriol. Rev.,* **41,** 47 (1977).
84. R. Knowles, "Denitrification," in E. A. Paul and J. N. Ladd, Eds., *Soil Biochemistry,* Vol. 5, Marcel Dekker, NY, 1981, pp. 323–369.
85. W. J. Payne, *Denitrification,* Wiley, New York, 1981.
86. W. J. Payne, *Bact. Rev.,* **37,** 409 (1973).
87. R. Conrad, *Microbiol. Rev.,* **60,** 609 (1996).
88. W. J. Payne and M. A. Grant, "Denitrification," in J. M. Lyons et al., Eds., *Genetic Engineering of Symbiotic Nitrogen Fixation and Conservation of Fixed Nitrogen,* Plenum Press, New York, 1981, pp. 411–428.
89. D. D. Focht, "Soil Denitrification," in J. M. Lyons et al., Eds., *Genetic Engineering of Symbiotic Nitrogen Fixation and Conservation of Fixed Nitrogen,* Plenum Press, New York, 1981, pp. 499–516.
90. S. J. Ferguson, *Antonie Leeuwenhock,* **66,** 89 (1994).
91. N. P. Revsbech and J. Sorensen, Eds., *Denitrification in Soil and Sediment,* Plenum Press, New York, 1990.
92. W. G. Zumpt, *Microbiol. Molec. Biol. Rev.,* **61,** 533 (1997).
93. W. H. Patrick Jr. and M. E. Tusneem, *Ecol.,* **53,** 735 (1972).
94. C. J. Smith and P.M. Chalk, *Soil Sci. Soc. Amer. J.,* **44,** 277 (1980).
95. J. O. Legg and J. J. Meisinger, "Soil Nitrogen Budgets," in F. J. Stevenson, Ed., *Nitrogen in Agricultural Soils,* American Society of Agronomy, Madison, Wisconsin, 1982, pp. 503–566.

96. G. Stanford, J. O. Legg, and T. E. Staley, "Fate of ^{15}N-Labelled Nitrate in Soils under Anaerobic Conditions," in E. R. Klein and D. P. Klein, Eds., *Proc. 2nd. Intern. Conf. on Stable Isotopes,* Argonne National Laboratory, Argonne, Illinois, 1975, pp. 667–673.

97. G. Stanford, J. O. Legg, S. Dzienia, and C. E. Simpson, *Soil Sci.,* **120,** 147 (1975).

98. A. R. Mosier, "Gas Flux Measurement Techniques with Special Reference to Techniques Suitable for Measurements over Large Ecologically Uniform Areas," in A. F. Bouwman, Ed., *Soils and the Greenhouse Effect,* Wiley, Chichester, 1990, pp. 289–302.

99. A. M. Galsworthy and J. R. Burford. *J. Soil Sci.,* **29,** 537 (1978).

100. A. D. Matthais, A. M. Blackmer, and J. M. Bremner, *J. Environ. Qual.,* **9,** 251 (1980).

101. G. L. Hutchinson, and A. R. Mosier, *Soil Sci. Soc. Amer. J.,* **45,** 311 (1981).

102. D. E. Rolston, F. E. Broadbent, and D. A. Goldhamer, *Soil Sci. Soc. Amer. J.,* **43,** 703 (1979).

103. D. E. Rolston, M. Fried, and D. A. Goldhamer, *Soil Sci. Soc. Amer. J.,* **40,** 259 (1976).

104. K. S. Smith, M. K. Firestone, and J. M. Tiedje, *Soil Sci. Soc. Amer. J.,* **42,** 611 (1978).

105. R. Knowles, *Appl. Environ. Microbiol.,* **38,** 486 (1979).

106. J. C. Ryden, L. J. Lund, and D. D. Focht, *Soil Sci. Soc. Amer. J.,* **43,** 104 (1979).

107. J. C. Ryden, L. J. Lund, J. Letey, and D. D. Focht, *Soil Sci. Soc. Amer. J.,* **43,** 110 (1979).

108. J. M. Bremner, A. M. Blackmer, and E. L. Schmidt, *Appl. Environ. Microbiol.,* **40,** 1060 (1980).

109. G. P. Robertson and J. M. Tiedje, *Soil Biol. Biochem.,* **19,** 187 (1987).

110. O. A. L. O. Saad and R. Conrad, *Biol. Fertil. Soils,* **15,** 21 (1993).

111. R. M. Engler, D. A. Antie, and W. H. Patrick, Jr., *J. Environ. Qual.,* **5,** 230 (1976).

112. Y. Avnimelech and A. Raveh, *Water Res.,* **8,** 553 (1974).

113. K. R. Reddy and W. H. Patrick, Jr., *Soil Biol. Biochem.,* **8,** 491 (1976).

114. E. G. Beauchamp, C. Gale, and J. C. Yeomans, *Commun. in Soil Sci. Plant Anal.,* **11,** 1221 (1980).

115. J. M. Bremner and A. M. Blackmer, *Science,* **199,** 295 (1978).

116. J. R. Burford and J. M. Bremner, *Soil Biol. Biochem.,* **7,** 389 (1975).

117. R. J. K. Myers and J. W. McGarity, *Plant Soil,* **35,** 145 (1971).

118. G. Stanford, R. A. Vander Pol, and S. Dzienia, *Soil Sci. Soc. Amer. Proc.,* **39,** 284 (1975).

119. D. K. Patten, J. M. Bremner, and A. M. Blackmer, *Soil Sci. Soc. Amer. J.,* **44,** 67 (1980).

120. E. McKenzie and L. T. Kurtz, *Soil Sci. Soc. Amer. Proc.,* **40,** 534 (1976).

121. M. S. Smith and J. M. Tiedje, *Soil Sci. Soc. Amer. J.,* **43,** 951 (1979).

122. M. B. Meek, L. B. Grass, and A. J. MacKenzie, *Soil Sci. Soc. Amer. Proc.,* **33,** 375 (1969).

123. I. Koike and A. Hattori, *Appl. Environ. Microbiol.,* **35,** 278 (1978).

124. J. Sørensen, *Appl. Environ. Microbiol.,* **35,** 301 (1978).

125. M.-O. Samuelsson and U. Rönner, *Appl. Environ. Microbiol.,* **44,** 1241 (1982).

126. M.-O. Samuelsson, *Appl. Environ. Microbiol.,* **50,** 812 (1985).

127. D. W. Nelson, "Gaseous Losses of Nitrogen Other Than Through Denitrification," in F. J. Stevenson, Ed., *Nitrogen in Agricultural Soils,* American Society of Agronomy, Madison, Wisconsin, 1982, pp. 327–363.

128. P. M. Chalk, D. R. Keeney, and L. M. Walsh, *Agron. J.,* **67,** 33 (1975).

129. M. M. Court, R. C. Stephen, and J. S. Waid, *Nature (London),* **194,** 1263 (1974).

130. R. D. Hauck and J. M. Stephenson, *J. Agric. Food Chem.,* **13,** 486 (1965).

131. R. Wetselaar, J. B. Passioura, and B. R. Singh, *Plant Soil,* **36,** 159 (1972).

132. V. L. Cochran, L. F. Elliott, and R. I. Papendick, *Soil Sci. Soc. Amer. J.,* **45,** 307 (1981).

133. F. E. Clark, *Trans. 4th Intern. Congr. Soil Sci.,* **IV–V,** 153 (1962).

134. D. W. Nelson and J. M. Bremner, *Soil Biol. Biochem.,* **1,** 229 (1969).

135. F. E. Broadbent and F. J. Stevenson, "Organic Matter Interactions," in M. H. McVickar et al., Eds., *Agricultural Anhydrous Ammonia,* American Society of Agronomy, Madison, Wisconsin, 1966, pp. 169–187.

136. D. F. Bezdicek, J. M. MacGregor, and W. P. Martin, *Soil Sci. Soc. Amer. Proc.,* **35,** 397 (1971).

137. F. Führ and J. M. Bremner, *Atompraxis,* **10,** 109 (1964).

138. C. J. Smith and P. M. Chalk, *Soil Sci. Soc. Amer. J.,* **44,** 288 (1980).

139. F. J. Stevenson, R. M. Harrison, R. Wetselaar, and R. A. Leeper, *Soil Sci. Soc. Amer. Proc.,* **34,** 430 (1970).

140. F. J. Stevenson and R. J. Swaby, *Soil Sci. Soc. Amer. Proc.,* **34,** 773 (1964).

141. A. T. Austin, *Sci. Prog.,* **49,** 619 (1961).

142. H. A. Mills, A. V. Barker, and D. N. Maynard, *Agron. J.,* **66,** 355 (1974).

143. D. S. Mikkelsen and S. K. DeDatta, "Ammonia Volatilization from Wetland Rice Soils," in W. G. Rockwood and C. Mendoza, Eds., *Nitrogen and Rice,* International Rice Institute, Laguna, Philippines, 1982, pp. 135–155.

144. G. L. Terman, *Adv. Agron.,* **31,** 189 (1976).

145. O. T. Denmead, J. R. Freney, and J. R. Simpson, *Science,* **185,** 609 (1974).

146. D. E. Kissel, H. L. Brewer, and G. F. Arkin, *Soil Sci. Soc. Amer. J.,* **41,** 1133 (1977).

147. O. T. Denmead, J. R. Simpson, and J. R. Freney, *Soil Sci. Soc. Amer. J.,* **41,** 1001 (1977).

148. O. T. Denmead, R. Nulsen, and G. W. Thurtell, *Soil Sci. Soc. Amer. J.,* **42,** 840 (1978).

149. Y. Avnimelech and M. Laher, *Soil Sci. Soc. Amer. J.,* **41,** 1080 (1977).

150. L. B. Fenn, *Soil Sci. Soc. Amer. Proc.,* **39,** 366 (1974).

151. L. B. Fenn and D. E. Kissel, *Soil Sci. Soc. Amer. J.,* **40,** 394 (1976).

152. L. B. Fenn and R. Escarzaga, *Soil Sci. Soc. Amer. J.,* **41,** 358 (1977).

153. R. G. Woodmansee, *BioScience,* **28,** 448, (1978).

154. G. Stanford, "Assessment of Soil Nitrogen Availability," in F. J. Stevenson, Ed., *Nitrogen in Agricultural Soils*, American Society of Agronomy, Madison, Wisconsin, 1982, pp. 651–688.

155. B. L. McNeal and P. F. Pratt, "Leaching of Nitrate from Soils," in P. F. Pratt, Ed., *Proceedings, National Conference on Management of Nitrogen in Irrigated Agriculture*, University of California, Riverside, 1978, pp. 195–230.

156. T. M. Addiscott, A. P. Whitmore, and D. S. Powlson, *Farming, Fertilizers and the Nitrate Problem*, CAB International, Wallingford, England, 1991.

157. R. Wetselaar and G. D. Farquhar, *Adv. Agron.*, **33**, 263 (1980).

158. J. G. Lipman and A. B. Conybeare, *Preliminary Note on the Inventory and Balance Sheet of Plant Nutrients in the United States*, New Jersey Agric. Exp. Station Bull. 607, 1936.

159. G. Stanford, C. B. England, and A. W. Taylor, *Fertilizer and Water Quality*, ARS 41-168, U.S. Department of Agriculture, U.S. Government Printing Office, Washington, DC., 1970.

160. E. A. Paul, "Nitrogen Cycling in Terrestrial Ecosystems," in J. O. Nriagu, Ed., *Environmental Biogeochemistry*, Vol. 1, Ann Arbor Science, Ann Arbor, Michigan, 1976, pp. 225–243.

6

THE INTERNAL CYCLE OF NITROGEN IN SOIL

A useful concept of recent origin is that an internal N cycle exists in soil that is distinct from the global cycle of N, but interfaces with it (see Figure 1 of Chapter 5). A key feature of the internal cycle is the biological turnover of N through mineralization–immobilization:

$$\text{Organic N} \underset{\text{immobilization}}{\overset{\text{mineralization}}{\rightleftharpoons}} \text{NH}_3, \text{NO}_3^-)$$

These processes result in the interchange of inorganic-N forms with organic N. A decrease in mineral-N levels with time indicates net immobilization, while an increase indicates net mineralization. The observation that levels of mineral N are often constant within a narrow range cannot be interpreted to mean that an internal cycling is not operating, but that mineralization–immobilization rates, even though vigorous, are equal.[1]

Several interrelated organic matter fractions must be considered when discussing N-organic matter interactions in soil.They include plant and animal residues, microbial biomass, partially stabilized dead organic matter (e.g., remains of microbial cells, melanins produced by fungi, and newly formed humic substances), and the stable humus fraction (Fig. 6.1). As plant residues decay in soil, inorganic N is incorporated into the microbial biomass, a portion of which is converted to newly formed humic substances and ultimately into stable humus. The mean residence time (MRT) of the N in any given pool ranges from a few days or weeks for some biomass components to 1000 or more years for the stable humus fraction (see Chapter 1). When steady-state levels of organic matter have been attained, mineralization of native humus is being compensated for by synthesis of new humus.

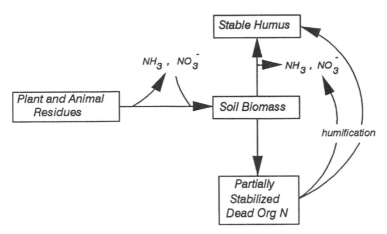

Fig. 6.1   Major pools of organic N in soils and their interactions. The decay of organic residues leads to incorporation of N into the biomass, part of which is converted into partially stabilized forms and ultimately into stable humus through a process called humification (see text).

Biochemical processes such as ammonification, nitrification, denitrification, and assimilation are the major biological transformations that occur within the soil. In addition, chemical processes such as fixation of NH_4^+ by clay minerals and of NH_3 by soil organic matter also play a prominent role.

BIOCHEMISTRY OF AMMONIFICATION AND NITRIFICATION

The process of organic residue decay in soil is complex and is brought about by the joint activities of a wide range of macro- and microfaunal organisms.[2] The overall decay process is discussed in Chapter 1, and only those aspects that relate directly to the conversion of organic N to available mineral forms will be discussed here. Release of C and N from residues differ in that part of the C is inevitably lost as CO_2, while N tends to be conserved, particularly when the residues have a high C/N ratio.

Conversion of organic N to available mineral forms (NH_4^+, NO_3^-) is mediated by microbial transformations and influenced by factors that affect microbial activity (temperature, moisture, pH, etc.). The first step (ammonification) is conversion of organic N to NH_3 and is carried out primarily by heterotrophic microorganisms, with minor participation of fauna. The subsequent conversion of NH_3 to NO_3^- occurs primarily through the combined activities of two groups of autotrophic bacteria, *Nitrosomonas,* which converts NH_4^+ to NO_2^-, and *Nitrobacter,* which converts NO_2^- to NO_3^-.

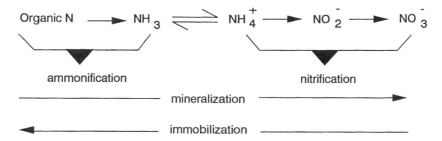

Because mineralization is nearly always accompanied by immobilization, results obtained for NH_4^+ and NO_3^- accumulations cannot be used to calculate a "mineralization rate," nor can increases in NO_3^- levels be used to determine a "nitrification rate."

Both aerobic and anaerobic microorganisms are involved in the ammonification process, whereas only aerobes oxidize NH_3 to NO_3^-. Thus, conditions that restrict the O_2 supply of soil permit NH_3 (as the NH_4^+-ion) to accumulate, such as in rice paddy fields. Nitrate is the predominant available form of N in well-aerated cultivated soils, except for a short time after addition of fertilizers containing NH_3, NH_4^+, or urea.

Ammonification

Ammonification is an enzymatic process in which the N of nitrogenous organic compounds is liberated as NH_3. The review of Ladd and Jackson[2] shows that a wide array of enzymes is involved, each acting on a specific compound or class of organic compound. The initial substrate is often a macromolecule (protein, nucleic acid, aminopolysaccharide), from which simpler N-containing biochemicals are formed (e.g., amino acids, purine and pyrimidine bases, amino sugars). The simpler compounds are usually taken up by microbial cells and subsequently degraded with formation of NH_3.

Breakdown of Proteins and Peptides Degradation of proteins and peptides in soil involves an initial cleavage of peptide bonds by proteinases and peptidases to form amino acids, from which NH_3 is released through the action of such enzymes as amino acid dehydrogenases and oxidases.

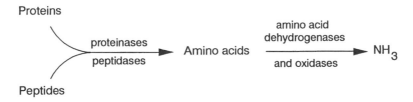

A typical reaction for the oxidative deamination of amino acids is as follows:

$$R-\underset{\underset{NH_2}{|}}{CH}-COOH + NAD \xrightarrow{\quad NADH + H^+ \quad} R-\underset{\underset{NH}{\|}}{C}-COOH \xrightarrow{\quad H_2O \quad} R-\underset{\underset{O}{\|}}{C}-COOH + NH_3$$

Degradation of Nucleic Acids   Conversion of the N of nucleic acids to NH_3 also requires the action of a variety of enzymes. The nucleic acids, which occur in all living organisms, consist of individual mononucleotide units (purine or pyrimidine base–sugar–phosphate) joined by a phosphoric acid ester linkage through the sugar. During degradation (Fig. 6.2), nucleic acids are converted to mononucleotides by the action of nucleases, which catalyze the hydrolysis of ester bonds between phosphate groups and pentose units. The mononucleotides, in turn, are converted to nucleosides (*N*-glycosides of purines or pyrimidines) and inorganic PO_4^{3-} by nucleotidases. The nucleosides are then hydrolyzed to the purine or pyrimidine bases and the pentose component by the nucleosidases, which are subsequently converted to NH_3 by reactions catalyzed by amidohydrolyases and amidinohydrolyases.

Decay of Amino Sugars   Amino sugars are common as structural components of cell walls of microorganisms in combination with mucopeptides and mucoproteins.

The formation of NH_3 from glucosamine (a common soil amino sugar) is catalyzed by the combined action of glucosamine kinase and glucosamine–6–phosphate isomerase; the products are ammonia and fructose-6-phosphate.

Urease in Soil

In addition to containing nitrogenous components of plant or animal tissues, soils may contain organic N as urea, which is a constituent of the urine of

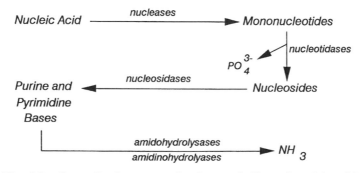

Fig. 6.2   Generalized sequence for the metabolism of nucleic acids.

grazing animals and is often added to the soil as fertilizer. Decomposition of urea to yield NH_3 (and CO_2) is catalyzed by urease, an enzyme found in practically all soils.

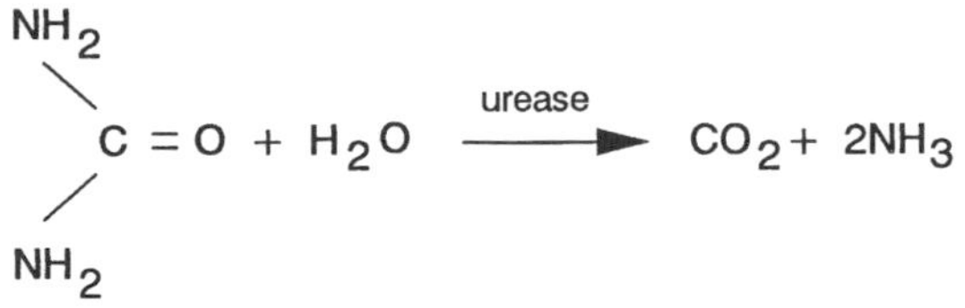

$$\underset{\underset{NH_2}{|}}{\overset{\overset{NH_2}{|}}{C}} = O + H_2O \xrightarrow{\text{urease}} CO_2 + 2NH_3$$

Nitrification

Except in poorly drained or submerged soils, the NH_3 formed through ammonification (see previous section) is readily converted to NO_3^-, as shown below. Energy produced by some of the reactions is utilized for biosynthesis and cell maintenance.

$$NH_3 \underset{-H^+}{\overset{+H^+}{\rightleftarrows}} NH_4^+ \xrightarrow{\textit{Nitrosomonas}} NO_2^- \xrightarrow{\textit{Nitrobacter}} NO_3^-$$

That nitrification is a two-step biological process was established in 1889 by Winogradsky,[3] who also found that *Nitrosomonas* and *Nitrobacter* were the organisms involved. Pertinent reviews on nitrification include those of Focht and Verstraete[4] and Schmidt.[5]

The nitrifying bacteria are Gram-negative chemoautotrophic (chemolithotrophic) archebacteria comprising the family Nitrobacteriaceae. Five genera of NH_3 oxidizers and three genera of NO_3^- oxidizers are generally recognized, although only a single species of each group is considered to be primarily responsible for NH_3 and NO_2^- oxidation in soil: *Nitrosomonas europaea* and *Nitrobacter winogradskyi,* respectively.

Nitrosomonas grows in pure culture as short rods that exhibit straight line motility; *Nitrobacter* typically occurs as large cells ($1.0–1.5 \times 1.0–3.0$ μm) with an irregular lobular shape and a tumbling motility. Practically all of the C for cell synthesis comes from an inorganic source, namely, HCO_3^- contained in the soil water.

Pathway of Nitrification   The six-electron transfer accompanying the oxidation of NH_3 (oxidation state of -3) to NO_2^- (oxidation state of -3) by *Nitrosomonas* is a two-step process:

$$NH_3 + 1/2\ O_2 + 2\ H^+ \longrightarrow NH_2OH$$

$$NH_2OH + H_2O + 2\,[e^- \text{ carrier(ox)}] \longrightarrow NO_2^- + 2\,[e^- \text{ carrier}(2e^-)]$$

The electron transport (cytochrome) system is:

$$2\,[e^- \text{ carrier}(2e^-)] + O_2 + 2ADP + 2Pi \longrightarrow$$

$$2\,[e^- \text{ carrier(ox)}] + 2H_2O + 2ATP$$

The first reaction (formation of NH_2OH, hydroxylamine) is not energy-yielding. The e^- carrier molecules ($2e^-$) and adenosine triphosphate (ATP) are used either for biosynthesis of cell components or fixation of CO_2 in the net reaction shown below:

$$3CO_2 + 6\,ATP + 6\,NADPH \longrightarrow$$

$$\text{glyceraldehyde-3-P} + 6\,ADP + 6\,Pi + 6\,NADP^+$$

To provide the energy for CO_2 fixation, 1 mole of NH_3 must be oxidized per mole of ATP and NADPH produced, for a total of 9 moles (153 grams) of NH_3 oxidized in order to convert 3 moles (36 grams) of C from CO_2 to organic forms.

Several studies have indicated that N_2O is a byproduct of NH_4^+ oxidation. This gas may arise by chemical dismutation of nitroxyl (NOH) and through the action of nitrite reductase, as shown in the following illustration. Some of the N_2O produced in soil, and subsequently evolved into the atmosphere, is generated during nitrification.[6,7]

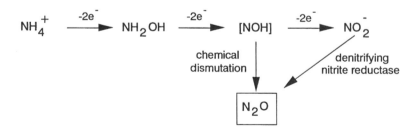

Nitrate is formed by *Nitrobacter* in a 2-electron change in the oxidation state of N (from +3 to +5), with release of 17.8 kcal/mole of energy:

$$O_2^- + e^- \text{ carrier(ox)} \xrightarrow{\text{nitrite oxidase}} NO_3^- + e^- \text{ carrier}(2e^-)$$

The electron transport (cytochrome) system generates ATP:

$$e^- \text{ carrier}(2e^-) + 1/2\,O_2 + ADP + Pi \longrightarrow e^- \text{ carrier(ox)} + H_2O + ATP$$

No intermediates are known. The reduced form of electron carrier and ATP are used for biomass formation as described for *Nitrosomonas.*

Isolation of Nitrifying Microorganisms   Approaches for the isolation of nitrifying bacteria in soil are limited by the complexities of the soil microhabitat and the physiology of the organisms involved. Isolation by plating methods is generally unsuitable because of problems associated with slow growth, small colony size, and overgrowth by heterotrophs. Most isolation procedures involve an enrichment step, which has distinct disadvantages in that a single isolate may achieve dominance during isolation, which precludes quantification of the initial population size. The most common approach is indirect enumeration by use of the most probable number (MPN) technique for statistical estimation of nitrifiers in an inoculum.[8]

Factors Affecting Nitrification   The main factors that affect nitrification in soil are temperature, moisture, pH, and concentrations of the substrates NH_3, O_2, and CO_2; CO_2 is rarely a limiting factor, since it is normally present at high concentrations in soil. Production of NO_3^- decreases with decreasing temperature below 30 to 35°C; below 5°C, very little NO_3^- is formed. As one might expect, nitrification proceeds at a very slow rate in cold, wet soils. For this reason, farmers in the northern areas of the United States are encouraged to wait until soil temperatures fall below 5°C (about 40°F) before applying ammoniacal fertilizers in the fall so as to limit nitrification and minimize losses of NO_3^- through denitrification and leaching.

The O_2 and CO_2 (as HCO_3^-) required by the organisms are contained in the solution phase of the soil; consequently, moisture content is of major importance. Depletion of O_2, with a decline in nitrification, is favored by: (1) the presence of easily decomposable organic matter, which increases O_2 demand by heterotrophs, (2) excess moisture, which saturates soil pores and restricts recharge of O_2 from the gaseous phase, and (3) high soil temperatures, which reduce the solubility of O_2 in water.

Under most soil conditions, oxidation of NO_2^- proceeds at a more rapid rate than the oxidation of NH_3; thus, NO_2^- is seldom found in more than trace amounts. Conditions that favor the presence of free NH_3 (high pH and low cation-exchange capacity) restrict nitrification due to NH_3 toxicity. *Nitrobacter* is somewhat more NH_3-sensitive than *Nitrosomonas;* consequently, use of NH_3 fertilizers or conditions that promote ammonification sometimes result in the temporary accumulation of NO_2^-.

A wide variety of organic compounds, including simple organic acids, some amino acids, and pyridine bases, inhibit the growth of nitrifiers in enrichment cultures. Evidence that nitrification is inhibited by decomposition products of organic residues in soil, or by metabolites excreted by plants or microorganisms, is still only suggestive.[9] Grassland soils consistently contain moderate amounts of exchangeable NH_4^+ but little if any NO_3^-, which has been attributed to inhibition of nitrification due to substances secreted by grass roots. Evidence for and against this hypothesis is discussed elsewhere.[5]

The nitrifiers are among the most sensitive of the common soil bacteria to herbicides, insecticides, and fungicides. However, any reduction in numbers appears to be of short duration. Various reviews on the subject (e.g., see Goring and Laskowski[10]) indicate that most pesticides, when applied at recommended rates, are unlikely to adversely affect nitrification on a long-term basis.

Several synthetic organic chemicals (e.g., nitrapyrin) have been proposed to retard nitrification in soil, the objective being to conserve fertilizer N and enhance its use by plants. The performance of these chemicals has been erratic, with good retention of N in some situations, but not others. Yield increases associated with their use are usually modest (<100 kg more grain produced per hectare).

Heterotrophic Nitrification   In addition to the autotrophic bacteria, several heterotrophs produce NO_2^- or NO_3^- from NH_3 and/or organic N compounds in pure culture. They include a number of bacteria, actinomycetes, and fungi, as noted in Table 6.1. Several algae also produce NO_2^- or NO_3^- from NH_4^+.

The ecological significance of heterotrophic nitrification has yet to be established with certainty. Nitrification occurs in soil under a broader range of environmental conditions than has been predicted from biochemical and physiological studies of the autotrophic nitrifiers, suggesting the participation of heterotrophs in the process. For example, nitrification proceeds in soil at pH values well below that observed for *Nitrosomonas* and *Nitrobacter* in pure culture.

NET MINERALIZATION vs. NET IMMOBILIZATION

Some, and often all, of the NH_3 and NO_3^- formed through ammonification and subsequent nitrification is rapidly utilized by the heterotrophic microflora and converted into microbial components; that is, the inorganic N is said to be immobilized. The biochemical pathways and enzymes involved are different than those described earlier for ammonification and nitrification in that NO_3^- is reduced to NH_3 (assimilatory reduction), which subsequently com-

TABLE 6.1   Examples of Heterotrophic Nitrifying Microorganisms

Bacteria	Actinomycetes	Fungi
Arthrobacter sp.	*Streptomyces*	*Aspergillus flavus*
Azotobacter sp.	*Nocardia*	*Neurospora crassa*
Pseudomonas fluorescens	*Penicillium* sp.	
Klebsiella aerogenes		
Bacillus megaterium		
Proteus sp.		

bines with a C substrate to form first simple N-containing biochemicals (amino acids, purine and pyrimidine bases, amino sugars, etc.), which are then polymerized into complex molecules (proteins, nucleic acids, aminopolysaccharides, etc).

The C/N Ratio

The decay of organic residues in soil is accompanied by conversion of C and N into microbial cells and other products. In the process, part of the C is liberated as CO_2 (see Chapter 1). As the C/N ratio is lowered, and as dead microbial cells are attacked (with synthesis of new biomass), a portion of the immobilized N is released through net mineralization.

A commonly used rule of thumb for the relationship between the C/N ratio of crop residues and mineral N levels is as follows:

C/N Ratio		
<20	20 to 30	>30
Net gain of NH_4^+ and NO_3^-	Neither gain nor loss	Net loss of NH_4^+ and NO_3^-

Residues with C/N ratios greater than about 30, equivalent to N contents of about 1.5% or less, lower mineral N reserves because of *net immobilization*. On the other hand, residues with C/N ratios below about 20, or N contents greater than about 2.5%, often lead to an increase in mineral N levels through *net mineralization*. The qualification of this rule is that the C and N must be in compounds with similar degradation rates. For example, sewage sludges (C/N ratios of 6:1 to 8:1) contain large amounts of C and N in slowly degraded forms, and therefore the organic N cannot serve as a rapidly available N-source. The N contents and C/N ratios of some organic materials are recorded in Table 6.2.

Changes in NO_3^- and CO_2 levels attending the decay of low-N crop residues in soil are illustrated in Fig. 6.3. Under conditions suitable for microbial activity, rapid decay occurs with concurrent liberation of considerable quantities of C as CO_2. To meet the N requirements of microorganisms, mineral N is consumed; that is, there is a net immobilization of N. However, when the C/N ratio of the decaying material has been lowered to about 20, NO_3^- levels once again increase due to net mineralization.

The time required for microorganisms to lower the C/N ratio of carbonaceous plant residues to the level where mineral forms of N accumulate will depend upon such factors as climate, application rate, lignin content, degree of comminution, and level of activity of the soil microflora. A reasonable estimate is that, under conditions favorable for microbial activity, net mineralization will commence after four to eight weeks of active decay. Accord-

TABLE 6.2   Typical C/N Ratios of Some Organic Materials

Material	C/N
Microbial biomass	6–12
Sewage sludge	5–14
Soil humus	10–12
Animal manures	9–25
Legume residues and green manures	13–25
Cereal residues and straw	60–80
Forest (woody) wastes	150–500
Composts	15–20

ingly, if crop residues with high C/N ratios are added to the soil immediately prior to planting, extra fertilizer N may be required to avoid N starvation of the crop.

Complicating factors in studying the mineralization–immobilization relationship are:

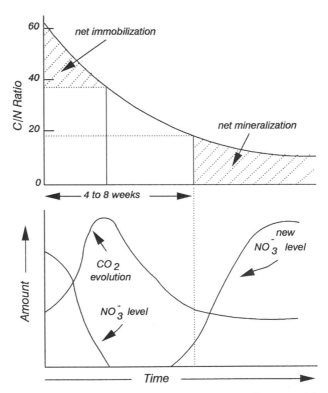

Fig. 6.3 Changes in NO_3^- levels attending the decay of crop residues in soil.

1. Ammonium is the preferred form of inorganic N used by soil microorganisms.
2. Part of the NH_4^+ is subject to nitrification. Under conditions where net immobilization occurs, microorganisms consume NO_3^- (as well as NH_4^+).
3. Substantial losses of N by denitrification can occur in organic matter-rich environments.

Following a change to net mineralization, part of the previously immobilized NO_3^-–N reappears in the NO_3^- pool, now diluted with N from other sources that participated in the turnover.

Mineralization and Immobilization Rates

A prime objective of studies on N transformations in soil is to determine quantitative values for mineralization and immobilization rates. Elaborate mathematical formulas based on ^{15}N data have been derived for this purpose and have been used in some studies. As Jansson[11] pointed out, the calculations are based on some assumptions that have not proven to be valid, and so they give only a rough approximation of the actual rates.

Mineralization–immobilization is an important component of many N-cycle models (discussed below). Several pools for organic C are often included in these models, such as native humus, the soil biomass, and components of plant residues. The various C substrates decay at different rates, the order for plant residue components being proteins > carbohydrates > cellulose and hemicellulose > lignin. Levels of NH_4^+ and NO_3^- in the soil increase or decrease, depending not only on the C/N ratio of the substrate but also on its chemical composition.

Transformations in Sediments

The ultimate product of mineralization is modified in zones where O_2 is absent, such as sediments. Ammonia is the end product of mineralization since nitrification does not occur in wet sediments, although N-oxidation may occur at the sediment–water interface where sufficient O_2 is present to meet the needs of nitrifying bacteria. In some cases, ammonification (conversion of organic N to NH_3) may also be curtailed or otherwise modified. Since anaerobic degradation (fermentation) provides less energy for biosynthesis than aerobic processes, larger amounts of substrate must be oxidized per unit of C assimilated, which means that less mineral N will be incorporated into microbial biomass than for aerobic degradation. As a result, competition of microorganisms with plants for available N may be less severe in aquatic environments than in aerobic soils.

Wetting and Drying

Cycles of wetting and drying have been shown to cause flushes (increases) in mineral N levels when compared to field-moist soils, with each successive cycle causing a slightly smaller flush. The magnitude of the flush, or "partial sterilization" effect, is related to organic matter content and the length of time the soil has remained dry. Several reasons can be given for the net release of mineral N, including the following:

1. Microorganisms are killed by the drying treatment (partial sterilization), and easily degradable nitrogenous constituents are released as dead cells undergo autolysis and decay.
2. Drying leads to conversion of organic N to more soluble compounds, which are utilized by microorganisms with release of mineral N.
3. Wetting and drying leads to the breakdown of water-stable aggregates, with exposure of new surfaces and substrates for microbial attack. Freezing and thawing affects mineralization in the same way as drying and wetting, but to a lesser degree.

MINERAL N ACCUMULATIONS

Only a small fraction of the N in soils, generally $<0.1\%$, exists in available mineral compounds at any one time (as NO_3^- and exchangeable NH_4^+).[12] Thus, only a few kg of N per hectare may be immediately available to the plant, even though as much as 6000 kg N/ha may be present in organic forms. Ammonium and NO_3^- are temporary byproducts of biological transformations involving organic and gaseous forms of N. Factors influencing levels of NO_3^- and exchangeable NH_4^+ are considered in detail in early reviews by Harmsen and van Schreven[13] and Harmsen and Kolenbrander.[14]

Levels of exchangeable NH_4^+ and NO_3^- vary from day to day and from one season to another and will depend upon a variety of environmental factors, including:

Seasonal variation: Levels of exchangeable NH_4^+ and NO_3^- are greatly affected by temperature and rainfall. The amounts found in the surface layer of soils of the temperate humid climatic zone are lowest in winter due to low mineralization rates and increased leaching losses, rise in spring as nitrification and mineralization of organic matter commences, decrease in summer because of plant uptake and transient inhibition of microbial activity in drier soils, and increase once again in the fall when plant growth ceases and crop residues start to decay. The level in winter seldom exceeds 10 mg N/kg soil, but may increase four- to sixfold during the spring.[13]

Mineralization and immobilization: Biological turnover leads to the inter-change of NH_4^+ and NO_3^- with N of the organic matter, as noted above. Accordingly, mineral N levels represent a delicate balance between mineralization and immobilization and are affected by the activities of soil microorganisms and the C/N ratios of plant residues.

Growing plants: Plants decrease mineral N levels in soil. As well as by direct uptake, NH_4^+ and NO_3^- levels may be altered by immobilization and denitrification in the root zone, and possibly by inhibition of nitrification by root excretion products.[13,14] The influence of biochemical processes in the rhizosphere on biochemical N transformation is reviewed by Rovira and McDougal.[15]

Leaching of NO_3^-: Nitrate is the most mobile N-form in soil and is susceptible to leaching and movement into water supplies. The magnitude of NO_3^- leaching is difficult to estimate and will depend upon a number of variables, including quantity of NO_3^-, amount and time of rainfall, infiltration and percolation rates, evapotranspiration, water-holding capacity of the soil, and presence of growing plants. Leaching is generally greatest during cool seasons when precipitation exceeds evaporation; downward movement in summer is restricted to periods of heavy rainfall.

Volatilization of NH_3: Rapid changes in NH_3 levels can occur as a consequence of volatilization. As noted in Chapter 5, losses are greatest on calcareous and saline soils, especially when NH_3-forming fertilizers are used. Only slight losses occur in soils with pH values less than 7, but losses increase markedly as the pH increases. For any given soil, losses increase with an increase in temperature, and they can be appreciable when neutral or alkaline soils containing NH_3 in the surface layer are dried out. The presence of adequate moisture reduces volatilization, even from alkaline soils.

Losses of NO_3^- through denitrification: Significant loss of NO_3^-–N can and does occur as a consequence of denitrification (Chapter 5). Under anaerobic conditions, such as occur frequently in soils following a heavy rain, NO_3^-–N can be lost in a comparatively short time, particularly when energy is available in the form of organic residues. Losses are generally negligible at moisture levels below about two-thirds of the water-holding capacity of the soil; above this value, the magnitude of loss is correlated directly with moisture regime. The process may also occur in anaerobic microenvironments of well-drained soils, such as in small pores containing water and in the vicinity of roots and decomposing residues.

Buildup of NH_4^- and NO_3^- by fertilizer applications: In soils of humid and semihumid regions, any fertilizer N added in excess of plant or microbial needs will be subject to loss through leaching and/or denitrification. Thus, mineral forms of N seldom carry over from one season to the next. However, where leaching and denitrification are minimal, such as in soils of arid and semiarid regions, some carryover occurs and repeated annual applications of N fertilizer can lead to a buildup of NO_3^- in the soil profile. Figure 6.4 shows the buildup in NO_3^-–N in the upper 2.44 meters of four different soil types following application of N fertilizer for seven years to continuous corn in Missouri.[16] The relatively lower levels in the silt loam and sandy soils can be ascribed to greater leaching of NO_3^-.

Increasing attention is now being given to NO_3^- accumulations in soil, particularly below the rooting zone, because of possible movement into water supplies. Extremely high levels of NO_3^-–N (2000–4000 kg/ha) have been observed in soil under cattle feedlots.[17]

NITROGEN AVAILABILITY INDEXES

Most temperate-zone soils will contain several thousand kg N/ha, but only a small fraction of this N, estimated at no more than 1 to 2% of the total, will be available to plants during a growing season. Various soil tests have been proposed from time to time in attempts to predict the soils' contribution of N to the crop, but in general, these tests have not attained the same success as those for available P and K. The subject of tests for available N and their limitations is covered in several reviews.[13,18,19]

Before a discussion of the various soil tests, it should be stated that there is usually a sound scientific basis for many fertilizer N recommendations.

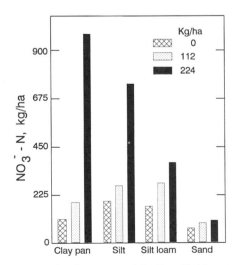

Fig. 6.4 Buildup of NO_3^-–N in the upper eight feet (2.44 meters) of four soil types after eight years of fertilizer N application (annual rates of 0, 112, and 224 kg N/ha) to continuous corn in Missouri. Adapted from Stevenson and Wagner.[16]

Hundreds of field experiments have been conducted throughout the world for the purpose of establishing the relationship between yield and N fertilizer rate for numerous crops under a broad range of soil, climatic, and management conditions. The results of these trials, along with information on crop N requirement, yield potential, level of management, and nature of the previous crop (legume or nonlegume), have been of immense value in making N fertilizer applications. Given the uncertainty in yields resulting from soil and weather variations, fertilizer N applications based on previous cropping history, as described above, may prove as reliable as, or more than, chemical or biological soil tests.

A variety of tests have been proposed as indexes of soil N availability, including soil NO_3^- levels, mineralizable N by aerobic and anaerobic incubations, hot-water or hot-salt extractable N or NH_3, amount of NH_3 recovered by alkaline $KMnO_4$ distillation, and total N or organic matter content. Some of the more promising indexes are discussed below.

Residual NO_3^- in the Soil Profile

In climates of low rainfall, and where very little leaching or denitrification occurs, residual NO_3^- in the root zone is approximately equivalent in availability to fertilizer N and can be taken into account when making fertilizer N recommendations.[18,19] The test has had wide application to soils of arid and semiarid regions, including large areas of western United States and Canada, where NO_3^- leaching below the root zone is minimal.

Incubation Methods

These methods typically involve incubation (7 to 25 days) of the soil under aerobic or anaerobic (waterlogged) conditions. Most soil test correlations (with yield or plant uptake of N) have been done under greenhouse conditions. Modifications in the test include preleaching to remove residual NO_3^-, using vermiculite or sand to improve aeration and leachability, and adding $CaCO_3$ or a nutrient solution. Advantages of anaerobic incubation methods are:

1. Only NH_4^+ needs to be measured.
2. Difficulties in establishing optimum moisture levels are eliminated.
3. More N is mineralized in a given period than under aerobic conditions.
4. Higher temperatures (which lead to more rapid mineralization) can be used.

In some early work, Stanford and his associates[20] developed an incubation approach designed to define the mineralizable (or labile) soil N pool. In this approach, measurements of N mineralized from the soil over an extended time period (up to 30 weeks) are made, with inorganic N being extracted at various time intervals. Incubation is continued for as long a time as is deemed

necessary to describe adequately the relationship between cumulative N mineralization (N_t) and time of incubation, as depicted in Fig. 6.5. The N mineralization potential (N_o) was assumed to follow first-order kinetics ($dN/dt =$
kN) and was estimated as log ($N_o - N$) = log $N_o - k_t/2.303$. The accepted
value of N_o was that which gave the best fit for the linear relationship between
log ($N_o - N_t$) vs. time (t). Both k and N_o were believed to be definitive soil
characteristics upon which N mineralization potential could be based.

Limitations of incubation methods are:

1. The results are affected by conditions prevailing in the soil at the time
   of sampling.
2. High results are attained with soils of poor structure.
3. Reliable results, sufficiently correlated with the N requirement of field
   crops, can only be expected when the technique is calibrated to a given
   soil type in a given climatic zone and when all samples are collected
   at the same time during the season.[13]

Chemical Methods

A number of chemical tests have been proposed for estimation of soil N
availability, but they are empirical in nature and, for the most part, are too
expensive and time-consuming for routine use. They range in severity of
extraction from drastic (e.g., total organic N and hydrolyzable N, to intermediate (e.g., alkaline $KMnO_4$ distillation) to mild (e.g., hot-water and hot-
salt extractions). The desirability of developing tests based on relatively mild
extractants has been emphasized.[19]

The mild extraction technique of Stanford and Smith[21] involves the determination of the NH_4^+–N that is formed when the soil is heated with 0.01 M
$CaCl_2$ in an autoclave at 121°C for 16 hours. The procedure, which was

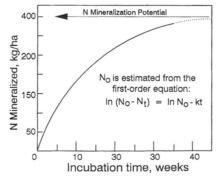

Fig. 6.5 Potentially mineralizable soil organic N based on cumulated amounts of N
mineralized during consecutive incubations. From Stanford.[19]

believed to release little N from stabilized organic matter, represents an attempt to measure N in the more biologically and chemically labile organic forms.

The ultimate value of any given chemical method of determining soil N availability will depend on the degree to which the test can be correlated with crop yield or N uptake under conditions existing in the field. Progress has been made in recent years in developing reliable and rapid methods of assessing soil N availability, but none of the proposed indexes have been adequately tested under a broad range of field conditions and none are in general use.[19] A critical part of the problem with these methods is that, when applied to soil samples collected at the beginning of the growing season, they provide an estimate of *potentially* available N, but that potential may not be realized for a variety of reasons, including dry conditions or atypically cool or hot weather (which reduce microbial degradation of the organic matter). The methods also cannot predict N losses, such as through leaching and/or denitrification during extremely wet weather.

STABILIZATION AND COMPOSITION OF SOIL ORGANIC N

Nitrogen entering the soil system through biological N_2 fixation, as fertilizer, or through other means is subjected to various chemical and biological transformations. Some of the N is utilized by the crop, some is lost by leaching, denitrification, or NH_3 volatilization, and some is incorporated into stable humus forms.[1] Results of studies using the stable isotope [15]N (discussed in Chapter 7) have shown that as a general rule, from 20 to 40% of the N applied as fertilizer remains behind in the soil in organic forms after the first growing season. Only a small portion of this immobilized N (<15%) becomes available to plants during the subsequent growing season, and availability decreases even further in subsequent years.

The Humification Process

The pathways whereby humic substances are formed are a key to understanding the process whereby fertilizer N is converted to biologically resistant forms. As noted in Chapter 1, one popular theory is that humic and fulvic acids are formed by a multiple-stage process that includes:

1. Degradation of all plant polymers and low-molecular-weight components, including lignin, into simple monomers

2. Metabolism of the monomers by microorganisms with an accompanying increase in the soil biomass

3. Repeated recycling of the biomass C (and N) with death of old cells and synthesis of new cells,

4. Concurrent polymerization of reactive monomers into high-molecular-weight polymers[22-24]

The concensus is that polyphenols (quinones) derived from lignin, together with those synthesized by microorganisms, polymerize in the presence or absence of amino compounds (amino acids, NH_3, etc.) to form brown-colored polymers.

The reaction between amino acids and polyphenols involves simultaneous oxidation of the polyphenol to the quinone form, such as by polyphenol oxidase enzymes. The addition product readily polymerizes to form brown nitrogenous polymers, according to the general sequence shown in Fig. 6.6.

The net effect of the humification process is conversion of the N of amino acids into the structures of humic and fulvic acids. Thus, whereas much of the N in microbial biomass occurs in known biochemical compounds (e.g., amino acids, amino sugars, purine and pyrimidine bases), most of the organic N in soil (>50%) occurs in unknown forms (see next section).

Chemical Distribution of the Forms of Organic N in Soils

Current information regarding organic N in soil has come largely from studies utilizing acid hydrolysis to convert the organic N to soluble forms.[23] In a

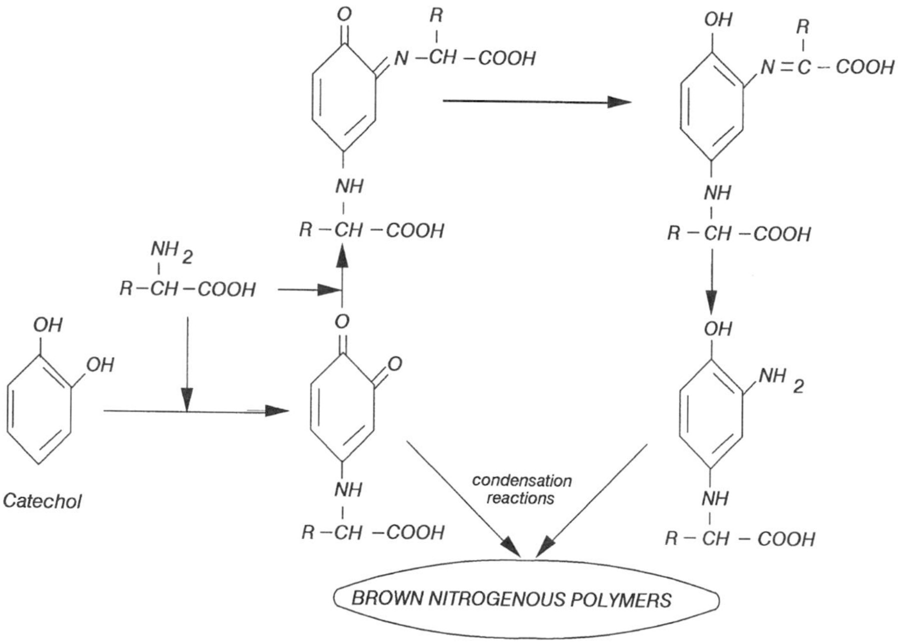

Fig. 6.6 Formation of brown nitrogenous polymers as depicted by the reaction between amino acids and polyphenols. Adapted from Kelly and Stevenson.[24]

typical procedure, the soil is heated with 3 N or 6 N HCl for 12 to 24 hours, after which the N is separated into the fractions outlined in Table 6.3. Methods have been developed that permit recovery of the different forms of N as NH_3, thereby facilitating the use of ^{15}N as a tracer.[25]

The main identifiable organic N compounds in soil hydrolysates are amino acids and amino sugars. Soils contain trace quantities of nucleic acids and other nitrogenous biochemicals, but specialized techniques are required for their separation and identification. Only one-third to one-half of the organic N in most soils can be accounted for in known compounds.

The N that is not solubilized by acid hydrolysis is usually referred to as acid-insoluble N; that recovered by distillation with MgO is NH_3–N. The soluble N not accounted for as NH_3 or in known compounds is the hydrolyzable unknown (HUN) fraction.

A rather large proportion of the soil N, usually on the order of 25 to 35%, is recovered as acid-insoluble N. Some of this N may be associated with the structures of humic and fulvic acids. The N of humic substances may exist in the following types of linkages:[22–24]

1. As a free amino (—NH_2) group
2. As an open chain (—NH—, =N—) group
3. As a part of a heterocyclic ring, such as an —NH— of indole and pyrrole or the —N= of pyridine

TABLE 6.3   Fractionation of Soil N by Acid Hydrolysis

Form	Definition and Method	% of Soil N (usual range)
Acid-insoluble N	Nitrogen remaining in soil residue following acid hydrolysis. Usually obtained by difference (total soil N– hydrolyzable N)	20–30
NH_3–N	Ammonia recovered from hydrolysate by steam distillation with MgO.	20–35
Amino acid N	Usually determined by the ninhydrin– CO_2 or ninhydrin–NH_3 methods. Recent workers have favored the latter.	30–45
Amino sugar N	Steam distillation with phosphate–borate buffer at pH 11.2 and correction for NH_3–N Colorimetric methods are also available.	5–10
Hydrolyzable unknown N (HUN fraction)	Hydrolyzable N not accounted for as NH_3, amino acids, or amino sugars. Part of this N occurs as non-α– amino–N in arginine, histidine, lysine, and proline.	10–20

4. As a bridge constituent linking aromatic rings or quinones together

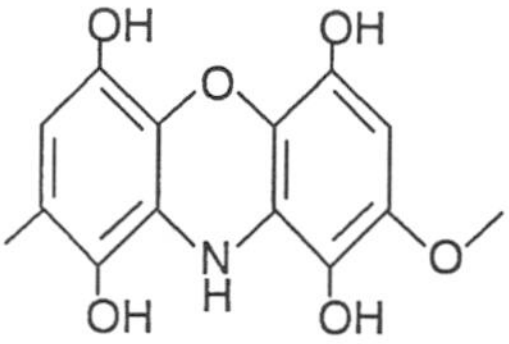

I

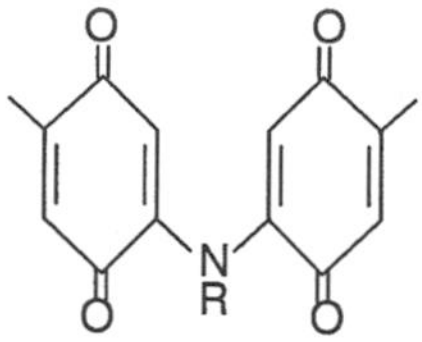

II

5. As amino acids or peptides attached to aromatic rings or quinones.

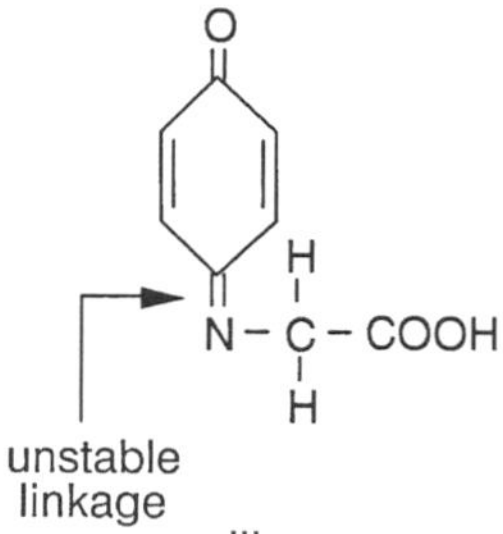

unstable
linkage

III

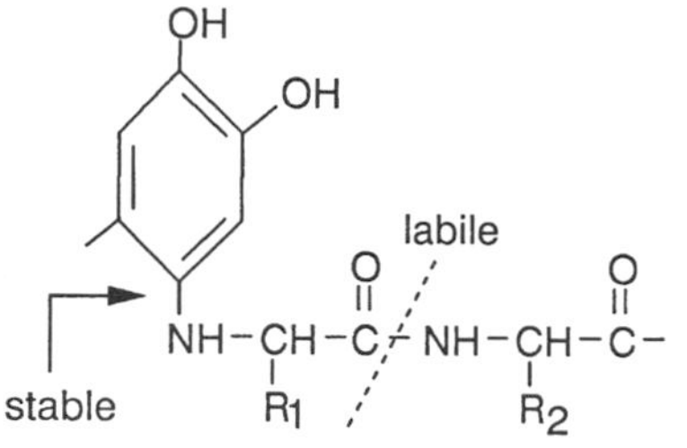

stable

IV

The N of heterocyclic rings (I), or that links aromatic rings or quinones together (II), would be expected to be recalcitrant to the acid treatment. On the other hand, acid hydrolysis would be expected to remove amino acids linked to quinones (III). For bound peptides, an amino acid directly attached to the aromatic ring may be less available to plants and microorganisms than amino acids of the peptide chain, as depicted in (IV).

During N fractionation, a large proportion of soil N (usually 20–30%) is recovered as NH_3. Some of the NH_3 is of inorganic origin; part comes from amino sugars and the amino acid amides, asparagine and glutamine. However, the origin of approximately one-half of the NH_3, equivalent to 10 to 12% of the total organic N, is still obscure.

The percentage of the soil N in the HUN fraction is also appreciable, often >20% of the total N. Some of this N, from one-fourth to one-half, may occur as the non-α-amino acid N of such amino acids as arginine, histidine, lysine, and proline. The non-amino N in these amino acids is not included with the amino acid N values as determined by the ninhydrin–NH_3 or ninhydrin–CO_2 methods (see Table 6.3).

Typical data showing the distribution of the forms of N in soils from different regions of the world are shown in Fig. 6.7. In general, differences in N composition within a similar group of soils are fully as great as for soils

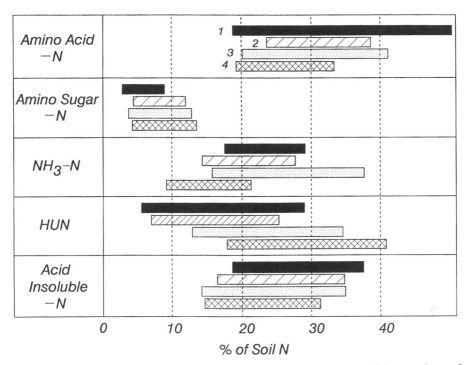

Fig. 6.7 Ranges in the distribution of the forms of N in the soils of four regions of the earth: 1 = United States (28 soils); 2 = Canada (34 soils); 3 = United Kingdom (20 soils); 4 = Africa (12 soils). Adapted from the summary of Stevenson.[23]

having contrasting chemical and physical properties. As noted below, changes have been observed through cropping.

It is a well-known fact that the N content of most soils declines when land is subjected to intense cultivation and attains a new equilibrium level characteristic of the cropping system used (see Chapter 2). This loss of N is not spread uniformly over all N fractions, although it should be pointed out that neither long-term cropping nor the addition of organic amendments to the soil greatly affects the relative distribution of the forms of N. All forms of soil N, including the acid-insoluble fraction, appear to be biodegradable.[27,28]

Typical data showing the effect of cultivation and cropping systems on the distribution of the forms of N in soil are recorded in Table 6.4. Cultivation generally leads to a slight increase in the proportion of the N as hydrolyzable NH_3, but this effect is due in part to an increase in the percentage of the soil N as clay-fixed NH_4^+. The proportion of the soil N as amino acid N generally decreases with cultivation, whereas the percentage as amino sugar N changes very little or increases. Climate has a major influence on N distribution, as seen in Table 6.5.

TABLE 6.4   Typical Data Showing the Effect of Cropping on the Distribution of the Forms of N in Soils

	% of Soil N				
Location[a]	Acid Insoluble	NH_3	Amino Acid	Amino Sugar	HUN
Illinois, U.S.A. (Morrow Plots)[b] Grass border and					
C–O–Cl rotation (2)	20.3	16.6	42.0	10.5	10.7
Continuous corn and C–O rotation (2)	20.2	16.7	35.0	14.4	13.9
Iowa, U.S.A.					
Virgin (10)	25.4	22.2	26.5	4.9	21.0
Cultivated (10)	24.0	24.7	23.4	5.4	22.5
Nebraska, U.S.A.					
Virgin (4)	20.8	19.8	44.3	7.3	7.8
Cultivated (4)	19.3	24.5	35.8	7.0	13.4

[a]Values in parentheses indicate number of soils analyzed. The data from Iowa and Nebraska are from Keeney and Bremner[27] and Meints and Peterson,[28] respectively.
[b]C–O–Cl = corn, oats, clover rotation with lime and P additions; C–O = corn–oats rotation.

Taken as a whole, results of N fractionations show that cultivation has only a negligible effect on the relative distribution of the forms of N, indicating that chemical fractionation of soil N following acid hydrolysis is of little practical value as a means of testing soils for available N or for predicting crop yields (see review of Stevenson[23]).

Application of ^{15}N–NMR Spectroscopy

A potentially useful technique for the characterization of soil N is CPMAS ^{15}N–NMR (described in Chapter 3). Due to the low natural abundance (and poor instrument sensitivity) of the ^{15}N isotope, the approach has had limited success for the analysis of natural soils, and much of the work has been done with ^{15}N-labeled substrates. With the recent development of new instrumental techniques and more powerful spectrometers, opportunities now exist for examining N cycle processes in soil.[29]

The technique of CPMAS ^{15}N–NMR has been applied in studies of (1) immobilized fertilizer N,[30,31] (2) the nature of the N in ^{15}N-enriched humic substances,[32] (3) NH_3 fixation by humic substances,[33] (4) nature of the N in composts,[34] and (5) the importance of the Maillard reaction in the formation of humic substances.[35] The method has also been used for the elucidation of oxygen-containing functional groups in humic and fulvic acids, namely, by reaction with ^{15}N-enriched hydroxylamine.[36]

In a study with ^{15}N-enriched $(NH_4)_2SO_4$, the soil was cropped for three years, following which humic and fulvic acids were extracted by conventional

TABLE 6.5 Nitrogen Distribution in Soils from Widely Different Climatic Zones[a]

Climatic Zone[b]	Total Soil N, %	Form of N, % of Total Soil N				
		Acid Insoluble	NH_3	Amino Acid	Amino Sugar	HUN
Arctic (6)	0.02–0.16	13.9 ± 6.6	32.0 ± 8.0	33.1 ± 9.3	4.5 ± 1.7	16.5
Cool temperate (82)	0.02–1.06	13.5 ± 6.4	27.5 ± 12.9	35.9 ± 11.5	5.3 ± 2.1	17.8
Subtropical (6)	0.03–0.30	15.8 ± 4.9	18.0 ± 4.0	41.7 ± 6.8	7.4 ± 2.1	17.1
Tropical (10)	0.24–1.61	11.1 ± 3.8	24.0 ± 4.5	40.7 ± 8.0	6.7 ± 1.2	17.6

[a]As recorded by Sowden et al.[26]
[b]Numbers in parentheses indicate number of soils.

techniques (see Chapter 1) and analyzed by CPMAS ^{15}N–NMR.[32] Most of the immobilized N (80–87%) was found in amide forms, with only a small portion as amino acids. All of the pyrrole N, representing 4 to 9% of the immobilized N, was accounted for in the HUN fraction following acid hydrolysis. Knicker et al.[34] found that essentially all of the N incorporated into composts (^{15}N-enriched KNO_3 was used) occurred in the form of peptides and free amino groups; signals that could be attributed to heteroaromatic structures (i.e., humic substances) were completely absent.

NONBIOLOGICAL REACTIONS AFFECTING THE INTERNAL N CYCLE

Not all transformations of N in soil are mediated by microorganisms; some are chemical in nature. These nonbiological reactions play a prominent role in the internal cycle of N in soils. Chemical reactions of inorganic forms of N are of three main types:

1. Fixation of NH_4^+ on interlamellar surfaces of clay minerals, notably vermiculite and the hydrated micas.
2. Fixation of NH_3 by the soil organic fraction, which supplements biological immobilization in that the fixed N is not readily available to plants or microorganisms.
3. Reactions of NO_2^- with organic constituents, including humic and fulvic acids. Part of the NO_2^-–N is converted to organic forms and part is lost from the soil system as N-containing gases.

The fate of inorganic forms of N in soils, including uptake by plants, is influenced by the magnitude of these processes, which vary from one soil to another and with cultural practices. Nitrite reactions (item 3) are discussed in Chapter 5.

Ammonium Fixation by Clay Minerals

It has long been known that many soils are capable of retaining considerable amounts of NH_4^+ in nonexchangeable forms. Fixation is the result of substitution of NH_4^+ for interlayer cations (Ca^{2+}, Mg^{2+}, Na^+) within the expandable lattice of clay minerals. According to one popular theory, NH_4^+ is fixed because the ion fits snugly into hexagonal holes or voids formed by oxygen atoms on exposed surfaces between the sheets of 2:1 lattice-type clay minerals. These voids have a diameter of about 2.8 nm, which is near the diameter of NH_4^+ (as well as K^+). When occupied by a fixable cation, the lattice layers contract and are bound together, thereby preventing hydration and expansion. Cations that have hydrated diameters greater than 2.8 nm, such as Na^+, Ca^{2+}, and Mg^{2+}, cannot enter the voids and are thereby able to move more freely in and out of the clay sheets. The subject of NH_4^+ fixation is reviewed by Nömmik[37] and Nömmik and Vahtras.[38]

Biological Availability of Fixed NH_4^+

The availability of NH_4^+ to microorganisms and higher plants can be reduced by fixation. However, Nömmik[39] suggests that fixation is usually not a serious problem under normal fertilizer practices. As noted below, K^+, being a fixable cation, is effective in blocking the release of fixed NH_4^+. Thus, the application of large amounts of K^+ simultaneously or immediately following NH_4^+ additions may reduce the availability of the fixed NH_4^+ to higher plants. Ammonium fixation cannot be considered entirely undesirable, since N losses through leaching may be reduced. In addition, fixation may ensure a more continuous supply of available N throughout the growing season.

Ammonium-Fixing Capacity

Individual soils vary in their capacity for fixing NH_4^+ from only a few kilograms to several hundred kilograms per hectare, depending on the factors outlined below.[37–39]

Type of Clay Mineral   Vermiculite, illite, and montmorillonite are the main minerals that retain NH_4^+ in a nonexchangeable form. These clays have a crystal lattice that is unique in that the individual sheets can expand and contract. Soils that contain significant amounts of vermiculite or weathered illite will fix 600 mmol NH_4^+/kg (1880 kg of NH_4^+–N/ha to plow depth); those soils where the clay fraction is dominated by montmorillonite will fix only 20 to 30 mmol/kg (62–108 kg of NH_4^+–N/ha). Practically no fixation occurs in soils where kaolinite is the predominant clay mineral.

The differential behavior of 2:1 type clay minerals in fixing NH_4^+ is due in part to the source of the negative charge, that is, whether the main part of the charge originates from the hexagonal Al-layer or the tetrahedral Si-layer. The attraction is greatest when the charge arises from isomorphous substitution of Al for Si in the tetrahedral layers, which gives a shorter distance between the interlayer cation and the negative site of the lattice. Substitution of Al by Si in the tetrahedral Si-sheet accounts for 80 to 90% of the substitution in vermiculites, about 65% or more in illites, and less than 20% in montmorillonites, a pattern that matches NH_4^+ fixation capacity.

Potassium Status of the Soil   Ammonium-fixing capacity of the soil depends to some extent on K^+ content. Fixation by micaceous minerals (such as illite) occurs only when the lattice has been impoverished of K^+ by weathering.

When the interlattice charge is balanced completely by a fixable cation (such as K^+), the sheets are fully contracted and NH_4^+ cannot enter. Soils containing large amounts of illite may or may not fix NH_4^+, depending on degree of weathering and saturation of the lattice by K^+. The capacity of many soils to fix NH_4^+ may be increasing because of removal of K^+ by cropping.

In general, surface soils have a much lower capacity for fixing NH_4^+ than subsoils, which may be due to a combination of the blocking effect of native K^+ and interference from organic matter.

Potassium fertilization can have a depressing effect on NH_4^+ fixation.[37-39] Addition of K^+ prior to or concurrently with fertilizer NH_4^+ reduces fixation of the NH_4^+ in proportion to the amount of K^+ applied, as shown in Fig. 6.8. Continuous fertilization with K^+ may reduce the ability of soil to fix NH_4^+. The blocking effect of K^+ can be avoided, if desired, by applying the K^+ well in advance of the NH_4^+.

Concentration of NH_4^+   The amount of NH_4^+ fixed by soil increases with rate of application. However, the percentage of the added NH_4^+ that is fixed decreases with an increase in the amount applied because fixation sites become saturated with NH_4^+. Although fixation can occur over a prolonged period, most fixation occurs within a few hours.

Moisture Conditions   Drying of the soil after addition of NH_4^+ increases both the rate and magnitude of fixation. Vermiculite and illite can fix NH_4^+ under moist conditions, but drying appears to be essential for fixation by montmorillonite. Alternate drying and wetting may be particularly effective in enhancing fixation. Freezing and thawing of the soil may influence NH_4^+ fixation in the same way as drying.

Soil pH   The relationship between soil acidity and NH_4^+-fixing capacity is

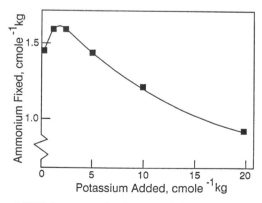

Fig. 6.8   Fixation of NH_4^+ by a vermiculite-containing clay soil as influenced by the simultaneous addition of K^+. Adapted from Nömmik.[37]

not pronounced. However, fixation tends to increase slightly with increasing pH, such as through liming. Highly acidic soils (<pH 5.5) generally fix little NH_4^+.

Naturally Occurring Fixed NH_4^+

Virtually all soils that contain K^+-bearing silicate minerals (e.g., illite) will also contain naturally occurring fixed NH_4^+.[40]

Data given in Table 6.6 show that a wide range of values has been recorded for fixed NH_4^+ in soils. The surface layer often contains 200 mg N/kg or more as fixed NH_4^+, equivalent to 448 kg N/ha. Clay and clay loam soils contain larger amounts of fixed NH_4^+ than silt loams, which in turn contain larger amounts than sandy soils. Spodosols contain rather low amounts of fixed NH_4^+, which can be attributed to their very low clay contents. Organic soils contain very little fixed NH_4^+.

The highest value thus far reported for fixed NH_4^+ in the surface layer of the soil appears to be that of Dalal,[41] who recorded a value of 1300 mg/kg in a Trinidad soil formed from micaceous schist and phyllite. Martin et al.[42] also recorded high values for fixed NH_4^+ in some subtropical soils derived from phyllite in Australia; in one soil, the content of fixed NH_4^+–N ranged from 415 mg/kg in the surface layer to over 1000 mg/kg at a depth of 120 cm. The amount of N in the soil as fixed NH_4^+ on an acre-profile basis can be appreciable, as illustrated by Fig. 6.9. The soil volume occupied by plant roots may contain over 1700 kg N/ha of fixed NH_4^+–N.

Such factors as drainage, type of vegetative cover, and extent of leaching of the profile by percolating water have little effect on the fixed NH_4^+ content of the soil; the amounts are more closely related to the kinds and amounts of clay minerals that are present.[43]

Results obtained for the vertical distribution of fixed NH_4^+ in representative forest and prairie grassland soils of the United States are given in Table 6.7. The rather high levels in the lower horizons of some of the profiles can be explained by the high illite content of the parent material.[43]

In contrast to the sharp decline in total N with depth, the soil's content of fixed NH_4^+ changes very little or increases. Thus, the proportion of the soil N as fixed NH_4^+ increases with depth (see Table 6.6). In general, less than 10% of the N in the plow layer of the soil occurs as fixed NH_4^+, although higher percentages are not uncommon. In some subsurface soils, over two-thirds of the N may occur as fixed NH_4^+. Narrowing of the C/N ratio with increasing depth in soil is due in part to fixed NH_4^+.

Changes occur in the absolute amount of fixed NH_4^+ in soil during long-time cropping, but they have been slight and variable. However, the proportion of the total N as fixed NH_4^+ tends to increase with cropping, because organic N levels decline when soils are placed under cultivation, while fixed NH_4^+ levels change very little. Walsh and Murdock[44] concluded that even under the most advantageous cropping conditions, very little native fixed NH_4^+ was available to crops, which was attributed to the blocking effect of K^+.

TABLE 6.6 Some Typical Values for Fixed NH_4^+ in Soils[a]

Location	Range, mg/kg	Comments
Australia	41–1076	From 221–1076 mg/kg (5–90% of N) in profiles developed on Permian phyllite and from 41–315 mg/kg (5–82% of N) in soils formed on other parent materials.
Canada		
Saskatchewan	158–330	From 7.7–13.3% of N in surface soil and up to 58.6% in subsoil. Cultivation did not effect fixed NH_4^+ content.
Alberta	110–370	From 7–14% of N in a wide variety of surface soils. Percentage increased with depth.
England	52–252	From 4–8% of N in surface soils and from 19–45% in subsoils.
Nigeria	32–220	From 2–6% of N in surface layers and from 45–63% in the subsoil.
Russia	14–490	From 2–7% of the N in surface soil but the percentage increases with depth.
Sweden	10–17	Values are for a Spodosol profile low in clay.
Taiwan	140–170	From 10.6–32.6% of the N in surface layer of 9 soils.
United States		
North Central	7–270	A wide range has been recorded, the lowest being in Spodosols and the highest in silt loams and soils rich in illite; from 4–8% of the N in the surface soils with the proportion increasing with depth.
Pacific Northwest	17–138	From 1.1–6.2% of N in surface soils with the proportion increasing with depth in some soils but not others.
Hawaii	0–585	Volcanic ash soils contained less (4–178 mg/kg) than soils derived from basalt (up to 585 mg/kg).

[a] For references see Young and Aldag.[40]

Fixation of NH_3 by Organic Matter

The chemical reaction of NH_3 with soil organic matter is frequently referred to as "NH_3 fixation," and this convention will be adopted herein. The term should not be confused with retention or "fixation" of the NH_4^+ ion by clay minerals (see previous section).

For complete coverage of NH_3 fixation by organic matter, the reader is referred to several reviews on the subject.[37,45] Results using ^{15}N-labeled an-

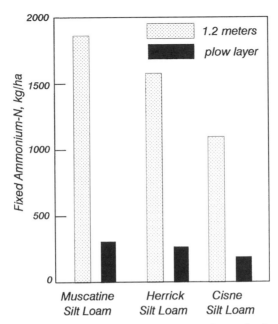

Fig. 6.9   Content of fixed NH_4^+–N in the plow layer and to a depth of 1.22 m in three agricultural soils of Illinois.

hydrous and aqueous NH_3 have confirmed the importance of organic matter in fixing NH_3 in soils.[46–49]

The ability of lignin and soil organic matter to react chemically with NH_3 has been known for more than 60 years, and numerous patents have been issued over this period for the conversion of peat, sawdust, lignaceous residues (corn cobs, etc.), and coal products into nitrogenous fertilizers by treatment with NH_3. Fixation is associated with oxidation (uptake of O_2) and is favored by an alkaline pH. Thus, the application of alkaline fertilizers to soil,

TABLE 6.7   Distribution of Fixed NH_4^+ in some Soil Profiles

Horizon	Mollisols (5)[a]		Alfisols (5)		Others[b]	
	mg/kg	% of N	mg/kg	% of N[c]	mg/kg	% of N
A_1	96–129	4.3–5.6	113–155	5.6	57–141	3.5–7.9
A_2	106–131	8.2–9.3	109–168	9.4	50–154	7.8–10.8
B_2	99–155	11.5–18.4	143–210	20.1	55–182	12.7–17.6
B_3	104–174	21.4–27.7	185–224	26.5	84–210	13.1–17.0
C	98–188	28.4–44.7	102–224	34.9	90–210	27.2–36.1

[a]Numerals refer to number of profiles examined.
[b]Includes an Ultisol and two Mollisols with argillic horizons.
[c]For one profile only.

such as aqueous or anhydrous NH_3, may result in considerable fixation. Injection of anhydrous NH_3, for example, results in a pronounced increase in soil pH, with the highest pH being along the injection line with a gradient extending outward from that line. Similarly, the highest concentration of NH_3 will be found in the injection zone. These conditions are highly favorable for NH_3 fixation by organic matter.

The relative importance of clay and organic colloids in retaining NH_3 (as NH_4^+) will depend upon the nature of the soil, but at pH values above 7, the organic fraction appears to be more reactive in relation to the amount present than is the clay.

Mechanisms of NH₃ Fixation

Very little is known regarding NH_3 fixation reactions, although several plausible mechanisms have been proposed. These are based on the observation that fixation proceeds most favorably at high pH's (Fig. 6.10) and that fixation is accompanied by the uptake of O_2 (Fig. 6.11). Fixation occurs rapidly under optimum conditions and can continue for a prolonged period, although at a diminishing rate. Burge and Broadbent[51] estimated the NH_3-fixing capacity of a series of organic soils and found that, under aerobic conditions, one molecule of NH_3 was fixed for every 29 C atoms. As one might expect, NH_3 fixation bears a close relationship to the organic matter content of the soil.

Flaig[52] suggested a mechanism for NH_3 fixation by phenolic compounds in which polymers containing N in heterocyclic linkages are formed. The initial step involves O_2 consumption by the phenol to form a quinone, which subsequently reacts with NH_3 to form a complex polymer. Catechol (V), for example, is converted by the action of O_2 under alkaline conditions to the o-quinone (VI), which is then hydrated to form benzenetriol (VII). Further oxidation was postulated to produce a mixture of o-hydroquinone (VIII) and p-hydroxy-o-quinone (IX), both of which are capable of reacting with NH_3.

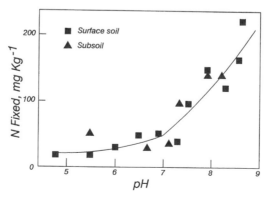

Fig. 6.10   Effect of pH on NH_3 fixation by an organic soil. Adapted from Broadbent et al.[50]

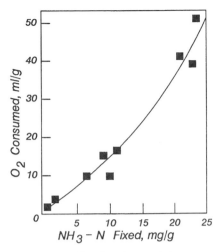

Fig. 6.11 Relationship between NH_3 fixation and O_2 consumption. Adapted from Broadbent and Stevenson.[45]

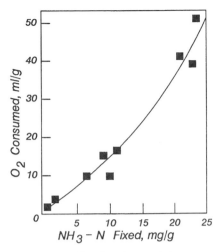

The incorporation of NH_3 into p-hydroxy-o-quinone (IX) was believed to produce structures of the types represented by X and XI.

An interesting feature of structure XI is that a relatively small amount of substrate will fix a considerable amount of NH_3. In this respect, it should be noted that results of ^{15}N–NMR spectroscopic studies have indicated only trace quantities of heterocyclic N in the reaction products between NH_3 and humic substances.[33]

As yet, there is no sound basis for selecting any one structure as the correct one, or for assuming that only one mechanism is involved. Fixation may involve the incorporation of NH_3 into a wide variety of compounds, including some of a refractory nature (such as lignin). Also, NH_3 may participate in polymerization or condensation reactions of small molecules to form polymers with properties similar to natural humic substances.

Availability of Chemically Fixed NH₃ to Plants

An important question with respect to NH_3 fixed by organic matter is whether the N is available to plants. Most research indicates that availability is relatively low.[37,45] In a greenhouse experiment using ^{15}N-labeled NH_3, Burge and Broadbent[51] observed that over 95% of the fixed NH_3 was no more available to plants than the indigenous soil N. He et al.,[49] also using ^{15}N-labeled NH_3, found that organic matter fixed NH_3 was initially more labile than the native soil N but became less labile with time.

Noticable solubilization of soil organic matter has been observed following application of anhydrous (or aqueous) NH_3 to soil,[46,53,54] a result that may enhance the availability of native humus N to plants and microorganisms.

Nitrite–Organic Matter Interactions

Under certain circumstances, NO_2^- can interact with soil humic substances to form nitrogenous gases. This subject is discussed in Chapter 5.

MODELING OF THE SOIL N CYCLE

Modeling is an attempt to describe the dynamic aspects of the soil N cycle in mathematical terms.[55–61] Many models are simulation models in that they attempt to forecast how a system will behave or perform without actually analyzing the physical system or its prototype. Mathematical models, on the other hand, utilize empirical or observational data to provide quantitative values for gains, losses, and transfers of N, as well as for the amounts of N contained in one or more pools as a function of time. Models can be local, regional, or global in scope.

Mathematical models are of three main types:

1. *Stochastic:* Based on the assumption that the processes to be modeled obey the laws of probability.

2. *Empirical:* Based on observational data. Input and output processes are expressed in terms of regression equations from soil-derived data.
3. *Mechanistic:* Based on well-established physical, chemical, and biological laws that describe the various processes.

Mechanistic models are more versatile than other types in that historical data are not required for their development, but data are required for validation of the model. Grant and coworkers[62–64] developed complex computer models by using a combination of fundamental principles of ammonification, nitrification, and denitrification and published data for the rates of various processes, and then validated the models experimentally. Their methods will probably serve as progenitors of similarly complex models in the future. Davidson et al.[58] have tabulated rate coefficients for mineralization, immobilization, nitrification, and denitrification in soil that may be used (with caution) for modeling purposes.

The primary purpose of modeling is to provide a method of determining the fate of fertilizer N in soil under a given set of conditions. Appraisals of plant uptake and NO_3^- losses through leaching or denitrification involve consideration of the many sources and sinks of N, as well as flow pathways of both NO_3^- and water. Other objectives of modeling are to:[55]

1. Obtain a better understanding and increased insight into complex problems
2. Evaluate available information and ascertain the adequacy of published data for problem evaluation
3. Test existing as well as new concepts and hypotheses
4. Estimate by difference an unknown output or input
5. Obtain a better evaluation or prediction of an observed phenomena
6. Identify research needs
7. Help develop guidelines for best management practices

Models are usually illustrated by a flow diagram showing one or more pools of N with inputs and/or outputs for each. Values are assigned to pool sizes and rates for inputs and outputs. The various models differ greatly in complexity, ranging from those designed to simulate a single process, such as leaching or denitrification, to those that include transformations occurring within the soil. Finally, there are models that involve interactions with other components of the ecosystem.

The degree of sophistication of any particular model is determined by the extent of understanding of the system to be modeled, the availability of reliable data to serve as inputs, and the intended application.[58] Those models involving transformations within the soil usually include mineralization, immobilization, nitrification, denitrification, and leaching, although one or more of these are sometimes ignored. The more complex models take into account differences in the decay rate of various constituents of plant residues (proteins, sugars, cellulose, and lignin), as shown in Fig. 6.12. Molina et al.[65] have developed and calibrated a model for the short-term dynamics of organic N, NH_4^+, and NO_3^- in which the soil organic phase is divided into two pools that degrade independently of each other. Each pool contained a labile and resistant component.

The absence of reliable data for modeling is a major obstacle in most modeling studies. Most processes are transient in nature; some occur simultaneously. Some pools and fluxes can be evaluated with a reasonable degree of accuracy, whereas others cannot. A major difficulty with models that include mineralization of the soil organic N is that this pool is very large compared to the other pools, and a slight mistake in estimating the mineralization rate introduces a major error in estimates for the amount of mineral N potentially available to the plant. Under certain circumstances, transformations occurring within the soil can be ignored and a "black box" approach to modeling can be used (see Chapter 7).

Some selected modeling approaches for soil N transformations are listed in Table 6.8. Data used for the various transformations (e.g., nitrification, mineralization, immobilization, denitrification, plant assimilation, etc.) are modeled in different ways: multiple regression, zero and first-order chemical kinetics, and Michaelis–Menten kinetics. A detailed discussion of the various approaches is beyond the scope of the present paper, and the reader is referred to several reviews for more detailed information.[55,57,58]

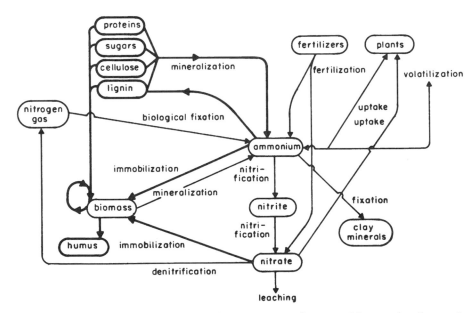

Fig. 6.12 Soil N transformations and components of crop residues and soil organic matter. From Frissel and van Veen,[56] reproduced by permission of Academic Press.

TABLE 6.8   Some Modeling Approaches for Soil N Transformations

Reference	Modeling Approaches[a]
Davidson et al.[58]	Nitrification, mineralization, and immobilization by first-order kinetics and modified by soil water pressure head; denitrification by first-order kinetics considering pressure head, water content, and organic matter content; NH_4^+ sorption by linear partition model; NH_4^+ and NO_3^- uptake by Michaelis–Menten kinetics.
Dutt et al.[59]	Nitrification, mineralization–immobilization, and urea hydrolysis modeled by regression equation; NH_4^+ sorption by equilibrium cation exchange equation; N plant uptake proportional to water uptake.
Frissel & van Veen[56]	Growth of *Nitrosomonas* and *Nitrobacter* by Michaelis–Menten kinetics including considerations of oxygen level; denitrification by a physical–biological model involving oxygen diffusion; mineralization–immobilization considers organic matter grouped into fresh applied organic matter such as animal manures, straw and waste water, and resistant biomass residues; NH_3 volatilization by physiochemical model including dissociation of NH_4OH; NH_4^+ clay fixation by reversible first-order kinetics.
Mehran & Tanji[60]	Nitrification, denitrification, mineralization, immobilization, and N plant uptake by irreversible first-order kinetics; NH_4^+ exchange by reversible first-order kinetics.
Molina et al.[65]	Computes short-term dynamics of organic N, NH_4^+, and NO_3^- that result from the process of residue decomposition, mineralization, immobilization, nitrification, and denitrification. Organic pools include the residues and two pools of the soil organic fraction, each containing a labile and resistant component.

[a]Derived in part from Tanji.[55]

SUMMARY

Nitrogen undergoes a wide variety of transformations in soil, most of which involve the organic fraction. Although considered individually, each process is affected by others occurring sequentially; in some cases, opposing processes operate simultaneously.

An internal "N cycle" exists in soil apart from the overall cycle of N in nature. Even if N gains and losses are equal, as may occur in a mature ecosystem, the "N cycle" is not static. Continuous turnover of N occurs through mineralization–immobilization, with incorporation of N into microbial cells, the so-called active fraction of organic matter. Whereas much of the newly immobilized N is recycled through mineralization, some is converted to stable humus forms. The processes involved in mineralization–

immobilization turnover are not fully understood, and as a result there is considerable confusion regarding the interpretation of tracer data about soil N transformations (see Chapter 7).

Fixation of NH_4^+ by clay minerals is of some importance in select soils, and significant amounts of the soil N often occur in the form of naturally occurring fixed NH_4^+. Ammonia fixation by organic matter will be particularly important when urea or anhydrous NH_3 is applied to soils rich in organic matter, an effect that has not been fully investigated under field conditions.

Ultimately, N-cycle processes in soil will be monitored by computer analysis of biological data for mineralization, mineralization, nitrification, biological N_2 fixation, and losses of N through leaching and denitrification. These analyses will provide precise estimates for the quantities of plant-available form of N in the soil at any one time. The internal N cycle will be manipulated to minimize N losses through leaching and denitrification, thereby protecting the environment. Conventional N fertilizers will continue to be used, but in a much more efficient manner than at present.

REFERENCES

1. S. L. Jansson and J. Persson, "Mineralization and Immobilization of Soil Nitrogen," in F. J. Stevenson, Ed., *Nitrogen in Agricultural Soils,* American Society of Agronomy, Madison, Wisconsin, 1982, pp. 229–252.

2. J. N. Ladd and R. B. Jackson, "Biochemistry of Ammonification," in F. J. Stevenson, Ed., *Nitrogen in Agricultural Soils,* American Society of Agronomy, Madison, Wisconsin, 1982, pp. 173–228.

3. S. Winogradsky, *Ann. Inst. Pasteur,* **4,** 213, 257, 760 (1890).

4. D. D. Focht and W. Verstrate, *Adv. Microbiol. Ecol.,* **1,** 135 (1977).

5. E. L. Schmidt, "Nitrification in Soil," in F. J. Stevenson, Ed., *Nitrogen in Agricultural Soils,* American Society of Agronomy, Madison, Wisconsin, 1982, pp. 253–288.

6. J. M. Bremner and A. M. Blackmer, *Science,* **196,** 295 (1978).

7. A. M. Blackmer, J. M. Bremner, and E. L. Schmidt, *Appl. Environ. Microbiol.,* **40,** 1060 (1980).

8. E. L. Schmidt and L. W. Belser, "Autotrophic Nitrifying Bacteria," in R. W. Weaver, et al., Eds., *Methods of Soil Analysis, Part 2: Microbiological and Biochemical Properties,* Soil Science Society of America, Madison, Wisconsin, 1994, pp. 159–177.

9. J. M. Bremner and G. W. McCarty, "Inhibition of Nitrification in Soil by Allelochemicals Derived from Plants and Plant Residues," in J.-M. Bollag and G. Stotzky, Eds., *Soil Biochemistry,* Vol. 8, Marcel Dekker, New York, 1993, pp. 181–218.

10. C. A. I. Goring and D. A. Laskowski, "The Effects of Pesticides on Nitrogen Transformations in Soils," in F. J. Stevenson, Ed., *Nitrogen in Agricultural Soils,* American Society of Agronomy, Madison, Wisconsin, 1982, pp. 689–720.

11. S. L. Jansson, "Use of ^{15}N in Studies on Soil Nitrogen," in A. D. McLaren and J. Skujins, Eds., *Soil Biochemistry 2,* Marcel Dekker, New York, 1971, pp. 129–166.

12. F. J. Stevenson, Ed., *Nitrogen in Agricultural Soils,* American Society of Agronomy, Madison, Wisconsin, 1982.

13. G. W. Harmsen and D. A. van Schreven, *Adv. Agron., 7,* 299 (1955).

14. G. W. Harmsen and G. J. Kolenbrander, "Soil Inorganic Nitrogen," in W. V. Bartholomew and F. E. Clark, Eds., *Soil Nitrogen,* American Society of Agronomy, Madison, Wisconsin, 1965, pp. 43–92.

15. A. D. Rovira and B. M. McDougall, "Microbiological and Biochemical Aspects of the Rhizosphere," in A. D. McLaren and G. H. Petersen, Eds., *Soil Biochemistry,* Dekker, New York, 1967, pp. 417–463.

16. F. J. Stevenson and G. H. Wagner, "Chemistry of Nitrogen in Soils," in T. L. Willrich and G. E. Smith, Eds., *Agricultural Practices and Water Quality,* Iowa State University Press, Ames, Iowa, 1970, pp. 125–141.

17. B. A. Stewart, F. G. Viets Jr., G. L. Hutchinson, and W. D. Kemper, *Environ. Sci. Tech., 1,* **736** (1967).

18. L. G. Bundy and J. J. Meisinger, "Nitrogen-Availability Indices," in R. W. Weaver, et al., Eds., *Methods of Soil Analysis, Part 2: Microbiological and Biochemical Properties,* Soil Science Society of America, Madison, Wisconsin, 1994, pp. 951–984.

19. G. Stanford, "Assessment of Soil Nitrogen Availability," in F. J. Stevenson, Ed., *Nitrogen in Agricultural Soils,* American Society of Agronomy, Madison, Wisconsin, 1982, pp. 651–688.

20. G. Stanford, J. N. Carter, and S. J. Smith, *Soil Sci. Soc. Amer. Proc., 38,* 99 (1974).

21. G. Stanford and S. J. Smith, *Soil Sci., 122,* 71 (1976).

22. F. J. Stevenson, *Humus Chemistry: Genesis, Composition, Reactions,* 2nd ed., Wiley, New York, 1994.

23. F. J. Stevenson, "Organic Forms of Soil Nitrogen," in F. J. Stevenson, Ed., *Nitrogen in Agricultural Soils,* American Society of Agronomy, Madison, Wisconsin, 1982, pp. 67–122.

24. K. R. Kelly and F. J. Stevenson, "Organic Forms of N in Soil," in A. Piccolo, Ed., *Humic Substances in Terrestrial Ecosystems,* Elsevier, Amsterdam, 1996, pp. 407–427.

25. F. J. Stevenson, "Nitrogen–Organic Forms," in D. L. Sparks, Ed., *Methods of Soil Analysis: Part 3,* Soil Science Society of America, Madison, Wisconsin, 1996, pp. 1185–1178.

26. F. J. Sowden, Y. Chen, and M. Schnitzer, *Geochim. Cosmochim. Acta, 41,* 1524 (1977).

27. D. R. Keeney and J. M. Bremner, *Soil Sci. Soc. Amer. Proc., 28,* 653 (1964).

28. V. W. Meints and G. A. Peterson, *Soil Sci., 124,* 334 (1977).

29. C. Steelink, "Application of N-15 NMR Spectroscopy to the Study of Organic Nitrogen and Humic Substances in Soil," in N. Senesi and T. M. Miano, Eds., *Humic Substances in the Global Environment and Implications on Human Health,* Elsevier, Amsterdam, 1994, pp. 405–426.

30. L. Benzing-Purdie et al., *J. Soil Sci.. 43,* 113 (1992).

31. L. Benzing-Purdie et al., *J. Agric. Food Chem.,* **34,** 170 (1986).

32. Z. Su-Neng and W. Qi-Xiao, *Pedosphere,* **4,** 307 (1992).

33. K. A. Thorn, J. B. Arterburn, and M. A. Mikita, *Environ. Sci. Technol.,* **26,** 107 (1992).

34. H. Knicker, R. Fründ, and H.-D. Lüdemann, "N-15 NMR Studies of Humic Substances in Solution," in N. Senesi and T. M. Miano, Eds., *Humic Substances in the Global Environment and Implications on Human Health,* Elsevier, Amsterdam, 1994, pp. 501–506.

35. L. Benzing-Purdie, J. A. Ripmeester, and C. M. Preston, *J. Agric. Food Chem.,* **31,** 913 (1983).

36. K. A. Thorn and M. A. Mikita, *Environ. Sci. Technol.,* **113,** 67 (1992).

37. H. Nömmik, "Ammonium Fixation and Other Reactions Involving a Non-enzymatic Immobilization of Mineral Nitrogen in Soil," in W. V. Bartholomew and F. E. Clark, Eds., *Soil Nitrogen,* American Society of Agronomy, Madison, Wisconsin, 1965, pp. 198–258.

38. H. Nömmik and K. Vahtras, "Retention and Fixation of Ammonium in Soils," in F. J. Stevenson, Ed., *Nitrogen in Agricultural Soils,* American Society of Agronomy, Madison, Wisconsin, 1982, pp. 123–171.

39. H. Nömmik, *Acta Agric. Scand.,* **7,** 395 (1957).

40. J. L. Young and R. W. Aldag, "Inorganic Forms of Nitrogen in Soil," in F. J. Stevenson, Ed., *Nitrogen in Agricultural Soils,* American Society of Agronomy, Madison, Wisconsin, 1982, pp. 43–66.

41. R. C. Dalal, *Soil Sci.,* **124,** 323 (1977).

42. A. E. Martin, R. J. Gilkes and J. O. Skjemstad, *Aust. J. Soil Res.,* **8,** 71 (1970).

43. F. J. Stevenson and A. P. S. Dhariwal, *Soil Sci. Soc. Amer. Proc.,* **23,** 121 (1959).

44. L. M. Walsh and J. T. Murdock, *Soil Sci.,* **89,** 183 (1960).

45. F. E. Broadbent and F. J. Stevenson, "Organic Matter Reactions," in M. H. McVickar et al., Eds., *Agriculture Anhydrous Ammonia,* American Society of Agronomy, Madison, Wisconsin, 1966, pp. 169–187.

46. R. J. Norman, L. T. Kurtz, and F. J. Stevenson, *Soil Sci. Soc. Amer. J.,* **51,** 235, 809 (1987).

47. X.-T. He, R. L. Mulvaney, F. J. Stevenson, and R. M. Vanden Heuvel, *Soil Sci. Soc. Amer. J.,* **11,** 54 (1990)

48. R. J. Norman, L. T. Kurtz, and F. J. Stevenson, *Soil Sci. Soc. Amer. J.,* **51,** 809 (1987).

49. X.-T. He, R. L. Mulvaney, and F. J. Stevenson, *Biol. Fert. Soils* **11,** 145 (1991).

50. F. E. Broadbent, W. D. Burge, and T. Nakashima, *Trans. 7th Intern. Congr. Soil Sci. (Madison),* **2,** 509 (1960).

51. W. D. Burge and F. E. Broadbent, *Soil Sci. Soc. Amer. Proc.,* **25,** 199 (1961).

52. W. Flaig, *Z. Pflanzenahr. Dung. Bodenk.,* **51,** 193 (1950).

53. D. J. Tomasiewicz and J. L. Henry, *Can. J. Soil Sci.,* **65,** 737 (1985).

54. R. G. Meyers and S. J. Thien, *Soil Sci. Soc. Amer. J.,* **52,** 516 (1988).

55. K. K. Tanji, "Modeling of the Soil Nitrogen Cycle," in F. J. Stevenson, Ed., *Nitrogen in Agricultural Soils,* American Society of Agronomy, Madison, Wisconsin, 1982, pp. 721–772.

56. M. J. Frissel and J. A. van Veen, "A Critique of Computer Simulation Modeling for Nitrogen in Irrigated Croplands," in D. R. Nielsen and J. G. MacDonald, Eds., *Nitrogen in the Environment,* Vol. 1, Academic Press, New York, 1978, pp. 145–162.

57. M. J. Frissel, Ed., *Cycling of Mineral Nutrients in Agricultural Ecosystems,* Elsevier, New York, 1978.

58. J. B. Davidson, D. A. Graetz, S. C. Rao, and H. M Selim, *Simulation of Nitrogen Movement, Transformation, and Uptake in Plant Root Zone,* USEPA, Athens, Georgia, 1978.

59. R. G. Dutt, M. J. Shaffer, and W. J. Moore, *Computer Simulation Model of Dynamic Biophysicochemical Processes in Soils,* Arizona Agric. Exp. Sta. Tech. Bull. 196, 1972.

60. M. Mehran and K. K. Tanji, *J. Environ. Qual.,* **3,** 391 (1974).

61. K. K. Tanji and S. K. Gupta, "Computer Simulation Modeling for Nitrogen in Irrigated Cropland," in D. R. Nielsen and J. G. MacDonald, Eds, *Nitrogen in the Environment,* Vol. 1, Academic Press, New York, 1978, pp. 79–130.

62. R. F. Grant, N. G. Juma, and W. B. McGill, *Soil Biol. Biochem.,* **25,** 1317 (1993).

63. R. F. Grant, *Soil Biol. Biochem.,* **26,** 305 (1994).

64. R. F. Grant, *Soil Biol. Biochem.,* **27,** 1117 (1995).

65. J. A. E. Molina, C. E. Clapp, M. J. Shaffer, F. W. Chichester, and W. E. Larson, *Soil Sci., Soc. Amer. J.,* **47,** 85 (1983).

7

DYNAMICS OF SOIL N TRANSFORMATIONS AS REVEALED BY ^{15}N TRACER STUDIES

Many facets of the soil N cycle can be examined (and understood) only by using the stable isotope ^{15}N. In a typical experiment, a known amount of ^{15}N-labeled fertilizer is applied to soil, and the relative amounts of the soil ^{14}N and fertilizer ^{15}N used by plants and the quantities of fertilizer N retained in mineral and organic forms are determined at intervals. Any fertilizer not accounted for in the crop or in the soil is assumed to have been lost through leaching, denitrification, or NH_3 volatilization. The use of ^{15}N in soil N studies is discussed in several reviews.[1-3]

Nitrogen-15 has been used in the following types of studies:

1. Nitrogen balance of the soil
2. Stabilization of N through immobilization
3. Uptake of soil and fertilizer N by plants and fate of residual fertilizer N in soil
4. Losses of soil and fertilizer N through leaching and denitrification
5. Biological N_2 fixation
6. Fixation of NH_4^+ by clay and of NH_3 by organic matter and the availability of the fixed N to plants and microorganisms
7. Relative use of NH_4^+ and NO_3^- by microorganisms and higher plants

Those aspects related to transformations occurring within the soil through mineralization-immobilization are emphasized herein, namely, items 1 through 5.

NITROGEN ISOTOPE TECHNIQUES

There are six known isotopes of N, but only those of mass numbers 14 and 15 are stable and occur naturally. The four radionuclides (mass numbers of

12, 13, 16, and 17) have very short half-lives and are thus not suitable for most studies on soil N transformations. The ^{13}N radionuclide, with a half-life of 10.05 minutes, has had restricted use in short-term studies (e.g., see Bremner and Hauck[4]). Most of Earth's N occurs as the stable isotope ^{14}N (99.634% of atmospheric N), and the natural abundance of ^{15}N (atmosphere) is 0.366%. Experiments with ^{15}N are conducted by using an N-source whose percentage of ^{15}N is either greater than natural abundance (^{15}N-enriched) or less than natural abundance (^{15}N-depleted).

The advantage of ^{15}N as a tracer stems from its nonradioactivity; consequently, the isotope can be used in experiments carried out over extended time periods. In contrast, short-lived isotopes like ^{13}N undergo radioactive decay so rapidly that much of the original activity is lost within a few days or weeks. Accounting for loss of ^{13}N due to radioactive decay or physiochemical processes such as volatilization or leaching is difficult and can lead to uncertainty in the interpretation of results. There are no health hazards associated with the use of the stable isotope ^{15}N, and a permit is not required for experiments carried out in the field or greenhouse. By comparison, radioactive isotopes like ^{13}N can only be used under confined and carefully controlled conditions in order to comply with regulations regarding use of radioactive materials.

Methods for the use of ^{15}N as a tracer for soil and fertilizer N research are described in several reviews[4-8] and in Knowles and Blackburn.[9]

Assumptions

Three key assumptions are made in the use of ^{15}N in soil N studies:

1. The isotopic composition of N (e.g., ^{14}N and ^{15}N) in the natural soil remains constant over time.
2. Living organisms, including soil microbes, cannot distinguish between the two N isotopes (i.e., ^{14}N and ^{15}N are used indiscriminately during N metabolism).
3. The chemical reactivity and response to physical factors of the two isotopes are identical and remain constant over time.

As noted below, these assumptions are not completely valid for all studies on soil N transformations. Other assumptions that relate to a particular soil N transformation, and that are not always valid, are discussed by Hauck and Bremner.[2]

Steps in the Determination of Isotopic ^{15}N

For isotope determination, all forms of N must be converted into a suitable gaseous form for instrumental analysis, elemental N_2 being preferred because it is chemically inert and of low molecular weight. Also, N_2 can easily be formed from a variety of organic and inorganic compounds. The following steps are involved in an N isotope analysis:

1. Conversion of all forms of N to NH_3 (recovered as NH_4^+)

Most early studies were restricted to isotopic analysis of total N in the sample, which was done by recovery of the N as NH_4^+ using some modification of the Kjeldahl method. Procedures are now available for isotopic analysis of specific N forms, such as exchangeable- and clay-fixed NH_4^+, NO_3^-, NO_2^-, and various organic N compounds (e.g., amino acids and amino sugars). These procedures are described in detail elsewhere.[2,4,9] Methods for recovery of the different organic N forms as NH_4^+ are outlined in Chapter 6 (see Table 6.3).

2. Distillation and recovery of NH_4^+–N

The NH_4^+ recovered in Step 1 must be separated from the reagents used for the conversion, as well as constituents in the experimental sample that would interfere with subsequent hypobromite oxidation. One popular method is steam distillation under alkaline conditions, using an apparatus of the type shown in Fig. 7.1.

With microdistillation procedures, the distillate volume typically exceeds 40 mL, which must be reduced to 2–5 mL for subsequent analysis. Evaporation can be done in open beakers on a hot plate, but precautions must be taken to prevent contamination from atmospheric NH_3. Acidification of the distillate (see Table 6.3) is essential to prevent volatile loss of N as NH_3.

3. Oxidation of NH_4^+–N to N_2 by alkaline hypobromite in the absence of air:

$$2 \ NH_3 + 3 \ OBr^- \rightarrow N_2 + 3 \ H_2O + 3 \ Br^-$$

The reaction is generally carried out in a Rittenberg "Y" tube (Fig. 7.2) that can be tilted to mix the hypobromite solution with the experimental sample.

4. Determination of the isotopic composition of the N_2.

The steps noted above are time-consuming, and considerable care is required to avoid errors such as those listed in Table 7.1.

Isotope-Ratio Analysis by Mass Spectrometry

Mass spectrometric (MS) analysis is the preferred method for the determination of the isotopic composition of the N_2 formed in the above reaction.

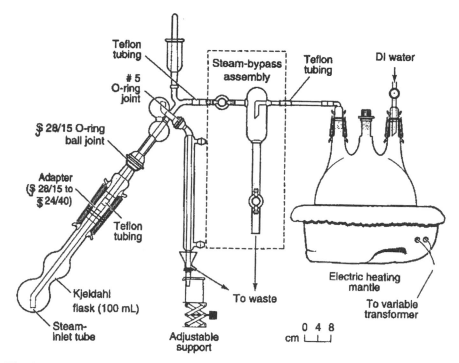

Fig. 7.1 Steam distillation apparatus. Provided through the courtesy of R. L. Mulvaney.

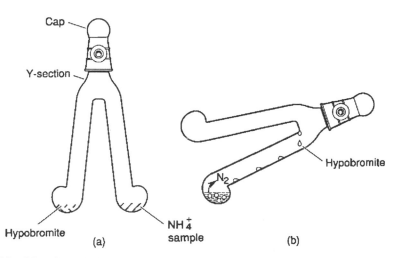

Fig. 7.2 Rittenberg "Y" tube: (*a*) Upright position for evacuation and degassing; (*b*) Tilted position for hyprobromite oxidation of NH_4^+ to N_2. From Mulvaney,[6] reproduced with permission of Academic Press Inc.

TABLE 7.1   Sources of Errors in the Isotopic Analysis of ^{15}N^a

Step	Source of Error
Conversion of labeled N to NH_4^+	Nonspecific and/or incomplete conversion
Distillation of NH_4^+	Incomplete distillation. Cross-contamination
Conversion of NH_4^+ to N_2	a. Failure to remove dissolved and gaseous N_2 prior to conversion
	b. Incomplete oxidation of NH_4^+ by hypobromite
	c. Incomplete recovery of oxidation product (i.e., N_2)
	d. Air leakage
	e. Contamination of N_2 by N_2O and other gaseous impurities during hypobromite oxidation

^{a}Adapted from Hauck and Bremner.[2]

As the name implies, a mass spectrometer separates ions into a spectrum according to their masses, or more precisely, mass to charge (m/e) ratio. A variety of instruments are available for this purpose, but the most common ones are magnetic deflection types consisting of three major components: an analyzer tube, ion source, and collectors. A schematic diagram of a 60°-sector mass spectrometer equipped with double collectors is shown in Fig. 7.3.

From the gas inlet, N_2 is admitted to the ion source. When an electron is lost, isotopes of masses $^{28}N_2^+$, $^{29}N_2^+$, and $^{30}N_2^+$ are formed, and then they are drawn out of the ion source and into the magnetic sector of the spectrometer. Due to the influence of the magnetic field, the paths of the three ions are curved (see Fig. 7.3), the radius decreasing in order of the mass of the ions. The intensities of the three beams, corresponding to m/e 28, 29, and 30, are directly related to the isotopic composition of the N_2 under analysis. The current generated by the three ion beams is amplified and recorded and subsequently converted into data of isotopic composition of the sample.

The mass spectrometer provides values for the ratio of the intensities of the currents produced by two or more of the ion beams having masses of 28, 29, and 30 (i.e., $^{28}N_2^+$, $^{29}N_2^+$, and $^{30}N_2^+$). These three ions correspond to the formulas $(^{14}N^{14}N)^+$, $(^{14}N^{15}N)^+$, and $(^{15}N^{15}N)^+$ and are are distributed according to the equation:

$$(a + b)^2 = a^2 + 2ab + b^2 \tag{1}$$

where a and b are ^{14}N and ^{15}N, respectively. Several equations can be used for calculating atom% ^{15}N from the experimental data, depending on the ratio that is measured (see Mulvaney[6]). When the measured ratio (r) is $^{29}N_2^+/^{28}N_2^+$, the equation is:

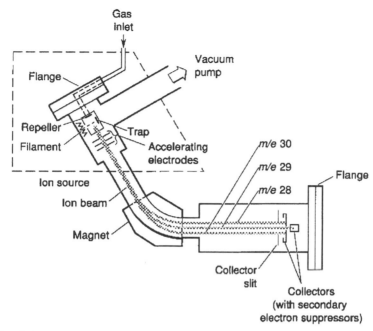

Fig. 7.3 Schematic diagram of a double-collector mass spectrometer for isotope ratio analysis of N_2. From Mulvaney,[6] reproduced with permission of Academic Press Inc.

$$\text{Atom}\% \ ^{15}\text{N} \ = \ 100r/(2 + r) \tag{2}$$

Nitrogen isotopic composition is often expressed in terms of atom% ^{15}N excess, which is obtained by subtracting the background concentration (natural abundance) of ^{15}N (conventionally $0.3663 \pm 0.0004\%$ for atmospheric N_2).

$$\text{Atom}\% \ ^{15}\text{N excess} = \text{atom}\% \ ^{15}\text{N} - 0.3663 \tag{3}$$

A flaw in this practice is that considerable variation exists in the natural abundance of ^{15}N in pedogenic materials (e.g., soils, sediments, organic deposits of various types), which can lead to uncertainties in the interpretation of the data. Natural variations in the isotopic composition of ^{15}N, as applied to soil N studies, are discussed below.

In concluding this section, it should be noted that the mass spectrometer is a complicated and expensive instrument that requires considerable skill to operate and maintain. In recent years, however, these problems have been minimized by the development of automated systems that are relatively easy to maintain and that permit isotope ratio analyses on samples containing microgram quantities of N, and at a rate of up to several hundred samples a day.[6]

NITROGEN BALANCE SHEETS

A critical problem relative to the cycling of N in agricultural soils and natural plant communities is the difficulty of preparing reliable balance sheets in which all gains and losses of N are accounted for. In those cases where fertilizer N has been applied, unexplained losses have occurred, and these have typically been attributed to leaching, denitrification, or NH_3 volatilization.

Two general approaches have been used in N balance studies. One involves a complete budget and documents the appropriate inputs and outputs, including crop removal. The second uses a specific ^{15}N-labeled input, from which a balance is calculated for the labeled input. The latter documents the fate of the labeled ^{15}N but provides little information about over-all gains and losses for the entire system.

Only limited success has been achieved thus far in obtaining a balance for total N in the soil–plant system under field conditions. The budget for any given ecosystem is the product of numerous complex transformations that interact with each other over time, and in most instances quantitative long-term data are lacking for one or more major processes.

Allison's[10] summary of N balance sheets for a large number of lysimeter experiments conducted in the United States is of historical interest. He reported that:

1. Crops commonly recovered only 50 to 75% of the N that was added or made available from the soil. Low recoveries were usually obtained where large additions of N were made, where the soils were very sandy, and when the crop did not consume all of the mineral N.

2. The N content of soils decreased regardless of how much N was added as fertilizer unless the soil was kept in uncultivated crops.

3. A large proportion of the N not recovered in the crop was found in the leachate, but substantial unaccounted for N losses occurred. Nitrogen gains were few. *Unaccounted-for N* is a widely used, but imprecise, term for N losses that are attributed to leaching, denitrification, and volatilization.

4. The magnitude of the unaccounted for N was largely independent of the form in which the N was supplied, whether as NH_4^+, NO_3^-, or organic N.

An advantage of the ^{15}N approach is the ability to trace a given N input (e.g., ^{15}N-labeled NH_4^+, NO_3^-, or urea) through the various pools, including plant uptake. In practice, precision is limited because of spacial variability of the soil and accompanying sampling errors. As Legg and Meisinger[3] pointed out, the use of ^{15}N can markedly improve tracing sensitivity in fertilizer N balance work, but only if the following conditions are met: (1) reliable estimates are obtained for the size of the various N pools, (2) representative samples are collected for analysis, and (3) the experiment is performed with adequate replication and proper local controls. A problem also arises because of biological discrimination in the use of ^{14}N and ^{15}N by microorganisms and because not all fractions of the native N contain the same background level of ^{15}N, as noted below.

Complete balance sheets have seldom been obtained for ^{15}N-labeled fertilizer N because of the difficulty of obtaining accurate estimates for leaching and denitrification losses. Typical values for such balance sheets based on early ^{15}N studies are given by Hauck[11] and Kundler.[12] Kundler's summary, which was based on a 10-year international study using ^{15}N-labeled fertilizers, gave these results:

Fertilizer N recovered in the crop	30 to 70%
N retained in organic matter the first year	10 to 40%
N lost from the system	10 to 30%

Most of the N losses were believed to occur as gaseous escape. Leaching losses under regular farming conditions were estimated to be not more than 5 to 10% of the fertilizer N.

The survey of Hauck[11] produced the following overall N balance sheet:

Fertilizer N recovered by the crop	55%
Immobilization (N tied up in organic matter)	up to 45%
Denitrification	15%
Leaching and runoff	negligible to considerable, depending on soil porosity and rainfall

The balance sheets above represent the range of recoveries for a variety of soils and environmental conditions. They apply in only a general way to a specific agricultural system—*each class of soil or type of ecosystem will have its own characteristic N balance sheet.* For example, a balance sheet for fertilizer N applied to a poorly drained, heavy-textured soil where conditions are suitable for denitrification will be entirely different from the balance sheet for an irrigated sandy loam soil, where extensive leaching of NO_3^- may occur. Results of specific N balance studies for a variety of crops (small grain, corn, rice, grassland systems, and forests) grown under a range of soil and environmental conditions were discussed at length by Legg and Meisinger.[3] Whereas N recoveries in the plant or soil vary considerably from one ecosystem to another, values for plant uptake and immobilized N generally follow the trends noted above. In this respect, it should be noted that the efficiency with which fertilizer N is used by plants, as well as the extent to which losses occur through denitrification and leaching, will be related to fertilizer application rate.[3,13–15]

Considering all of these studies, the following conclusions can be made regarding the efficiency of N fertilizer use.[3]

1. Nitrogen losses through leaching and denitrification are greatest when N inputs exceed crop requirements, a situation that leaves excess NO_3^-–N in the soil and permits leaching and denitrification after the growing season. Losses of N through leaching and denitrification can range from nil to over 50% in some cases.
2. Nitrogen can be used effectively and efficiently by plants, provided N inputs do not exceed crop assimilative capacity and the N is applied in phase with plant uptake.

EFFICIENCY OF FERTILIZER N USE BY PLANTS

The interchange between mineral and organic forms of soil N has a bearing upon the use of ^{15}N to determine the efficiency of fertilizer N use by plants and the capacity of the soil to provide available N. Due to mineralization–immobilization turnover, the conventional method of determining fertilizer use efficiency from the difference in crop uptake between the fertilized soil and the untreated plot gives higher recoveries of fertilizer N than does the tracer method. The basic reason for this effect can be seen by considering the following hypothetical case, as outlined by Clark.[16]

Assume that during the course of the growing season a total of 60 kg of N is mineralized from soil organic-N sources, but that over this same period 30 kg of N is immobilized by microorganisms into new biomass-N. A total of 30 kg of N will thereby be available to the plant (60 kg mineralized N − 30 kg immobilized N = 30 kg available N).

Assume now that 60 kg of ^{15}N-labeled fertilizer is applied and that microorganisms draw indiscriminately from the fertilizer and the mineralized soil N to meet their N requirements. Of the 30 kg of N needed, 15 kg will come from the soil (^{14}N) and 15 kg from the applied fertilizer (^{15}N). A total of 90 kg of N will thus be available to the plant (120 kg total mineral N − 30 kg immobilized N = 90 kg of plant available N). As shown below, 45 kg will come from the soil and 45 kg from the fertilizer.

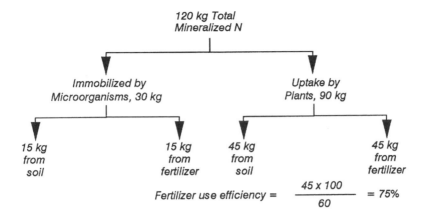

Even if the plants consume all of the available N, isotopic analysis for ^{15}N in the plant would indicate 75% recovery of applied fertilizer N. In contrast, the conventional method of measuring fertilizer N efficiency from the difference between the N uptake of plants growing on fertilized and unfertilized soil would have shown 100% recovery in this example.

The above line of reasoning is an oversimplification of the complex reactions that occur in soils, but it serves to emphasize that plant uptake of applied ^{15}N does not necessarily provide a true measure of fertilizer N efficiency. Because of turnover by mineralization–immobilization, some of the soil N not otherwise available is taken up by the plant and a corresponding amount of fertilizer N is immobilized.

The limitations outlined above for using ^{15}N to measure fertilizer N efficiency also apply when the ^{15}N isotope is used to evaluate leaching and denitrification losses. In this case, substitution of fertilizer N for mineralized soil N during turnover would lead to a low estimate for the *true* effect of the fertilizer in contributing to environmental pollution (NO_3^- through leaching and N_2 and N_2O by denitrification) because a larger amount of N would be lost than indicated by ^{15}N analysis alone.

INFLUENCE OF FERTILIZER N ON THE UPTAKE OF NATIVE SOIL N

A major objective of many ^{15}N experiments has been to determine the relative contribution of soil and fertilizer N to the N budget of plants. A unique feature of this work is that additions of fertilizer N invariably lead to increases in the amount of soil N taken up by the plant (Fig. 7.4). Explanations given for the increased consumption include:

1. Increased uptake is a special feature of the mineralization-immobilization process, as noted in the previous section.

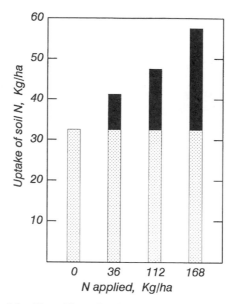

Fig. 7.4   Influence of fertilizer N application rate on the uptake of soil N by Sudan grass (*Sorghum suganensis*) in a pot experiment. The solid bars indicate the additional amount of soil N taken up by the plant in the presence of increasing amounts of fertilizer N. Adapted from Legg and Allison.[15]

2. Fertilizer N causes enhanced mineralization of native humus N through a "priming" action.
3. Plants growing in fertilized soil develop a more extensive root system, thereby permitting better utilization of soil ^{14}N by the plant.

The last explanation fails to account for results obtained in pot experiments, where the volume is limited and the soil is fully occupied by roots. Thus, most workers have attributed the increase to either turnover through mineralization–immobilization (item 1) or a priming action (item 2). We agree with the conclusion of Jansson[1] and others that the so-called priming action has been overemphasized and that the major cause of the increased uptake is a result of the interchange of fertilizer N for the native humus N (see previous section). For a detailed account of the priming effect, see the review of Hauck and Bremner.[2]

LOSSES OF FERTILIZER N THROUGH LEACHING AND DENITRIFICATION

Although adequate data are not available for a firm conclusion, there is evidence to indicate that losses of fertilizer N through leaching and denitrification are not proportional to the amount of N applied, but will be greatest when the amount of added N exceeds the optimum rate for maximum yield.[3] Field data obtained by Broadbent and Carlton,[13] using ^{15}N-depleted fertilizer, illustrate this point (Fig. 7.5). According to Broadbent and Carlton,[13] the key to minimizing excess $NO_3^- - N$ in the soil (and subsequent losses) is to adjust N fertilizer rates to reflect both crop N requirements and the soil's ability to provide available N. While this recommendation is scientifically sound, it is difficult to implement because of the impossibility of estimating in advance how much N will be supplied by the soil (see Chapter 6).

Losses of fertilizer N at low application rates (percentage basis) will be minimal because of net immobilization by microorganisms that bring about the decay of plant residues from the previous crop. Any N added in excess of microbial requirements will be highly susceptible to leaching and denitrification unless it is used by plants. The mineralization–immobilization sequence provides a natural safeguard against N losses because of net immobilization during periods when the potential for leaching and denitrification will be highest.

The magnitude and consequences of fertilizer N loss are currently subjects of considerable debate. Because of environmental concerns and the need to conserve energy, a rational approach to fertilizer N use is required. Since the point of maximum economic return from fertilizer N is generally below the point of maximum yield, it may be possible to adjust application rates for

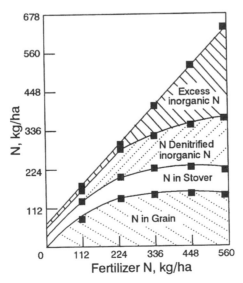

Fig. 7.5 Average annual distribution of total N inputs on an irrigated sandy loam soil. Adapted from field data of Broadbent and Carlton[13] as depicted by Legg and Meissinger.[3]

optimum efficiency while at the same time reducing losses of N to the environment.

Nitrogen losses from soil cannot be eliminated completely, but they can be minimized by proper management practices. As our understanding of the N cycle in soil increases, better ways will be found to manage N for maximum efficiency and minimal pollution.

COMPOSITION AND AVAILABILITY OF IMMOBILIZED N

Several approaches using ^{15}N have been applied in attempts to characterize immobilized N in soil, including fractionations based on acid hydrolysis (see Chapter 6), extraction of humic and fulvic acids by classical alkali extraction, and extractability with chemical reagents proposed as indexes of plant-available soil N. Other approaches have included (1) release of the organic N to mineral forms (NH_4^+ + NO_3^-) by incubation under ideal conditions in the laboratory and (2) Uptake by plants grown in the greenhouse or field.

Results of studies on uptake or extractability of applied ^{15}N are often expressed in terms of an availability or extractability ratio. For plant uptake, the equation is:

$$\text{Availability ratio} = \frac{^{15}\text{N in plant/total N in plant}}{^{15}\text{N in soil/total N in soil}} \tag{4}$$

A similar equation is used to express results of incubation experiments for mineralizable N (NH_4^+ and NO_3^-).

$$\text{Availability ratio} = \frac{\text{residual }^{15}\text{N mineralized/total N mineralized}}{\text{residual }^{15}\text{N in soil/total N in soil}} \tag{5}$$

For chemical extraction, the ratio becomes:

$$\text{Extractability ratio} = \frac{\text{tagged }^{15}\text{N in extract/total N in extract}}{\text{tagged }^{15}\text{N in soil/total N in soil}} \tag{6}$$

A ratio of unity (1.0) indicates that the residual fertilizer N and total soil N are equally available to plants, microorganisms, or chemical extraction. Values greater or less than 1.0 represent enhanced or reduced susceptibility of the residual fertilizer N to biological utilization or chemical extraction.

Microbial Biomass N

As noted above, the decay of organic residues in soil is accompanied by conversion of N (and C) into microbial components. The biomass, along with the decaying residues, represent the active phase of the soil organic matter. As a general rule, from 1 to 6% of the soil organic N resides in the microbial biomass at any one time.[17-20]

Biomass N has been determined using incubation and extraction procedures similar to those used for biomass C (see Chapter 3). In the incubation method for biomass N, measurements are made of the amount of N (inorganic + organic) released following incubation of soil fumigated with chloroform, the extractant being 0.5 M K_2SO_4.[17] Biomass N is obtained from the relationship:

$$\text{Biomass N} = F_n/k_n \tag{7}$$

where F_n is the difference between the amount of N released by incubation of fumigated and unfumigated soil and k_n is the fraction of the biomass N that is released as soluble forms during incubation. Estimates for k_n have been somewhat variable, but a value for k_n of 0.68 has been recommended.

A limitation of the fumigation–incubation method is that denitrification and/or immobilization during incubation may alter the amount of N found in soluble forms, a problem that can be eliminated by extraction and chemical analysis of the fumigated sample (i.e., incubation step eliminated).[18] An additional advantage of direct extraction is that there is no need for complete removal of fumigant or for prolonged incubation of the soil. Joergensen and Brookes[19] described a colorimetric method for biomass N based on the extra amount of ninhydrin-reactive N (e.g., amino acids and NH_4^+) released during fumigation.

Information on the composition and availability of newly immobilized N has been provided by studies in which the microbial biomass was labeled by short-term incubation of soil with an ^{15}N-labeled N source and a suitable C source. The variables have included form of applied N, type of substrate, soil properties, and effects of moisture content and soil drying. Under optimum conditions for microbial activity, net immobilization of added ^{15}N proceeds rapidly and reaches a maximum in as little as three days with a simple substrate (e.g., glucose) to as much as two months or more for a complex substrate such as mature crop residue.[21,22]

Mineralization studies have shown that the turnover of N from dead microbial cells is on the order of five times higher than for the native soil organic N.[20] Cytoplasmic constituents are easily broken down, while cell wall components are slowly mineralized. Fungal melanins resist attack by microorganisms and may persist in soil for a long time.

Drastic changes in soil moisture or temperature (i.e., drying and rewetting, freezing and thawing) result in a flush of mineral N during subsequent incubation (see Chapter 6). The organic N made more available to biological attack is probably derived in part from microbial cells killed during the treatment.

Nature of Newly Immobilized N

Results of a study[23] on the extractability of labeled ^{15}N from newly immobilized N (soil incubated with ^{15}N-labeled NH_4^+ over a seven-day period using glucose as a C source) are given in Table 7.2. The milder the extractant, the greater was the extractability ratio (defined earlier). Except for acid $KMnO_4$, the extractants removed more of the newly immobilized N (up to 11%) than the native soil N (2.4 to 3.8%). Regardless of extractant, most of the extracted N was derived from the soil.

In a follow-up study,[24] the ^{15}N-labeled soil was extracted sequentially with 0.15 M $Na_4P_2O_7$ and 0.1 M KOH, following which the N was partitioned into the classical humic fractions (see Chapter 1). A high percentage of the newly immobilized N (46%) was found in humin; an additional 16% was recovered in the humic acid fraction. The explanation given for these results was that much of the newly immobilized N existed in insoluble cellular components of microorganisms, including fungal melanins that were not extractable under the conditions used.

Short-term incubations, as used in the above study, result in an initial increase in bacterial numbers, followed by an increase in fungi. A subsequent decline in the fungal population would be expected to be followed by a second increase in bacterial numbers. On a mass basis, the fungal biomass greatly exceeds that of bacteria. It should be noted also that fungal melanins are not completely solubilized by dilute alkali reagents, and the insoluble residues have properties similar to those of the humin fraction of soil organic matter.

In any event, these studies suggest that, for the soil used, a significant portion of the newly immobilized N is no more available to plants than the native soil N, an observation that agrees with other findings (below).

Kai et al.[21] determined the distribution of the forms of immobilized N in a soil incubated for up to 20 weeks with [15]NO_3^- and glucose, straw, or cellulose (C/N = 32). For all C substrates, an initial net immobilization of [15]N

TABLE 7.2   Extractability Ratio and Sources of N in Some Extractants Proposed as Indexes of Plant Available Soil N[a,b]

Extractant	Extractability Ratio[c]	% of Extracted N Present in	
		Biomass	Native Humus
0.01 M $NaHCO_3$	3.0	41.2	58.8
Hot water	2.4	33.7	66.3
Hot 0.005 M $NaHCO_3$	2.4	32.3	67.8
Hot 0.01 M $CaCl_2$	2.2	32.1	67.9
Acid $KMnO_4$	0.9	13.2	86.8

[a]From Kelley and Stevenson.[23]
[b]The soil was one in which the biomass had been labeled with [15]N.
[c]The extractability ratio is defined in Equation (6).

was followed by a period of net mineralization. At the point of maximum incorporation of N into the biomass, there was a distinct difference in the percentage distribution of the forms of organic N between the newly immobilized ^{15}N and the native humus N. As shown in Fig. 7.6, the immobilized N was higher in amino acid N and in hydrolyzable unknown forms (HUN fraction) but lower in acid-insoluble N; percentages of the N as NH_3–N and as amino sugar N were approximately the same. The results of Kai et al.[21] also demonstrated that, following maximum tie-up of N, there was a net release of immobilized ^{15}N to available mineral forms, particularly from the amino acid fraction.

Availability ratios of immobilized N as calculated from mineralization data have ranged from 0.4 to 10.0, depending upon source and rate of applied ^{15}N, type and amount of residue added, number of successive crops grown, and length of incubation period.[20,25,26] The findings indicate that applied fertilizer ^{15}N is rapidly converted to relatively stable forms during decay of organic residues in soil.

The high stability of immobilized ^{15}N has been confirmed by plant uptake studies. In general, relatively little of the immobilized ^{15}N is taken up by the first crop, and availability decreases during consecutive cropping.[25,27,28]

Several attempts have been made to determine the availability of immobilized N in soils using chemical extraction techniques. Legg et al.[27] repeatedly extracted ^{15}N-labeled soil by autoclaving with 0.01 M $CaCl_2$. Extractability ratios were compared with availability ratios based on N mineralization during laboratory incubation and by N uptake by a series of oat (*Avena sativa L.*) crops. Whereas the biologically based availability ratios indicated that the immobilized ^{15}N was twice as available to plants as the total soil N, the extractability ratios indicated equal susceptibility for both forms. This result suggests that chemical extraction is not selective in recovering plant-available N, a conclusion also reached by Stanford et al.[29]

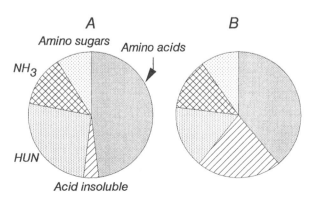

Fig. 7.6 Comparison of the distribution of the forms of organic N in soil: (*a*) newly immobilized N; (*b*) native humus N. Adapted from data of Kai et al.[21]

Availability of Residual Fertilizer N

As noted above, from 10 to 40% of the fertilizer ^{15}N applied to the soil is present in organic forms after the first growing season. Not more than about 15% of this residual N is available to plants during the second growing season, and availability decreases even further for succeeding crops.[30-35] In the study conducted by Westerman and Kurtz,[35] the percentage of the residual fertilizer ^{15}N recovered in the plant tops (three harvests) of a sorghum–Sudan hybrid (*Sorghum sudanense*) was 13 to 18%, equivalent to 4 to 6% of the ^{15}N originally applied. The residual ^{15}N remaining after the second growing season, representing approximately 24% of the initial fertilizer, was even less available (1.5%) and was assumed to be in equilibrium with the native soil N.[36]

The low availability of residual fertilizer N has been demonstrated for a number of cropping systems, as noted in Fig. 7.7 for the uptake of labeled $^{15}NH_4^+$ by rice (*Oryza sativa L.*).[37] In this case, approximately 50% of the applied ^{15}N was recovered in the crop (grain plus stover) during the first growing season (left side of Fig. 7.7); approximately 26% was found in the soil (roots + soil). Following harvest, the rice straw was returned to the soil and a second rice crop was grown, with some plots receiving supplemental unlabeled fertilizer N. Less than 10% of the residual labeled ^{15}N was recovered in the second rice crop (right side of Fig. 7.7). The addition of unlabeled N had only a slight influence on the uptake of residual fertilizer ^{15}N.

Chemical Characteristics of Residual Fertilizer N

Clues as to the high stability of residual fertilizer N in soils have come from fractionations based on acid hydrolysis (see Table 6.3). In the case of the field study of Allen et al.,[36] an average of one-third of the initially applied fertilizer N was present in the surface soil after the end of the first growing

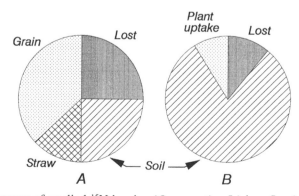

Fig. 7.7   Recovery of applied ^{15}N by rice (*Oryza sativa L.*) in a flooded soil; (*a*) first cropping season; (*b*) residual ^{15}N (straw and soil) during the second cropping season. Adapted from Reddy and Patrick.[37]

season, the remainder having been used by plants or lost through leaching and denitrification. Isotope-ratio analyses revealed that the residual N had been incorporated into the organic matter. Comparison of the distribution pattern for the fertilizer-derived N with that of the native humus N (Fig. 7.8) shows that a considerably higher proportion of the fertilizer N occurred in the form of amino acids (59.0 vs. 36.0%) and amino sugars (9.9 vs. 8.0%); lower proportions occurred as hydrolyzable NH_3 (10.6 vs. 18.1%), acid-insoluble N (10.3 vs. 21.7%), and the HUN fraction (10.2 vs. 16.2%). When the plots were resampled four years later, the fertilizer-derived N that remained in the soil, about one-sixth of that initially applied, had a composition very similar to that of the native humus N.

Similar results were obtained by Smith et al.,[38] who also found that fertilizer N was initially incorporated into such compounds as amino acids and subsequently into more stable forms. As compared to the native soil N, more of the residual fertilizer N occurred in an amino acid-containing fraction, with lower amounts being accounted for as hydrolyzable NH_3 and insoluble N. Equilibrium with the native soil N had not been achieved in three years.

Residual fertilizer-derived N has been partitioned into the classical humus fractions (e.g., humic acid, fulvic acid, and humin) by alkali extraction.[34,39] The relative amounts of N recovered in the humic and fulvic acids were found by Wojcik-Wojtkowiak[39] to depend on a variety of factors, including form of applied N. McGill and Paul[40] and McGill et al.[41] isolated humic and fulvic acids from ^{15}N-labeled soils and observed a higher degree of labeling in the fulvic acid fraction.

Long-Term Effects

The postulated long-term fate of residual fertilizer N in Mollisol soils of the Corn Belt Region of the United States is shown in Fig. 7.9. The mean residence time (MRT), or average lifespan, for the N retained after the first sea-

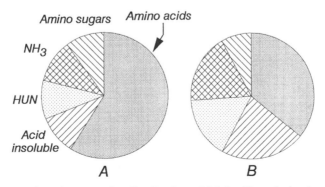

Fig. 7.8 Comparison between the distribution of (*a*) fertilizer-derived organic N and (*b*) the native humus N. From Allen et al.[36]

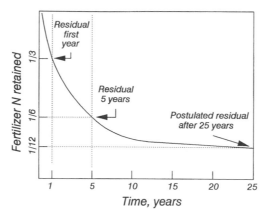

Fig. 7.9 Postulated long-time fate of residual fertilizer N in Mollisol soils of the Corn Belt Region of the United States. The amount retained is compensated for by mineralization of an equivalent amount of humus N, thereby maintaining steady-state levels of organic N in the soil.

son, is about 5 years. Because of increased humification, any N retained after this period will have an MRT of about 25 years. Thereafter, the MRT will be the same as for the native humus, estimated at from 200 to 800 years. From these results, it can be seen that a small amount of the fertilizer N will remain in the soil for a long time, perhaps centuries.

LABELED BIOLOGICALLY FIXED (LEGUME) N

In many parts of the world, adequate N for sustained crop production depends upon N$_2$ fixation by legumes. In the wheat belt of South Australia, for example, the conventional practice is to grow legumes in rotation with the cereal crop; the legume is used as grazed pasture or for grain production.

Ladd[42] and Ladd et al.[43] summarized results of an elaborate field study in which doubly labeled ^{14}C- and ^{15}N-labeled legume material (e.g., *Medicago littoralis*) was mixed with the topsoil at three field sites in South Australia and allowed to decay for about eight months before the soils were sown to wheat (*Triticum aestivum L.*). As shown in Fig. 7.10, only 11 to 17% of the ^{15}N was taken up by the wheat, with most of the remainder (72 to 78%) being recovered in the organic phase of the soil. Even after four years, nearly one-half of the added ^{15}N was still present in the soil in stable organic forms.[42,43] The conclusion was reached that the value of legumes as a source of N was due not so much to their capacity to provide relatively large amounts of immediately available N but rather to long-term benefits whereby soil organic N levels are maintained or increased, thereby ensuring an adequate supply of N by slow decay of the stable organic N. Based on these results, the cereal–

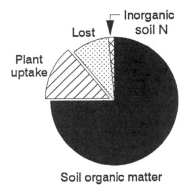

Fig. 7.10   Recovery in wheat (*Triticum aestivum L.*) of ^{15}N in decomposing legume residue. Most of the residue N was retained in the soil in organic forms (see text). Adapted from Ladd[42] and Ladd et al.[43]

legume rotational system is not only economically productive (a crop with cash value is produced each season), but also progressively builds reserves of organic matter and stable N in the soil.

Laboratory incubation studies using ^{15}N-labeled leguminous plant materials[44-46] have confirmed that a portion of the biologically fixed N is transformed to stable organic N forms. Obviously, more data are required regarding the fate of biologically fixed N in field soils.

HUMUS AS A SOURCE OF N

The assumption is often made that 1 to 3% of the soil organic N is mineralized during the course of a growing season, and presumably available to plants. However, as explained in Chapter 5, his statement must be accepted with considerable reservation, because the humus content of most soils is in a state of quasi-equilibrium (i.e., the total amount does not change significantly from one year to another).

At steady-state, the amount of N mineralized during the year, and recovered in either the harvested portion of the crop or lost through leaching and denitrification, is balanced by incorporation of N from other sources into newly formed humus. A net annual release of N occurs only when organic matter levels are declining, a condition that is to be avoided as exploitation of this type leads to a reduction in the productive capacity of the soil unless large amounts of fertilizer are used.

When the organic N level of the soil is at equilibrium, and the cropping sequence does not include a legume, transformations occurring within the soil can be ignored when modeling the N cycle (see Chapter 6), and only inputs and outputs of N need to be considered. A simplified "black-box" approach for modeling the soil N cycle is shown in Fig. 7.11. The approach has been used by Tanji et al.[14] to predict N losses from soil. More complex approaches for modeling the soil N cycle are discussed in Chapter 6.

For most soils, the quantity of N returned in crop residues, in rain water, or through biological N_2 fixation is sufficient to meet the needs of microorganisms in synthesizing new humus. Thus, a reasonably good estimate of the quantity of fertilizer N required by the crop can be taken as equal to the amount removed in the harvested portion divided by the efficiency with which the fertilizer N is used.

$$N_{\text{fertilizer}} = \frac{N_h}{E_f} \tag{8}$$

where N_h is the amount of N expected to be removed in the harvested portion and E_f is the efficiency factor. The assumptions are made that the amount of N contained in the stover will be compensated for by that returned in residues from the previous crop, and that any excess mineral N will have been lost through leaching, denitrification, or NH_3 volatilization. When the stover is not returned to the soil, or is burned (e.g., sugar cane production or wheat in some areas), additional fertilizer N must be added to compensate for the extra N when the stover is burned; nearly all of the N is lost to the atmosphere as nitrogenous gases. The efficiency factor will vary with the crop and environmental conditions but will generally range from 0.50 to 0.75. Better ways are needed to increase the efficiency of fertilizer N use by plants.

The use of Equation (8) to estimate fertilizer N requirements can be illustrated by considering N removal in the grain for a good yield of corn (*Zea mays L.*). From 120 to 150 kg N/ha will be removed. For an assumed efficiency factor of 0.66, from 180 to 230 kg of fertilizer N/ha will be required to attain the desired yield when corn follows corn in a rotation. In the Corn

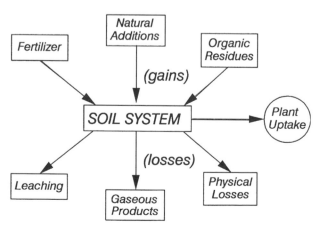

Fig. 7.11   "Black-box" approach for modeling the soil N cycles.

Belt region of the United States, corn often follows soybeans (*Glycine max L.*) in a rotation, in which case the recommended fertilizer rate is reduced by 25 to 45 kg N/ha to account for N in the soybean residues. For those soils where leaching and denitrification is not a problem (e.g., many unirrigated arid and semiarid zone soils), an adjustment in fertilizer N rate can be made for residual mineral N in the rooting zone.[47]

A more elaborate equation based on soil test results for mineralizable N (laboratory incubation) is given by Parr[48] for predicting the amount of fertilizer N to be applied to corn:

$$N_{fertilizer} = \frac{N_{crop} - (N_{om} + N_r)}{E_f} \qquad (9)$$

where N_{crop} is the total amount of N removed (grain + stover), N_{om} is the N mineralized from the soil organic matter in a laboratory incubation soil test, and N_r is the residual mineral N (exchangeable $NH_4^+ + NO_3^-$). The assumption is made that the mineralized N will be utilized by the crop at the same efficiency as the fertilizer N. Limitations of incubation tests for assessing soil N availability are discussed in Chapter 6.

When crop residues are returned to the soil, only the N in the harvested portion represents a net loss from the soil–plant system. In general, more N is contained in the harvested portion than in the stover, vines, straw, or roots, as noted in Chapter 5. Grasses or forages harvested for hay will remove large quantities of N. However, not all of the N removed by leguminous crops will come from the soil, but will be derived from the air through biological N_2 fixation. In sugar cane production, very little N is removed in the portion of the plant from which sugar is extracted, but huge losses occur when the tops and leaf trash on the cane field are burned to facilitate harvest. Over 100 kg N/ha is removed in the harvested portion of many nonleguminous crops, thereby accounting for the large amounts of fertilizer needed for sustained production.

Crop yields on continuously cropped arable land are often related to the N supply. Since surplus inorganic N is susceptible to losses through leaching and denitrification, efficiency in the utilization of soil and fertilizer N is dependent to some extent on the effective management of crop residues. The ideal situation is to have adequate mineral N in the soil during periods of active uptake by plants and to avoid excesses that will be lost through leaching and denitrification, or to use an additional crop such as a winter cover crop of grass to capture the excess N before it is leached.

NATURAL VARIATIONS IN ^{15}N ABUNDANCE

Slight variations occur in the N isotope composition of soil. These variations result from isotopic effects during biochemical and chemical transformations (see review of Hauck and Bremner[2]). An isotope effect occurs when two isotopes of the same element (e.g., ^{14}N and ^{15}N) are not processed at the same rates or efficiencies by biological systems or if chemical reactions favor one isotopic form over another. The basis for the effect is the small difference in mass of the two isotopes and differences in bond energies for compounds containing ^{14}N vs. [15]N. For example, the oxygen–nitrogen bonds of ^{14}N–NO_3^- are slightly weaker than the bonds of ^{15}N–NO_3^-, and denitrification of ^{14}N–NO_3^- occurs more readily than denitrification of ^{15}N–NO_3^-. The overall effect of these isotope effects in soil is a slight increase in the average ^{15}N content of soil N and its fractions, as compared to the ^{15}N content of atmospheric N_2.

Natural variations in N isotopic abundance are usually expressed in terms of the per mil excess ^{15}N, or delta [15]N ($\delta^{15}N$). The equation is:

$$\delta^{15}N = \frac{\text{atom\% } {}^{15}\text{N in sample} - \text{atom\% } {}^{15}\text{N in standard}}{\text{atom\% } {}^{15}\text{N in standard}} \times 1{,}000 \quad (10)$$

Thus, a $\delta^{15}N$ value of $+10$ indicates that the experimental sample is enriched by 1% compared with the atom% [15]N of the standard (i.e., 0.3663 for atmospheric N_2). A negative value indicates that the sample is depleted in ^{15}N relative to the standard.

Several studies[49,50] have shown that the $\delta^{15}N$ value of the soil generally falls within the range of $+5$ to $+12$, although higher and lower values are by no means rare. For any given soil, variations exist in the $\delta^{15}N$ value of the various N fractions.

Natural variations in ^{15}N content have been used to determine the fate of fertilizer N in soils, including utilization by plants and movement within the soil and into surface waters.[49–56] The approach has also been used to obtain evidence for N_2 fixation by plants and transfer of legume-fixed N_2 to nonlegumes.

In some early work, Kohl et al.[56] used natural variations in ^{15}N content to estimate the relative contribution of soil and fertilizer N to the NO_3^- in surface waters, the basis being that the ^{15}N enrichment of soil-derived NO_3^- will be higher than that of NO_3^- originating from the fertilizer. Since fertilizer N is obtained by chemical fixation of atmospheric N_2, it is usually depleted in ^{15}N as compared to the soil N. The conclusions of Kohl et al.[56] regarding the contribution of fertilizer N to surface waters generated considerable interest and were challenged by some soil scientists;[57] for a response to these criticisms, the reader is referred to Kohl et al.[58] One problem in evaluating the contribution from the soil is that the ^{15}N content of mineralized N derived from humus is not constant and can be lower than that for the total soil N. Data obtained by Feigin et al.[59] show that the ^{15}N content of soil-derived NO_3^- increases with incubation time; long-term incubation often gave $\delta^{15}N$ values twice as large as those measured after a short-term incubation. In evaluating the relative contribution of soil and fertilizer N to the NO_3^- in drainage waters, the question naturally arises as to the proper $\delta^{15}N$ value to select for soil-derived NO_3^-.

Measurements of natural ^{15}N abundance require high precision and are subject to variable and significant errors due to variations in the ^{15}N content of different organic and inorganic N components of the soil. Hauck and Bremner[2] concluded that methods based on the use of variations in natural N abundance were capable of giving only qualitative (or at best roughly quantitative) information.

SUMMARY

Data obtained using the stable isotope ^{15}N have shown that only a portion of the fertilizer N added to soil, estimated at 30 to 70% of the total application, is used by plants and that a significant fraction (up to 40%) is retained in soil organic matter after the first growing season. This residual N is relatively unavailable to plants during the second growing season, and availability decreases even further in subsequent years due to conversion to stable humus forms. A similar effect has been noted for the N from crop residues. The value of keeping a rotating fund of easily decomposable organic materials in the soil through the frequent and periodic return of crop residues (including legumes) is due to a large extent on their effect in maintaining or increasing organic matter levels, thereby insuring a continuous supply of N through mineralization–immobilization turnover.

When the ^{15}N isotope was introduced into soil–plant studies nearly five decades ago, it was anticipated that the new tracer technique would provide an expedient and accurate method for evaluating N fertilizers. This has not been found to be the case. Instead, tracer studies have revealed that the soil is a highly dynamic system in which a change of one phase directly affects all other phases. Despite these complications, the ^{15}N tracer technique is a valuable approach for obtaining basic information about soil N transformations. Such knowledge will be indispensable in the long run for improving present-day empirical methods of managing fertilizer N as well as biologically fixed N.

Some goals for research using ^{15}N to follow N transformations in soil are as follows:

1. Develop methods for distinguishing between N in the various soil N pools (e.g., microbial biomass, active fraction of humus, stable fraction of humus).
2. Define and quantify reaction mechanisms whereby N becomes stabilized.
3. Establish the relationship between mineralization–immobilization turnover and gaseous loss of soil and fertilizer N.

4. Quantify the relationship between soil organic and inorganic N pools as influenced by soil type, cropping system, climate, and residue management practice.

5. Determine the long-time fate of immobilized N under different crop management practices and climates.

6. Devise management strategies.

   a. For effective use of biologically and chemically fixed N.

   b. For efficient use of N from decaying plant and animal residues under conditions existing in the field.

   c. To maintain and efficiently utilize N resources of the soil for efficient crop production.

REFERENCES

1. S. L Jansson, "Use of ^{15}N in Studies on Soil Nitrogen," in A. D. McLaren and J. Skujins, Eds., *Soil Biochemistry,* Vol. 2, Marcel Dekker, New York, pp. 129–166.

2. R. D. Hauck and J. M. Bremner, *Adv. Agron.,* **28,** 219 (1976).

3. J. O. Legg and J. J. Meisinger, "Soil Nitrogen Budgets," in F. J. Stevenson, Ed., *Nitrogen in Agricultural Soils,* American Society of Agronomy, Madison, Wisconsin, 1982, pp. 503–566.

4. J. M. Bremner and R. D. Hauck, "Advances in Methodology for Research in Nitrogen Transformations in Soils," in F. J. Stevenson, Ed., *Nitrogen in Agricultural Soils,* American Society of Agronomy, Madison, Wisconsin, 1982, pp. 467–502.

5. R. Fiedler, "The Measurement of ^{15}N," in M. F. L'Annunziata and J. O. Legg, Eds., *Isotopes and Radiation in Environmental Research,* Academic Press, London, 1984, pp. 233–282.

6. R. L. Mulvaney, "Mass Spectrometry," in R. Knowles and T. H. Blackburn, Eds., *Nitrogen Isotope Techniques,* Academic Press, New York, 1993, pp. 11–57.

7. R. D. Hauck, J. J. Meisinger, and R. L. Mulvaney, "Practical Considerations in the Use of Nitrogen Tracers in Agricultural and Environmental Research," in R. W. Weaver et al., Eds., *Methods of Soil Analysis: Part 2, Microbiological and Biochemical Properties,* Soil Science Society of America, Madison, Wisconsin, 1994, pp. 907–950.

8. D. C. Wolf, J. O. Legg, and T. W. Boulton, "Isotopic Methods for the Study of Soil Organic Matter Dynamics, in R. W. Weaver et al., Eds., *Methods of Soil Analysis: Part 2, Microbiological and Biochemical Properties,* Soil Science Society of America, Madison, Wisconsin, 1994, pp. 865–906.

9. R. Knowles and T. H. Blackburn, Eds., *Nitrogen Isotope Techniques,* Academic Press, New York, 1993.

10. F. E. Allison, "Evaluation of Incoming and Outgoing Processes That Affect Soil Nitrogen," in W. V. Bartholomew and F. E. Clark, Eds., *Soil Nitrogen,* American Society of Agronomy, Madison, Wisconsin, 1965, pp. 573–606.

11. R. D. Hauck, "Quantitative Estimates of Nitrogen-Cycle Processes: Concepts and Review," in *Nitrogen-15 in Soil Plant Studies. Proc. Research Coordination Meeting, Sofia, Bulgaria,* International Atomic Energy Agency, Vienna, 1971, pp. 65–80.

12. P. Kundler, *Albrecht Thaer Arch.,* **14,** 190 (1970).

13. F. E. Broadbent and A. B. Carlton, "Field Trials with Isotopically Labeled Nitrogen Fertilizer," in D. R. Nielsen and J. G. MacDonald, Eds., *Nitrogen in the Environment,* Vol. 1, Academic Press, New York, 1978, pp. 1–41.

14. K. K. Tanji, M. Fried, and R. M. Van De Pol, *J. Environ. Qual.,* **6,** 155 (1979).

15. J. O. Legg and F. E. Allison, in *Transactions of 7th International Congress of Soil Science* (Madison, 1960), Vol. 1, Elsevier, Amsterdam, 1961, 545.

16. F. E. Clark, "A Reevaluation of Microbial Concepts Concerning Nitrogen Transformations in the Soil," in *Soil and Fertilizer Nitrogen Research: A Projection into the Future,* Tennessee Valley Authority, Muscle Shoals, Alabama, 1964, pp. 18–23.

17. S. M. Shen, G. Pruden, and D. S. Jenkinson, *Soil Biol. Biochem.,* **16,** 437 (1984).

18. P. C. Brookes, A. Landman, G. Pruden, and D. S. Jenkinson, *Soil Biol Biochem.,* **17,** 837 (1985).

19. R. G. Joergensen and P. C. Brookes, *Soil Biol. Biochem.,* **22,** 1023 (1990).

20. M. Amato and J. N. Ladd, *Soil Biol. Biochem.,* **12,** 405 (1980).

21. H. Kai, Z. Ahmad, and T. Harada, *Soil Sci. Plant Nutr.,* **19,** 275 (1973).

22. E. A. Paul and J. A. van Veen, *Transactions of 11th International Congress of Soil Science* (Edmonton), Vol. 3, 1978, 61.

23. K. R. Kelley and F. J. Stevenson, *Soil Biol. Biochem.,* **17,** 517 (1985).

24. X.-T. He, F. J. Stevenson, R. L. Mulvaney, and K. R. Kelley, *Soil Biol. Biochem.,* **20,** 75 (1988).

25. F. E. Broadbent and T. Nakashima, *Soil Sci. Soc. Amer. Proc.,* **31,** 648 (1967).

26. F. E. Broadbent and T. Nakashima, *Soil Sci. Soc. Amer. Proc.,* **38,** 313 (1974).

27. J. O. Legg, F. W. Chichester, G. Stanford, and W. H. DeMar, *Soil Sci. Soc. Amer. Proc.,* **35,** 373 (1971).

28. F. E. Broadbent and T. Nakashima, *Soil Sci. Soc. Amer. Proc.,* **29,** 55 (1965).

29. G. Stanford, J. O. Legg, and F. W. Chichester, *Plant Soil,* **33,** 425 (1970).

30. S. L. Jansson, *Soil Sci.,* **93,** 31 (1963).

31. F. E. Broadbent, *Agron. J.,* **72,** 325 (1980).

32. J. O. Legg and F. E. Allison, *Soil Sci. Soc. Amer. Proc.,* **31,** 403 (1967).

33. E. A. Paul and N. G. Juma, "Mineralization and Immobilization of Soil Nitrogen by Microorganisms," in F. E. Clark and T. Rosswall, Eds., *Terrestrial Nitrogen Cycles: Processes, Ecosystem Strategies and Management Impacts,* Ecol. Bull. 33, Stockholm, 1981, pp. 179–195.

34. Y. V. Rudelov, *Soviet Soil Sci.,* **14,** 40 (1982).

35. R. L. Westerman and L. T. Kurtz, *Soil Sci. Soc. Amer. Proc.,* **36,** 91 (1972).

36. A. L. Allen, F. J. Stevenson, and L. T. Kurtz, *J. Environ. Qual.,* **2,** 120 (1973).

37. K. R. Reddy and W. H. Patrick, *Soil Sci. Soc. Amer. J.,* **42,** 316 (1978).

38. S. J. Smith, F. W. Chichester, and D. E. Kissel, *Soil Sci.,* **125,** 165 (1978).

39. D. Wojcik-Wojtkowiak, *Plant Soil,* **49,** 49 (1978).
40. W. B. McGill and E. A. Paul, *Can. J. Soil Sci.,* **56,** 203 (1976).
41. W. B. McGill, J. A. Shields and E. A. Paul, *Soil Biol. Biochem.,* **7,** 57 (1975).
42. J. N. Ladd, *Plant Soil,* **58,** 401 (1981).
43. J. N. Ladd, J. M. Oades and M. Amato, *Soil Biol. Biochem.,* **13,** 119 (1981).
44. F. Azam, K. A. Malik, and M. I. Sajd, *Plant Soil,* **86,** 3 (1985).
45. F. Azam, K. A. Malik, and M. I. Sajd, *Plant Soil,* **95,** 97 (1986).
46. F. Azam, R. L. Mulvaney, and F. J. Stevenson, *Biol. Fert. Soils,* **8,** 54 (1989).
47. R. A. Olson and L. T. Kurtz, "Crop Nitrogen Requirements, Utilization, and Fertilization," in F. J. Stevenson, Ed., *Nitrogen in Agricultural Soils,* American Society of Agronomy, Madison, Wisconsin, 1982, pp. 567–604.
48. J. F. Parr, *J. Environ. Qual.,* **2,** 75 (1973).
49. D. A. Rennie, E. A. Paul and L. E. Johns, *Can. J. Soil Sci.,* **56,** 43 (1976).
50. G. Shearer, D. H. Kohl, and S.-H. Chien, *Soil Sci. Soc. Amer. Proc.,* **42,** 899 (1978).
51. R. E. Karamanos and D. A. Rennie, *Soil Sci. Soc. Amer. J.,* **44,** 57 (1980).
52. D. H. Kohl, G. B. Shearer, and B. Commoner, *Soil Sci. Soc. Amer. Proc.,* **37,** 888 (1973).
53. V. W. Meints, L. V. Boone, and L. T. Kurtz, *J. Environ. Qual.,* **4,** 486 (1975).
54. R. E. Farrell, P. J. Sandercock, D. J. Pennock, and C. Van Kessel, *Soil Sci. Soc. Amer. Proc.,* **60,** 1410 (1996).
55. G. Shearer and J. O. Legg, *Soil Sci. Soc. Amer. Proc.,* **39,** 896 (1975).
56. D. H. Kohl, G. B. Shearer, and B. Commoner, *Science,* **174,** 1331 (1971).
57. R. D. Hauck et al., *Science,* **177,** 453 (1972).
58. D. H. Kohl, G. B. Shearer, and B. Commoner, *Science,* **177,** 454 (1972).
59. A. Feigin, D. H. Kohl, G. Shearer, and B. Commoner, *Soil Sci. Soc. Amer. Proc.,* **38,** 90 (1974).

8

IMPACT OF NITROGEN ON
HEALTH AND THE ENVIRONMENT

There are numerous reasons for concern regarding the integrity of the soil N cycle and the fate of N that does not remain in the soil–plant system. They include ecological damage and health hazards associated with the presence of nitrate (NO_3^-) in natural waters, excess NO_3^- and nitrite (NO_2^-) in human food and animal feed, and the potentially adverse effects of N from soil and fertilizer in increasing the concentration of nitrous oxide (N_2O) in the atmosphere, thereby contributing to the greenhouse effect and possibly to ozone (O_3) depletion in the upper atmosphere. The potential effect of N on promoting unwanted growth of algae in lakes and streams (eutrophication) has been debated for decades, while concern over depletion of the O_3 layer is a more recent development. Potential adverse health and environmental effects of N are listed in Table 8.1.

The well-publicized need for using greater amounts of fertilizer N for worldwide production of food and fiber is, to some extent, in conflict with the necessity for controlling levels of NO_3^- in water supplies and limiting N_2O in the atmosphere. The environmental impact of NO_3^- leaching from agricultural land has been discussed from many different perspectives and from divergent points of view, as attested by several volumes and reviews on the subject,[1-15] to which the reader is referred. The subject of N_2O emissions from agricultural sources is reviewed by Granli and Bockman.[16]

HEALTH ASPECTS

Concerns about NO_3^- in drinking water and food arise from the fact that when consumed in large amounts, NO_3^- has the potential to cause methemoglobi-

TABLE 8.1   Potential Adverse Health and Environmental Impacts of N[a]

Impact	Causative Agents
Human health	
Methemoglobinemia in infants	Excess NO_3^- and NO_2^- in water and food
Cancer	Nitrosamines formed from HNO_2 and secondary amines
Respiratory illness	Peroxyacyl nitrates, alkyl nitrates, NO_3^- aerosols, NO_2^-, and HNO_3 vapor in urban atmospheres
Animal health	
Loss of livestock	Excess NO_3^- in feed and water
Crop production	
Stunted growth	High levels of NO_2^- in soil
Excessive vegetative growth	Excess available N
Environmental quality	
Eutrophication	Inorganic and organic N in surface waters
Stratospheric ozone depletion	Nitrous oxide from nitrification, denitrification, stack emissions, and auto exhausts.
Damage to materials and ecosystems	HNO_3 aerosols in rainfall

[a] Adapted from Keeney.[1]

nemia and stomach cancer. Opinion is divided, however, over the incidence of these health effects and the relationship between fertilizer N use and the NO_3^- content of surface waters.[6,7] The problem is perceived as serious enough to warrant establishing limits for the NO_3^- content of drinking water; U.S. Public Health Service Standards specify that the NO_3^-–N content of safe drinking water must not exceed 10 mg NO_3^-—N/liter (ppm), or 45 mg/liter on a NO_3^- basis, while international standards specify a limit of 11 mg NO_3^-–N/liter. Some streams and reservoirs in the United States that are sources of drinking water approach or exceed the 10 mg NO_3^-–N/liter value on a regular basis. Rural residents who use well water need to be particularly aware of the potential danger of high NO_3^- levels, from the standpoint of poisoning of both infants and cattle. It has long been known that well water in many localities of the United States contains NO_3^- in excess of 100 mg/liter.

Nitrate in vegetables (especially with fertilizer N application) may be an equal or greater source of NO_3^- ingestion than water (see Burt et al.[7]). The U.S. Public Health Service has established NO_3^- limits for certain prepared meat products, as well as meat and fish.

Methemoglobinemia

Methemoglobinemia is a condition that impairs O_2 transport in the blood due to the presence of NO_2^-, or indirectly from NO_3^-. Nitrate becomes harmful (i.e., causes methemoglobinemia) through microbial reduction to NO_2^- in the digestive tract. The NO_2^- is then absorbed into the blood stream, where it oxidizes oxyhemoglobin to methemoglobin, which is incapable of transporting O_2. Hence, a victim of nitrate poisoning actually suffers from effects that are equivalent to suffocation.

The reaction of NO_2^- with hemoglobin to produce methemoglobinemia is of little consequence in adults but, although relatively rare, can be fatal in infants. Nitrate poisoning can also affect cattle, particularly ruminants. The NO_3^- problem as it relates to animal and human health is covered in several reviews.[2,3,17-20]

The first report of infant death due to methemoglobinemia was recorded in Iowa in 1945 by Comly,[21] and other reports soon followed (see Burt et al.[7] and NRC[3]). The greatest risk of methemoglobinemia is in rural agricultural areas where water quality (i.e., water from private wells) is not closely monitored.

According to a USEPA report (quoted by Follett et al.[6]), approximately 2000 cases of methemoglobinemia were reported in North America and Europe for the period 1945–1985; about 7 to 8% of the affected infants died. The World Health Organization reported 2000 cases of methemoglobinemia worldwide between 1945 and 1986; 160 of the infants died.[7]

Despite these statistics, methemoglobinemia is a rare event in developed countries in comparison to other causes of death. Burt et al.[7] indicate that only 14 confirmed cases of methemoglobinemia (one death) have been reported in Britain during the last 35 years. One preventive measure for infant methemoglobinemia is use of an alternative source of water when the usual water supply is known to have a high NO_3^- content. For example, in the United States, public water suppliers are required to supply bottled water to households with infants, without charge, when the NO_3^- content of the tap water exceeds legal limits.

Situations under which NO_3^- poisoning can occur include:

1. In the stomach of infants under about three months of age, where high stomach acidity permits the growth of microorganisms capable of reducing NO_3^- to NO_2^-. From time to time, deaths from this cause have been reported, as noted above.
2. In the rumen of cattle, as well as the secum and colon of the horse.
3. In vegetables or prepared foods that contain high amounts of NO_3^- and that are stored under conditions that permit microbial proliferation.[18] Most cases have dealt with spinach.
4. In damp forage materials of high NO_3^- content. Ingestion by livestock

has been shown to be toxic. See the review of Deeb and Sloan[17] for additional information on NO_3^- toxicity to farm animals.

Stomach Cancer

The relationship between NO_3^- and stomach cancer is not well understood but is believed to be due to the formation of nitrosamines, which are carcinogenic and may be formed through the reaction of amines with NO_2^- (discussed below) or from the reaction between food-derived amines and NO_2^- in saliva.

Attempts to link the incidence of stomach cancer with NO_3^- levels in drinking waters have been largely unsuccessful. As noted earlier, the high NO_3^- levels in some vegetables are of particular concern (i.e., many leafy vegetables, cultivated grasses and feeds are naturally high in NO_3^-).

Other Effects

A further hazard of excess NO_3^- arises when forage crops are ensiled. During ensiling, NO_3^- is converted to NO_2^- by denitrifying bacteria, from which poisonous yellow and brown gases (NO, NO_2 and N_2O_4) are formed. Under certain circumstances, these N oxides accumulate in silo chutes and attached buildings and are lethal when inhaled by animals and humans.

NITRATE IN WATER SUPPLIES

Impact zones for the NO_3^- in soil leachates are aquifers (ground water supplies), streams, surface waters (lakes and reservoirs), and wells.

Because 95% of rural inhabitants and substantial livestock populations consume ground water, the NO_3^- content of these waters is particularly critical. In contrast to ground water, stream flow tends to dilute incoming NO_3^-, thereby leading to lower and more stable NO_3^- concentrations, particularly in receiving lakes and reservoirs. There is evidence that the elevated levels of NO_3^- in the shallow aquifers underlying major agricultural regions in the United States are due to agricultural activities.[9]

Agricultural soils, particularly heavily fertilized ones, are known sources of NO_3^- in natural waters. Land that receives N fertilizer and irrigation water and/or precipitation in excess of evapotranspiration is especially susceptible to NO_3^- leaching. In drier regions, input of irrigation water must exceed evapotranspiration so that soluble salts do not accumulate in the surface soil and reduce plant growth. Hence, irrigated soils of arid zones are very vulnerable to NO_3^- leaching. An evaluation of NO_3^- contamination of ground water in North America is given by Power and Schepers.[10]

Additional sources of NO_3^- in water supplies include municipal and rural sewage, feedlots or barnyards, food processing wastes, septic tank effluents, natural NO_3^- accumulation (caliche of semiarid regions), application of organic wastes in excess of plant requirements, sanitary facilities of recreational areas, landfills, miscellaneous industrial wastes, and biological N_2 fixation. The relative contribution of each source will depend upon conditions existing at a particular location or ecosystem under consideration. The contribution from municipal sewage can be appreciable when raw or digested sewage is discharged directly into lakes or streams, as was frequently done in the past. For example, at one time domestic sewage accounted for about one-half of the N loading of San Francisco Bay and its tributaries; fertilizer N from irrigated agricultural land accounted for less than 2%.[3]

Natural variations occur in the ratio of $^{15}N/N^{14}$ in nature, and attempts have been made from time to time to use the natural abundance of ^{15}N as a means of estimating the contribution of fertilizer N to the NO_3^- in surface waters (see Chapter 7).

Factors Affecting NO_3^- Levels in Natural Waters

Critical factors affecting NO_3 contamination of water supplies are:

Soil properties: Leaching of NO_3^- from soil is affected by soil texture, depth, permeability, water storage capacity, the need for artificial drainage, and the extent to which NO_3^- has been depleted through bacterial denitrification and/or immobilization during decay of plant remains (see Chapter 6).

Instream cycling: Substantial amounts of NO_3^- in drainage waters can be removed during movement and transport to lakes and reservoirs. In addition to losses by denitrification, NO_3^- can be stripped from water by algae and other microorganisms and higher plants, and it can be immobilized during decay of dead plant remains in drainage channels and streams. Much of the NO_3^- entering shallow aquifers, especially those with fluctuating water tables, has a good chance of being removed through denitrification and/or uptake by deeply rooted plants.[6]

Historically, man has increased the NO_3^- content of surface waters by removing native vegetation, tilling the soil, increasing livestock numbers, and, more recently, practicing heavy fertilization with N. As noted in Fig. 8.1, deforestation can have a marked effect on the NO_3^- concentration of surface waters; for this ecosystem, by the end of the first summer after clearing, NO_3^-–N in surface waters from the cleared site had increased to about 60 mg/liter, in contrast to 2 mg/liter in the uncut watershed.[11]

A world-wide inventory of NO_3^- concentrations in river waters[6] has shown that NO_3^- pollution is particularly serious in Western Europe. The global median NO_3^-—N concentration of NO_3^-–N in surface water, excluding Europe, was found to be 0.25 mg/liter; the European mean concentration was 4.5 mg/liter. Recent increases in NO_3^- were noted in Belgium, France, the Neth-

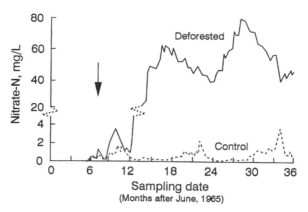

Fig. 8.1 Influence of deforestation on NO_3^- levels in drainage waters. The arrow indicates when deforestation occurred. Adapted from Likens et al.[11]

erlands, Italy, and Sweden. For 9 out of 25 rivers in the United Kingdom, mean NO_3^- concentrations for the period 1981–1985 were higher than in any previous period. Long-term trends in mean annual NO_3^- concentrations in several European rivers are illustrated in Fig. 8.2.

Trends for U.S. rivers follow closely those noted in Fig. 8.2 for the rivers of Europe, with some showing increases with time, some decreases, and some no change. Smith and Alexander's[13] review of water quality in 298 North

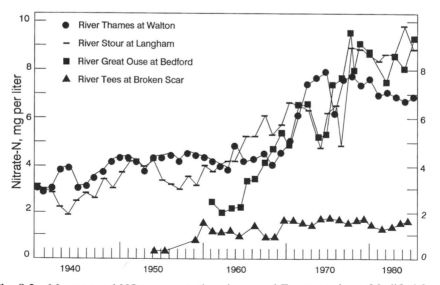

Fig. 8.2 Mean annual NO_3^- concentrations in several European rivers. Modified from Burt et al.,[7] as reproduced from Roberts and Marsh.[12]

American rivers shows that widespread increases have been observed for rural streams of agricultural regions. On the other hand, Aldrich[2] concluded that, for the 1960–1975 year period, there was no consistent trend between the NO_3^- content of the 13 major U.S. river waters and increased fertilizer N use over this same period.

The NO_3^- concentration of surface waters within any given watershed depends upon the supply of available NO_3^-, rainfall, and the extent to which NO_3^- is removed by instream processes. In the central United States, NO_3^- levels tend to be highest in spring, lowest in mid- to late summer, and intermediate in fall and winter, as shown in Fig. 8.3.

Rivers showing recent increases should be carefully monitored so as to establish future trends that might indicate an adverse effect on the environment.

Management of Fertilizer N for Minimal Pollution and Maximum Efficiency

In some instances, high NO_3^- levels of domestic and community water supplies have been attributed to NO_3^- leaching from the soil and into water systems. This has led to an evaluation of fertilizer N practices and ways of minimizing NO_3^- pollution, a subject that is covered in detail in Follett et al.[6] and Addiscott et al.[22]

The soil has its greatest influence on NO_3^- levels in streams during late

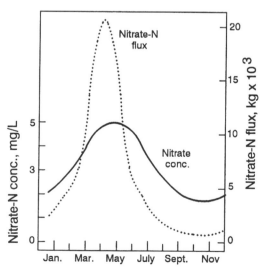

Fig. 8.3 Seasonal changes in NO_3^- concentrations of a typical Midwestern river in the United States. From Aldrich.[2]

winter and early spring, when crop growth is at a minimum and when precipitation exceeds the evaporation rate. As noted above (see Fig. 8.3), seasonal variations in the NO_3^- content of drainage waters and streams are relatively large, due in part to climatic factors that affect the amount of residual NO_3^- remaining in the soil following harvest.

Conditions conducive to leaching of NO_3^- from soil include the following:

1. Application of N fertilizer far in excess of plant requirements, as is often the case with horticultural crops on irrigated light textured soils or sands (see item 3) and at golf courses in regions of predominately sandy soils, like Florida in the United States. In well-drained soils of humid and semihumid zones, much of this unused N will be leached (as NO_3^-) beyond the rooting zone and will thus be subject to movement into water supplies.

2. Reduced consumption of N by crops as a result of low yields brought about by adverse climatic conditions (e.g., low rainfall during the growing season), thereby permitting unused NO_3^- to accumulate in the soil following harvest, ultimately to be leached into water supplies with the advent of winter or spring rains.

3. From coarse-textured and well-drained soils of humid or semihumid zones (or irrigated arid-zone soils) that have been well fertilized with N. The major environmental threat of movement of NO_3^- into water supplies is in areas of intensive crop production on sandy soils.

In considering the contribution of fertilizer N to the NO_3^- in surface waters, and hence lakes and reservoirs, two relationships are important: plant uptake, or yield, and leaching losses as influenced by application rate. The quantity of fertilizer N lost through leaching (or denitrification) cannot be regarded as being directly proportional to the amount applied, since the first increment of applied N will be used by microorganisms and incorporated into organic forms, particularly when residues with high C/N ratios have been applied to the soil.

A hypothetical relationship among crop yields, fertilizer application rate, and the NO_3^-–N content of drainage waters is illustrated in Fig. 8.4 for a crop with a high N requirement (corn). There is reason to believe that the NO_3^- content of drainage waters will follow line B rather than line A. Thus, leaching losses would be minimal when fertilizer N is applied in amounts below that required by plants. Should restrictions in fertilizer N use become necessary in order to keep the NO_3^- concentration in lakes and streams below some desired upper limit, the consequences will be substantially less severe if a type B relationship is followed, as compared to type A. It should be noted that type A describes a situation in which leaching losses would be directly related to the quantity of fertilizer N applied.

Two key strategies for reducing NO_3^- levels in leachate waters are as follows:[6]

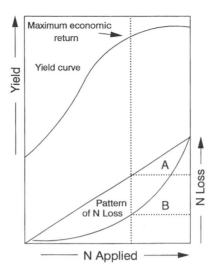

Fig. 8.4 Hypothetical relationship between crop yields, fertilizer application rate, and the NO_3^- content of drainage waters.

Through establishment of realistic yield goals: The potential for NO_3^- leaching from the root zone is relatively low with N rates less than those required to achieve maximum yield (N_{max}) and increases substantially as N rates increase beyond N_{max}. However, this relationship is easy to describe, but difficult to practice. A major problem that exists in approximating fertilizer N application rates is the inability to predict seasonal weather patterns and the amount of N that will be released from soil organic reserves, as noted above.

As Follett et al.[6] pointed out, farmers have a tendency to be optimistic in setting yield goals and will often apply more N than is needed for the yield that they can produce. In developed countries, the cost of fertilizer N is often only a small part of total production costs; accordingly, the grower will opt for applying more fertilizer N than recommended when, economically, there is little to lose and much to gain. Wiese et al.[14] cite examples in Nebraska where failure to achieve the yield goal resulted in N applications that averaged 44.8 kg of N/ha more than the level recommended by the University of Nebraska soil testing laboratory. Fig. 8.5 shows the influence of deviation from the recommended N rate on ground water NO_3^- concentrations. Wiese et al.[14] suggest that rather than basing a yield goal on N_{max}, a more realistic (and environmentally safe) approach can be chosen by using an average yield for a given field for a four- or five-year period and then adding 5% to that average.

Through appropriate timing of fertilizer N applications: There is little doubt that the efficiency of fertilizer N use by the crop could be in-

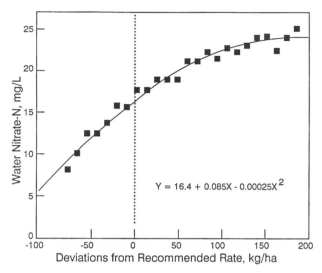

Fig. 8.5 Influence of deviation from recommended N rate on groundwater NO_3^-–N concentrations in Nebraska. From Pierzynski et al.,[8] as adapted from Schepers et al.[15]

creased substantially by more widespread adoption of fertilization practices that provide available N in synchrony with uptake by the crop. To minimize leaching of NO_3^-, application of N fertilizer must be timed to match the plant's peak N-use period. Efficiency can usually be improved substantially by split applications.

Pierzynski et al.[8] describe a program for increasing the efficiency of fertilizer N use by certain agricultural crops (e.g., corn). In brief, a small amount of preplant (starter) N is applied and the remainder is sidedressed when the crop is beginning its period of maximum uptake, with the amount applied being based on soil and climatic factors and tissue analysis for sufficiency of N in the plant. An idealized diagram for the application of fertilizer N in synchrony with N uptake by the crop (corn) is shown in Fig. 8.6.

Unfortunately, limitations are encountered by the farmer in improving the timing of N fertilization. Applications well in advance of planting (late fall and winter) are often preferred, from the standpoint of both efficient use of time (and equipment) and avoidance of potential problems associated with wet soil conditions at or following planting. Because of the increased risk of leaching (and losses through denitrification), this practice is environmentally unsound, although admittedly economically desirable under some existing farming conditions.

Other management practices for minimizing NO_3^- levels include:[8]

1. Modifying cropping patterns, such as crop rotations that include legumes, thereby reducing the need for N fertilizers.

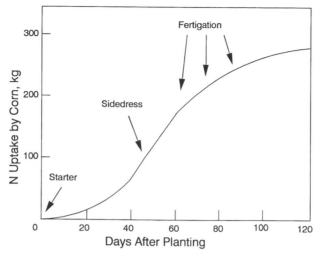

Fig. 8.6 Schematic representation of fertilizer N applications in synchrony with N uptake by corn. Adapted from Pierzynski et al.[8]

2. Introducing genetic alterations to provide N from biological N_2 fixation to nonlegumes such as corn and wheat (see Chapter 5).

3. Reducing N application rates at the expense of higher crop yields (see next section).

4. Using slow-release N fertilizers, which are of two main types: low-solubility compounds that depend on microbial activity for N release (e.g., urea formaldehyde) and coatings over a soluble fertilizer (e.g., sulfur-coated urea). Because of their high costs, slow-release fertilizers have not been used extensively under normal farming conditions.

5. Developing and using nitrification inhibitors. These chemicals may delay production of NO_3^- from urea and NH_4^+-based fertilizers and offer the potential for increasing efficiency of fertilizer N use, but the products available thus far (e.g., nitrapyrin, dicyandiamide) are restricted to specialty crops (mainly horticultural) due to lack of consistent benefits derived from their use.

6. Developing better models for predicting available soil N, thereby leading to more efficient use of fertilizer N. The dynamic nature of soil N and its sensitivity to unpredictable climatic factors such as rainfall and temperature have made it difficult to use chemical soil tests to estimate N availability, as is commonly done for other plant nutrients. Tests for residual NO_3^- have had some success for arid-zone soils, but not for those of humid or semihumid regions. For a discussion of soil N tests, the reader is referred to Chapter 6.

An approach that has the potential for reducing residual fertilizer NO_3^-

levels is the growth of a winter cover crop, established as soon as practical after the growing season, but ideally in early September (for temperate-zone soils). In this case, the cover crop (sometimes called a "catch crop" because it "captures" excess N) assimilates and converts the residual NO_3^-—N into plant organic N, which is returned to the soil in the spring when the cover crop is turned under and another crop is planted.

From time to time, the suggestion has been made that use of N fertilizers should be severely restricted. If their use is completely prohibited, yields of some crops, such as corn, would decrease 40 to 50% (see Hoeft[23]). Man, however, has the knowledge to *minimize* the movement of soil and fertilizer NO_3^- into water supplies, as others have pointed out.[23]

Nitrate in Wells

Attention has frequently been given to the NO_3^- content of shallow wells, and numerous examples can be cited where NO_3^-—N levels have exceeded by 10-fold or more the upper limit of 10 mg/liter regarded as safe for human consumption. Concern for NO_3^- in shallow wells can be traced to early studies indicating a high incidence of deaths to infants as a result of drinking high-NO_3^- water, as discussed above.

Data for NO_3^-–N levels in the well waters of several states are recorded in Tables 8.2 and 8.3. As can be seen from Table 8.3, NO_3^-–N levels are particularly high in shallow wells. These wells are of particular concern because most of them are drilled specifically as sources of private or public drinking water and the water is frequently used without treatment, whereas the water

TABLE 8.2   Nitrate Content of Rural Wells in Several States (mg N/liter)[a]

State	Type of Well	No. of Samples	Nitrate Content, mg N/liter			
			0–10	11–20	21–50	50+
			Samples Within Range, % of Total			
California	Domestic	168	69	13	12	6
	Domestic and irrigation	67	55	13	16	16
	Irrigation	306	75	10	11	4
Iowa	Dug	454	66	11	14	19
	Drilled	647	96	2	2	<1
	Bored	278	80	8	7	5
Michigan	—	2,847	95	—	7	—
Nebraska	Farm	275	78	—	22	—
	Nonmunicipal	2,687	80	—	20	—
Ohio	Rural	10,500	90	5	4	1

[a]From Aldrich.[2] For Illinois see Table 8.3.

TABLE 8.3   Relationship of Well Depth to NO_3^-–N Concentration in 8844 Illinois Wells[a]

Depth, m	No. of Wells Sampled	No. Exceeding 10 mg N/liter	% Exceeding 10 mg N/liter
< 8	480	134	28
8–15	926	185	20
15–30	1568	78	5
30–60	2042	61	3
> 60	3828	23	0.6

[a]Adapted from NRC.[3]

in lakes and streams may or may not be used for human consumption and it is often treated before consumption.

Hoeft[23] noted that as of September 1989, the U.S. Environmental Protection Agency (EPA) had completed the analysis of 295 wells, consisting of 180 community and 115 domestic wells. Of those, eight (less than 3%) had levels in excess of the 10 mg/liter NO_3^-–N standard; 145 wells (49%) tested positive for NO_3^-.

As for surface waters, the NO_3^- in wells can be derived from numerous sources. For rural wells, the direct contribution of soil is often secondary because the NO_3^- content of percolating waters is normally too low to account for the observed high NO_3^- contents. Where high NO_3^- levels have been observed, a source other than agricultural soil is suspected, such as a barnyard, livestock feeding area, septic tank, or other source of animal or human contamination. High NO_3^- levels in shallow wells of certain counties in Illinois and Missouri have been attributed to proximity of the wells to livestock feeding areas or septic tank tile fields. For private wells and wells used for municipal water in Long Island, New York, infiltrated sewage and domestic waste disposal systems are regarded as major sources. When possible, shallow wells should not be located within or adjacent to areas where contamination can occur.

Nitrate in Animal Feed

Nitrate often accumulates in such high amounts in plants that they constitute a hazard when fed to cattle and sheep.[20] Annual weeds, grasses, and the cereal crops generally contain higher amounts of NO_3^- than perennial grasses and legumes.

Factors affecting NO_3^- levels in plants are discussed in detail elsewhere.[18,20] They include a high content of available N in the soil and adverse conditions for plant growth, such as drought, shade, and cloudy weather. Nitrogen fertilization at high rates does not constitute a major problem so long as conditions are favorable for plant growth. Areas where toxicities have been reported include the Northern Plains and the Corn Belt Region of the United States.

The biochemical basis for NO_3^- accumulation in plants is a loss of nitrate reductase activity under growth-limiting conditions. Nitrate reductase is the enzyme responsible for converting NO_3^- to NH_3 in plants (see Chapter 6), and if its activity is low, plants accumulate NO_3^- and store it internally.

Nitrate in Food Crops

Nitrate is often present in very high concentrations in certain vegetables, notably lettuce, celery, beets, spinach, radishes, and turnip greens.[17,18,24] Nitrate contents as high as 3000 mg/kg have been reported. The NO_3^- content of vegetables varies greatly, depending upon sampling date, variety, and environmental conditions. Factors that favor high accumulations include:

1. High levels of NO_3^- in the soil as a result of heavy fertilization
2. Reduced light intensity during maturation
3. Soil moisture deficiency
4. Nutrient deficiencies in the plant.

Nitrate is added to many prepared meat products, but fresh vegetables are the major source of dietary nitrate.[19] Another potential hazard arising from NO_3^- additions to meat products is the consumption of nitrosamines that are formed by reaction of NO_3^- (as NO_2^-) and protein degradation products in the food.[3]

POSSIBLE CONNECTION BETWEEN SOIL AND FERTILIZER N AND OZONE DEPLETION

Ozone (O_3) exists high up in the atmosphere (i.e., the stratosphere), where it provides a protective shield against excessive ultraviolet radiation reaching the earth's surface. The O_3 layer exists as a 16.3-km (10-mile) band between 16.7 and 33 km (10 and 20 miles) above the earth. Over 99% of the incoming ultraviolet radiation in the most hazardous wavelengths (< 300 nm) is absorbed by the O_3 layer.

Ozone is continuously being formed and destroyed in the stratosphere, and a decrease in the average amount present may cause a variety of problems for plant and animal life, including skin cancer in man. It should be noted that although the O_3 layer is quite thick (about 16 km), the layer would only be a millimeter or so thick at sea-level pressure and temperature. According to an NRC report,[3] each 1% reduction in the O_3 level of the stratosphere might lead to a 2% increase in skin cancer as a result of additional ultraviolet radiation reaching the Earth's surface. The conclusion was also reached that reduction in O_3 content might cause changes in global temperature and rainfall distribution patterns.

The basic reactions leading to the formation of O_3 in the stratosphere are as follows:[25]

$$O_2 \xrightarrow[< 242 \text{ nm}]{\text{UV light}} O\cdot + O\cdot \tag{1}$$

$$O\cdot + O_2 \rightarrow O_3 \tag{2}$$

This process is in dynamic equilibrium with processes that destroy O_3, as shown by the following reactions:

$$O_3 + NO \rightarrow NO_2 + O_2 \tag{3}$$

$$O_3 + \cdot OH \rightarrow O_2 + \cdot OOH \tag{4}$$

$$O_3 + Cl\cdot \rightarrow O_2 + ClO\cdot \tag{5}$$

where $Cl\cdot$ is a free radical formed by UV degradation of chlorofluorocarbons (CFCs). Processes that prevent O_3 formation by destroying singlet oxygen, $O\cdot$, are also involved.

Nitrous oxide (N_2O), which diffuses into the stratosphere over long time periods from the earth's surface, is a natural source of NO for reaction (3), as shown by the following sequence:

$$N_2O \rightarrow O\cdot + N_2 \tag{6}$$

$$O\cdot + N_2O \rightarrow 2 NO \tag{7}$$

From these reactions, it is evident that an increase in the amount of N_2O entering the atmosphere has the potential for decreasing O_3 levels. In addition to reacting with O_3 by reaction (3), the NO formed by reaction (7) can react with NO_2 to form HNO_3, which is returned to the troposphere (see Fig. 8.7).

The O_3 layer is vulnerable to a variety of products introduced into the atmosphere through the activities of man, including spray can propellants (CFCs), products from the burning of fossil fuels, and exhaust gases from supersonic airplanes. Over decades, these constituents may reach the upper troposphere (layer immediately below the stratosphere) and enhance the destruction of O_3. Since exhaust fumes of supersonic aircraft are injected at the lower boundary of the stratosphere, they can affect the O_3 layer rather quickly.

As noted above, among the many gases that tend to deplete the O_3 layer is N_2O, which is produced in soil by bacterial denitrification and nitrification. The fear has been expressed that an increase in fertilizer N use will be ac-

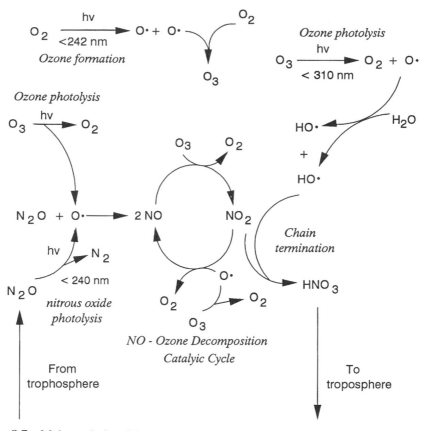

Fig. 8.7 Main cycle involving N₂O and other N oxides in the stratosphere. Adapted from NRC.[5]

companied by a concomitant increase in the amount of N_2O released into the atmosphere, ultimately causing a reduction of O_3 in the stratosphere and of the ability to absorb ultraviolet rays.[3,26–29]

Analysis of ice core samples (Fig. 8.8) suggests that the concentration of N_2O in the atmosphere remained rather constant for the period 1 to about 1770 A.D., but increased sharply thereafter.[30] For the three sites shown in Fig. 8.9, the rate of increase over the past two decades has remained rather constant, being equivalent to an annual increase of 0.2 to 0.3%. Accurate estimates cannot be made for the amounts of N_2O derived from various sources, but agriculture is suspected to be a major source.[16,30]

Prediction of the potential impact of soil-derived N_2O on O_3 in the stratosphere is fraught with uncertainties, including a lack of information as to the fraction of the denitrification product that occurs as N_2O (versus N_2) and the residence time of N_2O in the troposphere. Factors that affect the amount of N_2O emitted from soil (see Chapter 5) are:

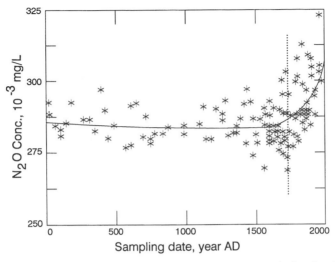

Fig. 8.8 Changes in N_2O concentrations in ice core samples, indicating increases in atmospheric levels starting in about 1770 A.D. From Robertson.[30]

1. Soil content of inorganic N, notably NO_3^-. Applications of N fertilizers, incorporation of residues rich in N (e.g., manures), and growing legumes can lead to an increase in N_2O flux, depending on soil conditions.

2. Presence of highly degradable organic materials that promote microbial activity and thereby a depletion of O_2, with enhanced N_2O emissions due to denitrification.

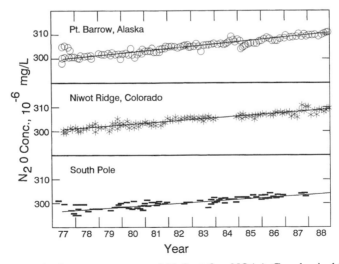

Fig. 8.9 Atmospheric measurements of N_2O at five NOAA-Geophysical Monitoring sites as recorded by Robertson.[30]

3. Soil aeration status and moisture content. Emissions of N_2O are greatest when soils are wet, but not waterlogged.
4. Soil temperature and pH. Emissions of N_2O increase with an increase in soil temperature.

A summary of the probable effects of stratojets, chlorofluorocarbons (spray can propellants), and fertilizer N on O_3 in the stratosphere has been prepared by NRC[3] and is summarized in Table 8.4. The estimate for fertilizer N contribution is a 1.5 to 3.5% reduction in O_3 by the year 2100. Aldrich[2] concluded that this estimate is high, since the soils' contribution has been declining over the years due to a reduction in mineralization losses.

The long response time for fertilizer N to reach the stratosphere (decades to centuries, as shown in Table 8.4) can be attributed to the long residence time of the fertilizer in soil or water, followed by many additional years in the troposphere. In contrast, chlorofluorocarbons are added directly to the atmosphere and can reach the stratosphere within a decade or so.

With regard to fertilizer N use, the conclusion reached by NRC[3] was that "the current value to society of those activities that contribute to global N fixation far exceeds the potential cost of any moderate . . . postponement of action to reduce the threat of future ozone depletion of N_2O." Research will ultimately provide information from which valid conclusions can be reached. It should be noted that production and use of chlorofluorocarbons have been greatly reduced because they were recognized as being major contributors to ozone-depleting volatile chemicals and gases.

Eutrophication

Eutrophication is the process of enrichment of water in lakes, ponds, and streams with nutrients, thereby giving rise to increased growth of aquatic plants, including phytoplankton and algae. Eutrophication detracts from the

TABLE 8.4   Postulated Decrease in Atmospheric O_3 by Three Different Mechanisms[a]

Mechanism	Ozone Response Time	Steady-State Global O_3 Reductions	
		Long-Term Use at Current Level, %	Long-Term Use at Possible Future Level, %
Stratospheric aviation	Years	<0.1	6.5
Chlorofluorocarbons	Decades	14	Large
Fertilizer N[b]	Decades to centuries	1.5	2.5

[a] From NRC.[3]
[b] Current level is 79×10^9 kg/yr. Projected level is 200×10^9 kg/yr.

recreational value of lakes, and it can lead to depletion of O_2 so that fish and other water-dwelling animals cannot survive. Odor and taste of water are often adversely affected, thus increasing the cost of providing high-quality water for human use.

Eutrophication is not new; it has taken place since early geological time. For example, the vast peat bogs in cooler regions of the earth are a direct manifestation of eutrophication. There is little doubt, however, that the process has in many instances been enhanced by man's activities.

The suggestion has been made that NO_3^- in drainage waters from agricultural lands leads to man-induced eutrophication of lakes and streams, a claim that has yet to be substantiated. Many investigators believe that other factors outweigh any possible effect of NO_3^- in drainage waters.

Phosphorus, rather than N, is usually the limiting nutrient for growth of aquatic plants. Also, N_2 fixation by cyanobacteria, along with N in atmospheric precipitation, may provide sufficient N to meet the needs of the common algae and aquatic macrophytes. It should be noted, however, that the productivity of estuaries is often limited by N.[1]

Ecosystem Damage due to Acid Rain

Acid rain (defined as rainfall of pH <5.7) occurs over broad areas of northern Europe, eastern Canada, and the northeastern United States. The trend is toward a decrease in pH and an expansion of the affected area with time. Acidity is due mainly to the mineral acids H_2SO_4 and HNO_3 in precipitation, with a minor contribution from HCl.

The most noticeable effect from acid rain has been a lowering of pH in numerous lakes in eastern North America and in Scandinavia. Accompanying the decrease in pH has been an increase in dissolved Al, which is toxic to plants. As a consequence, some lakes are virtually lifeless; others are approaching this state.

Ecological effects of acid rain are difficult to evaluate but may include disruption of biogeochemical cycling, nutrient turnover, organic matter degradation, damage to growing plants, a decline in soil fertility, and acidification of lakes, with accompanying effects on algae, macrophytes, invertebrates, and fish.[31-33]

A combination of many human activities is responsible for the H_2SO_4 and HNO_3 in acid rain, including stack gases from the burning of coal, gases from oil-burning facilities, and emissions from motor vehicles. The soil is not considered to be a significant contributor to the acid-forming constituents in rain water.

Formation of Nitrosamines

The potential hazard of nitrosamines as toxicants in certain foodstuffs, or formed following ingestion, has been the subject of considerable interest.

Nitrosamines, which are carcinogenic, mutagenic, and acutely toxic at very low concentrations, are formed through the reaction of NO_2^- with amines.

Primary aliphatic amines react with NO_2^- to form diazonium salts, which are unstable and break down to yield N_2 and a complicated mixture of organic compounds (e.g., alcohols and alkenes):

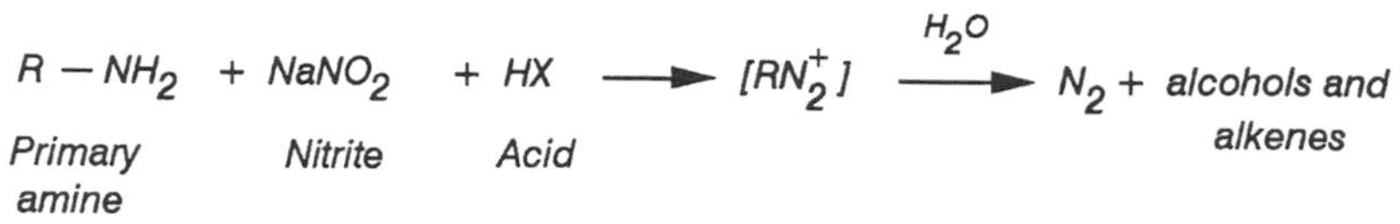

$$R-NH_2 + NaNO_2 + HX \longrightarrow [RN_2^+] \overset{H_2O}{\longrightarrow} N_2 + \text{alcohols and alkenes}$$

Primary   Nitrite   Acid
amine

In contrast, secondary aliphatic and aromatic amines react with NO_2^- to form nitrosamines, according to the following reaction:

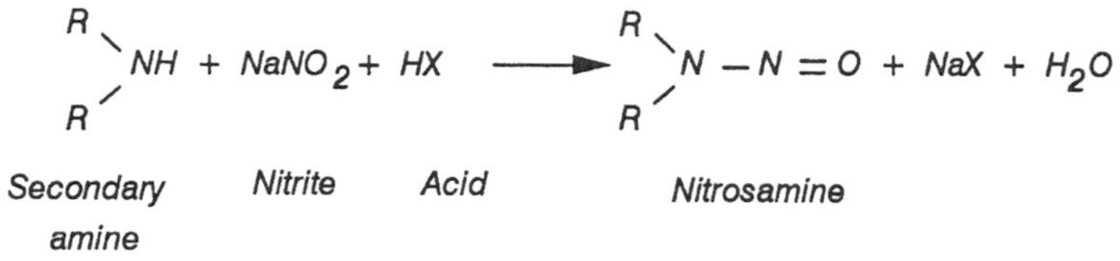

$$\begin{matrix} R \\ \diagdown \\ \diagup \\ R \end{matrix} NH + NaNO_2 + HX \longrightarrow \begin{matrix} R \\ \diagdown \\ \diagup \\ R \end{matrix} N-N=O + NaX + H_2O$$

Secondary   Nitrite   Acid   Nitrosamine
amine

As was the case for methemoglobinemia (discussed earlier), NO_2^- constitutes a problem only when the NO_3^- is reduced to NO_2^-, as can occur in vegetables or prepared foods that are stored under conditions that permit microbial growth. The subject of nitrosamines in foods is discussed elsewhere.[3,19] Wolff and Wasserman[19] point out that evidence for the presence or formation of nitrosamines in foods is limited and that some earlier reports indicating the occurrence of nitrosamines in foods may have resulted from inadequacies in analytical procedures for determining nitrosamines.

A potential hazard to the health of man and animals would exist if nitrosamines were formed in soil from pesticide degradation products or from precursors present in manures and sewage sludge. A health hazard would become a reality, however, only if the nitrosamines thus formed were leached into water supplies and subsequently consumed by human or animals or taken up by plants used as food by livestock or humans.

Trace quantities of nitrosamines have been detected in soils amended with known amines (dimethylamine, trimethylamine) and NO_3^- or NO_2^-, but for the most part, this work has been done under ideal conditions for nitrosamine formation in the laboratory, such as high additions of reactants and a strongly acidic pH.[34-37] Evidence is lacking that the synthesis of nitrosamines in field soils represents a threat to the environment. Mosier and Torbit[37] were unable to detect N-dimethylnitrosamine and N-diethylnitrosamine in manures even though the necessary precursors were known to be present.

SUMMARY

Several real and potential health and environmental impacts of N are known to exist. The soil N cycle affects primarily those that involve excess NO_3^- in drinking water, eutrophication, and possibly O_3 depletion of the stratosphere. Where high NO_3^- levels are observed in shallow wells, a source other than the soil is suspect, such as an animal production facility or domestic waste-disposal system. The contribution of fertilizer N to the NO_3^- in lakes and storage reservoirs can be kept at acceptable levels by proper management techniques.

The world-wide demand for food and fiber will require continued use of fixed N in agriculture, much of which will need to come from fertilizer. Better management practices are desired in order to maximize production of crops while at the same time minimizing adverse health effects and pollution of the environment.

REFERENCES

1. D. R. Keeney, "Nitrogen Management for Maximum Efficiency and Minimum Pollution," in F. J. Stevenson, Ed., *Nitrogen in Agricultural Soils,* American Society of Agronomy, Madison, Wisconsin, 1982, pp. 605–649.

2. S. R. Aldrich, *Nitrogen in Relation to Food, Environment, and Energy,* Illinois Agr. Exp. Sta. Special Publ. 61, Urbana, Illinois, 1980, pp. 1–452.

3. National Research Council (NRC), "Nitrates: an Environmental Assessment," *National Academy of Sciences,* Washington, D.C., 1978.

4. R. J. Haynes, Ed., *Mineral Nitrogen in the Plant–Soil System,* Academic Press, London, 1986.

5. J. A. Hauser and K. Hendrickson, Eds., *Nitrogen in Organic Wastes Applied to Soil,* Academic Press, New York, 1989.

6. R. F. Follett, D. R. Keeney, and R. M. Cruse, Eds., *Managing Nitrogen for Groundwater Quality and Farm Profitability,* Soil Science Society of America, Madison, Wisconsin, 1991.

7. T. P. Burt, A. L. Heathwaite, and S. T. Trudgill, Eds., *Nitrate: Processes, Patterns, and Management,* Wiley, New York, 1993.

8. G. M. Pierzynski, J. T. Sims, and G. F. Vance, *Soils and Environmental Quality,* Lewis, Boca Raton, Florida, 1994

9. D. R. Keeney, *CRC Crit. Rev. Environ. Control.* **16,** 257 (1986).

10. J. F. Power and J. S. Schepers, *Agric. Ecosystems Environ.,* **226,** 165 (1989).

11. G. E. Likens, F. H. Bormann, M. Johnson, D. W. Fisher, and R. S. Pierce, *Ecolog. Monogr.* **40,** 23 (1970)

12. G. Roberts and T. Marsh, *Intern. Assoc. Hydrol. Eng. (IAHS),* **164,** 365 (1989).

13. R. A. Smith and R. B. Alexander, "Trends in Concentrations of Dissolved Solids, Suspended Sediments, Phosphorus and Inorganic Nitrogen in U.S. Geological Sur-

vey National Streams Quality Accounting Network Stations (NASQAN)," in *National Water Summary 1984—Hydrologic Perspectives, U.S. Geological Survey,* 1985, pp. 66–73.

14. R. A. Wiese, R. B. Ferguson, and G. W. Hergert, "Fertilizer Nitrogen Best Management Practice," in *Nebraska Coop. Ext. Serv. Nebraska Guide,* G87-829, 1987.

15. J. S. Schepers, M. G. Moravek, E. E. Alberts, and K. D. Frank, *J. Environ. Qual.* **20,** 12 (1991).

16. T. Granli and O. C. Bockman, "Nitrous Oxide from Agriculture," in *Norwegian J. Agric. Sci.: Supplement 12,* 1994, pp. 8–86.

17. B. S. Deeb and K. W. Sloan, *Nitrates, Nitrites, and Health,* Illinois Agr. Exp. Sta. Bull. **750,** 1975.

18. O. A. Lorenz, "Potential Nitrate Levels in Edible Plant Parts," in D. R. Nielsen and J. G. MacDonald, *Nitrogen in the Environment: Soil-Plant-Nitrogen Relationships,* Academic Press, NY, 1978, pp. 201–219.

19. I. A. Wolff and A. E. Wasserman, *Science,* **177,** 15 (1972).

20. M. J. Wright and K. L. Davidson, *Adv. Agron.,* **16,** 197 (1964).

21. H. J. Comly, *J. Amer. Med. Assoc.,* **129,** 112 (1945).

22. T. M. Addiscott, A. P. Whitmore, and D. S. Powlson, *Farming, Fertilizers, and the Nitrate Problem,* CAB International, Wallingford, England, 1991.

23. R. G. Hoeft, *Better Crops with Plant Food,* **74**(4), 4 (1990).

24. W. J. Corre and T. Breimer, *Nitrate and Nitrite in Vegetables,* Unipub, New York, 1979.

25. S. E. Manahan, *Environmental Chemistry, 6th ed.,* Lewis, Boca Raton, Florida, 1994, pp. 294–297.

26. Council for Agricultural Science and Technology (CAST), *Effect of Increased Nitrogen Fixation on Stratospheric Ozone: Report No. 53,* Iowa State University Press, Ames, Iowa, 1976.

27. P. J. Crutzen and D. H. Emhalt, *Ambio,* **6,** 112 (1977).

28. H. S. Johnson, *Proc. National Acad. Sci.,* **69,** 2369 (1972).

29. M. B. McElroy and J. C. McConnell, *J. Atmos. Sci.,* **28,** 1095 (1971).

30. G. P. Robinson, "Fluxes of Nitrous Oxide and Other Nitrogen Trace Gases from Intensively Managed Landscapes: A Global Perspective," in D. E. Rolston et al., Eds., *Agricultural Ecosystem Effects on Trace Gases and Global Climate Change,* Soil Science Society of America Special Publication 55, Madison, Wisconsin, 1993, pp. 95–108.

31. J. N. Galloway, G. E. Likens, and M. E. Hawley, *Science,* **226,** 829 (1984).

32. G. E. Likens, *Chem. Eng. News.,* **54,** 29 (1976).

33. R. Patrick, V. P. Binetti, and S. G. Halterman, *Science,* **211,** 446 (1981).

34. A. L. Mills and M. Alexander, *J. Environ. Qual.,* **5,** 437 (1976).

35. S. K. Pancholy, *Soil Biol. Biochem.,* **10,** 27 (1978).

36. R. L. Tate and M. Alexander, *Soil Sci.,* **118,** 317 (1974).

37. A. R. Mosier and S. Torbit, *J. Environ. Qual.,* **5,** 465 (1976).

9

THE PHOSPHORUS CYCLE

The P cycle in soil is a dynamic system involving soil, plants, and microorganisms. Major processes include uptake of soil P by plants, recycling through return of plant and animal residues, biological turnover through mineralization–immobilization, fixation reactions at clay and oxide surfaces, and solubilization and formation of mineral phosphates through chemical reactions and activities of microorganisms. In natural systems, essentially all of the P utilized by plants is returned to the soil in plant and animal residues; under cultivation, some P is removed in the harvest and only part is returned. Most losses of soil P arise through erosion; smaller losses occur as a result of leaching.

The general cycle of P in soils has been diagrammed in a number of ways, depending on the particular aspect to be emphasized (e.g., biological transformations, soil–plant interrelationships, phosphate reactions). In Fig. 9.1, emphasis is given to plant–soil–microbiological relationships in which the P is partitioned into "pools" that vary in availability to plants. Soil solution P is shown to be in equilibrium with a given quantity of *labile* inorganic P such that as P is taken up by plants, or immobilized by microorganisms, additional inorganic P is solubilized. Stewart[1] has depicted microbial activity as a "wheel" that rotates in the soil, simultaneously consuming and releasing P to the soil solution. Studies using the radioactive isotope ^{32}P to follow changes in the forms of P in soils have shown that tranfers between the pools are relatively rapid.

In many respects, the P cycle in soil is analogous to the N cycle. However, the former is less spectacular in that valency changes do not occur during transformations by microorganisms, which is the case for N (see Chapter 5). Next to N, P is the most abundant nutrient contained in the microbial biomass, making up as much as 2% of the dry weight. Partly for this reason, P is the second most abundant nutrient in soil organic matter.

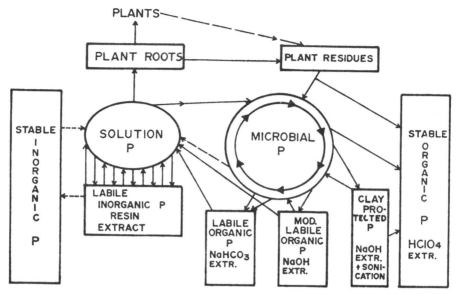

Fig. 9.1 The P cycle in soils, showing the partition of organic and inorganic forms of P into pools based on availability to plants. A flow diagram for the sequential extraction of P in the various fractions (pools) is given in Fig. 9.4. From Stewart,[1] reproduced by permission of the Potash & Phosphate Institute, Atlanta, Georgia.

Chemical and biochemical aspects of the P cycle have been reviewed from several standpoints, including fluxes of P on a global scale,[2,3] interactions with the C, N, and S cycles,[1] pedogenesis,[4] plant nutrition,[5] inorganic forms and fixation reactions,[5] soil organic P and associated transformations,[7-10] and environmental quality.[11,12] A soil and plant computer model has been developed for simulating P transformations in soil.[13]

GLOBAL ASPECTS OF THE P CYCLE

Major reserves of P in the earth (Table 9.1) are marine sediments (840,000 $\times 10^{12}$ kg), terrestrial soils (96 to 160 $\times 10^{12}$ kg), dissolved inorganic phosphate (mostly HPO_4^{2-} and PO_4^{3-}) in the ocean (80 $\times 10^{12}$ kg), crustal rocks as apatite (19 $\times 10^{12}$ kg), and the biota or biomass (2.7 $\times 10^{12}$ kg). The amount of P in the terrestrial biota (2.6 $\times 10^{12}$ kg) far exceeds that of the marine biota (0.05–0.12 $\times 10^{12}$ kg).

An overview of fluxes of P on a global scale is illustrated in Fig. 9.2. The annual uptake of P by terrestrial plants amounts to 200 $\times 10^9$ kg, a small portion of which is removed in the harvested portion of the crop (4 to 7 $\times 10^9$ kg). Losses of terrestrial P to fresh waters, and ultimately to the oceans, are about 17 $\times 10^9$ kg/yr; a smaller amount (4.3 $\times 10^9$ kg/yr) is carried by wind into the atmosphere. Unlike the C, N, and S cycles, the P cycle does not have a gaseous component; accordingly, movement of P to and from the atmo-

TABLE 9.1   Major Reservoirs of P in the Earth[a]

Reservoir	Total P $\times$ 10^{12} kg
Marine	
Sediments	840,000
Dissolved (inorganic)	80
Deritus (particulates)	0.65
Biota	0.050–0.12
Terrestrial	
Soil	96–160
Mineable rock phosphate	19
Biota	2.6
Fresh water (dissolved)	0.090

[a]From Richey[2]

sphere is of minor importance due to the small amounts that are circulated as atmospheric particulates. Inspection of Fig. 9.2 shows that P eventually finds its way to the sea, where it is deposited as ocean sediment and thereby removed from the cycle. Man has intervened in the P cycle through the mining of phosphates for fertilizer use, as well as by release of P into the environment in domestic and industrial effluents.

In natural ecosystems, the P cycle is virtually closed, and most plant P is recycled by microbial breakdown of litter and organic debris. A typical example of this is in tropical rain forests, where essentially all of the P exists as living and dead organic matter and where the underlying soil generally contains such a low amount of P that good yields of crops cannot be maintained following clearing of the forest. As noted below, much of the P in grassland soils resides in the biomass.

Chemical Properties of Soil P

Phosphorus is an element with a mass of 30.98. Six isotopes are known, one of which (^{32}P) has a sufficiently long half-life to be of value in soil investigations. Although P exhibits coordination numbers of 1, 3, 4, 5, and 6, the vast number of P compounds have coordination numbers of 3 and 4. In the pedosphere, P is found largely in its oxidized state as orthophosphate, mostly as complexes with Ca, Fe, Al, and silicate minerals. Some secondary minerals of phosphate are wavellite $[Al_3(PO_4)_2(OH)_3 \cdot 5H_2O]$, vivianite $[Fe_3(PO_4)_2 \cdot 8H_2O]$, dufrenite $[FePO_4 \cdot Fe(OH)_3]$, strengite $[Fe(PO_4) \cdot H_2O]$, and variscite $[Al(PO_4) \cdot 2H_2O]$. Small amounts of phosphine gas, PH_3, often occur in lakes or marshes under highly reducing conditions.

The form of the phosphate ion in solution varies according to pH. In dilute solution, phosphoric acid dissociates as follows:

$$H_3PO_4 \underset{+H^+}{\overset{-H^+}{\rightleftharpoons}} H_2PO_4^- \underset{+H^+}{\overset{-H^+}{\rightleftharpoons}} HPO_4^{2-} \underset{+H^+}{\overset{-H^+}{\rightleftharpoons}} PO_4^{3-}$$

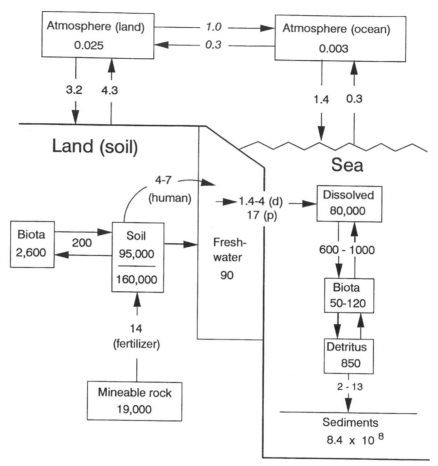

Fig. 9.2   Pools and annual fluxes of P for the global P cycle. Reservoirs are expressed in 10^9 kg and fluxes are in 10^9 kg/year: d = dissolved; p = particulate. Adapted from Richey.[2]

Successive pK_a's for the reactions are 2.15, 7.20, and 12.37, respectively. Thus, in the pH range of most soils (5 to 8), the amounts of the undissociated H_3PO_4 and trivalent PO_4^{3-} are negligible, and essentially all of the phosphate consumed by plants is in the $H_2PO_4^-$ and HPO_4^{2-} forms. At pH 6, about 94% of the phosphate occurs as $H_2PO_4^-$, but the percentage drops to 60% at pH 7.

Origin of P in Soils

The native P in soils was derived from the apatite of soil-forming parent materials. From a mineralogical point of view, apatite is a complex compound of tricalcium phosphate having the empirical formula $3[Ca_3(PO_4)_2] \cdot CaX_2$, where X can be either Cl^-, F^-, OH^-, or CO_3^{2-}. The most common minerals are the chloro-, fluor-, hydroxy-, and carbonate-apatites. These minerals are highly insoluble in water, and the P in them is not readily available to plants.

During centuries of weathering and soil development, the P of apatite is liberated and subsequently (1) absorbed by plants and recycled, (2) incorporated into the organic matter of soils and sediments, and (3) redeposited as sparingly soluble mineral forms, such as Ca–, Fe–, and Al–phosphates and the occluded P of Fe- and Al-oxides. An interesting account of changes in P forms along a chronosequence of soils in Hawaii (ages ranging from 300 years to 4.1×10^6 years) is given by Crews et al.[14]

Reserves of Mineable Phosphate Rock

Deposits rich in apatite (e.g., rock phosphate) are found in sediments deposited at the bottom of ancient seas, such as those in extensive areas of North America (5×10^{12} kg P). Other phosphate-rich deposits occur in tropical Africa (6×10^{12} kg P) and the Kola Peninsula in Russia (1×10^{12} kg P). Known reserves of rock phosphate are on the order of 5×10^{13} kg, representing a total of about 12×10^{12} kg of P based on an average P content of 10% due to inclusion of low-phosphate-containing rocks.[15] Results of recent surveys dealing with known world reserves, distribution of resources by geological deposit type, and trends in worldwide phosphate production rates can be found in Khasawneh et al.[5] Rock phosphates from various sources vary widely in chemical composition, but those rich in carbonate-apatite are the most commonly used fertilizer materials.

Treatment of rock phosphate with inorganic acids (e.g., sulfuric and phosphoric) is used to produce the more soluble phosphate fertilizers. Superphosphate is a product resulting from the mixing of approximately equal quantities of 60 to 70% H_2SO_4 and phosphate rock. The overall reaction is:

$$3[Ca_3(PO_4)_2] \cdot CaF_2 + 7\ H_2SO_4 + 3\ H_2O \rightarrow \begin{array}{c} 3\ Ca(H_2PO_4)_2 \cdot H_2O \\ \text{Ca–dihydrogen phosphate} \\ +7\ CaSO_4 + 2\ H_2O + 2\ HF \\ \text{Gypsum} \end{array}$$

The fertilizer obtained in this manner is referred to as *ordinary or normal superphosphate* and contains about 10% P. Reaction of rock phosphate with excess H_2SO_4 and removal of much of the gypsum gives *concentrated superphosphate*, a product containing about 20% P. The principal reaction leading to the production of H_3PO_4 by the so-called "wet" process is as follows:

$$3[Ca_3(PO_4)_2] \cdot CaF_2 + 10\ H_2SO_4$$

$$+ n\ H_2O \rightarrow 10\ CaSO_4 \cdot nH_2O + 6\ H_3PO_4 + 2\ HF$$

Three environmental problems are encountered with the mining of phosphate ores and the production of P fertilizers, namely, emission of fluorine as the highly volatile and poisonous HF gas, disposal of gypsum, and accumulation of Cd and other heavy metals in soils and possibly crops as a result of repetitive use of P fertilizers. The last topic is discussed in more detail in Chapter 11.

Although total reserves of phosphate rock would appear to be appreciable, phosphate is a scarce commodity when considered in terms of world-wide requirements. Accordingly, problems with phosphate deficiencies are likely to intensify in the not too distant future due to shortages of mineable phosphates, at which time crop yields will be limited by the rate at which phosphate is released from insoluble forms in the soil. Forecasts as to when shortages will develop vary considerably, depending on estimates that are made for unknown reserves and the P content of the rock at which mining will be profitable. At the present mining rate (about 7×10^{10} kg of rock phosphate annually), reserves will be depleted in about 700 years. However, should usage of phosphate fertilizers continue to grow at recent rates, known reserves will be used up in a much shorter time (about 100 years). Ultimately, ways may be found to exploit the soil's reserve of "fixed" phosphate, such as by the development of phosphate-efficient plants through genetic engineering. Another possibility is to recycle the large amounts of P that are present in waste materials, such as sewage sludges (about 4% P).

Phosphorus Content of Soils

The P content of soils in their natural state varies considerably, depending on the nature of the parent material, degree of weathering, and extent to which P has been lost through leaching. The P contents of the common soil-forming rocks vary from as little as 0.01% (100 mg/kg) in sandstones to over 0.2% (2,000 mg/kg) in high-phosphate limestones.

The usual range of P in soils is on the order of 500 to 800 mg/kg (dry-weight basis). A soil containing 500 mg P/kg will contain about 1120 kg P/ha to the plow depth. Total P is usually highest in the upper A horizon and lowest in the lower A and upper B horizons due to recycling by plants.

Soils formed from acidic igneous rocks are generally low in P; those derived from basic rocks contain moderate to high amounts. Unweathered calcareous soils of dry regions often have high P contents due to the general lack of leaching and the presence of appreciable amounts of P in the form of apatite. Dark-colored soils, such as the Mollisols, tend to be high in P (and other nutrients as well). It should be noted that the amount of P in plant-available forms does not necessarily bear a direct relationship to total P.

The content of native P in surface soils of the United States is shown in Fig. 9.3. The highest amounts are found in the soils of a large area in the Northwest; low amounts are found in coarse-textured (sandy) soils of the Atlantic and Gulf Coastal Plains. As noted below, the P content of many soils has been altered by cropping, additions of animal manure, and fertilization.

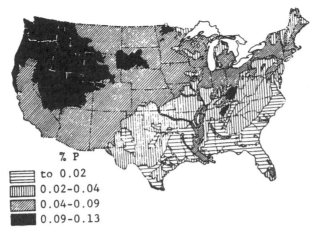

Fig. 9.3 Phosphorus content in the surface layer of U.S. soils. The original data were expressed in archaic terms (i.e., as P_2O_5). Adapted from Sauchelli.[3]

The amounts of P in subsurface horizons are generally on the same order as those found in the surface layer, as indicated by Table 9.2.[16] The profiles represent two major soil types, and the data serve to illustrate the effects of both parent material and climate on P content. The Mollisols are representative of the dark-colored, prairie soils of the upper Midwest of the United States, where weathering has been slight to moderate. These soils tend to be high in P. In contrast to the Mollisols, the Ultisols have very low P contents. The parent material from which the sandy loam Ultisol in Georgia was formed contained high amounts of P (1700 mg/kg), and the relatively low amounts in the profile (400 mg/kg or less) suggest that losses through leaching were extensive during weathering and soil formation.

Most of the P in soils occurs in inorganic forms, the main exception being peat (Histosols), where essentially all of the P occurs in organic forms. Prairie grassland soils, forest soils, and certain tropical soils also contain relative high amounts of P in organic forms, as shown below.

Effect of Cropping and Fertilization on Soil P

Long-time cropping of the soil without fertilizer or manure additions invariably leads to a reduction in P content. For a soil containing 500 mg/kg P, equivalent to 1120 kg P/ha to plow depth, continuous cropping for a period of 10 years under a management system where 10 kg of P is removed annually in the harvested portion of the crop will result in a reduction of total P of nearly 10%. Organic matter is lost at a rather rapid rate when soils are first placed under cultivation (see Chapter 2), and much of the initial loss of P can be attributed to mineralization of organic P, uptake by plants, and removal of P during harvest (e.g., grain, forages).

TABLE 9.2 Distribution of P in the Horizons and Parent Materials of Some Soil Profiles[a]

Type	Description	Horizon and Parental Material	Depth, cm	P, mg/kg
Mollisol	Tama silt loam, Iowa	A1	15–30	790
		A3	36–61	610
		B	64–102	440
		C	127–178	390
		Underlying loess	229+	440
Mollisol	Crete silt loam, Nebraska	A11	0–4	600
		A12	4–10	570
		A3	10–51	520
		B2	51–97	900
		C_{Ca}	97–152	920
		Underlying loess	152–213	440
Ultisol	Sandy loam, Georgia	A1	0–3	400
		A	3–23	200
		B1	23–36	200
		B2	36–71	400
		C	71–152	100
		Fresh granite		1,700
Ultisol	Silt loam, Virginia	A1	0–5	240
		A2	5–28	70
		B1	28–41	140
		B2	41–91	100
		C	91–142	140
		Fresh limestone		140

[a]From Simonson.[16]

In a study of soil P dynamics in cropping systems managed for 13 years by conventional and organic farming methods, Oberson et al.[17] observed only slight differences in the distribution of P between organic and inorganic forms, as well as in the percentage of the organic P that was extractable with 0.5 M $NaHCO_3$ and 0.1 M NaOH. They concluded that these parameters are inadequate as a means of predicting the long-term P fertility of the soil.

Data obtained by Haas et al.[18] for the soils of 15 dryland experiment stations in the U.S. Great Plains show that total P was reduced by an average of 8% by cropping with wheat over a period of 30 to 48 years, with most of the loss being organic P. Somewhat higher losses (from 12 to 34% after 60 to 90 years of cultivation) were obtained by Tiessen et al.[19] for the soils of three grassland soil associations of the Canadian prairies. Their data, given in Table 9.3, show that, for two of the associations, essentially all of the P

TABLE 9.3   Changes in Organic C and P by Long-Time Cultivation of Soils from Three Grassland Soil Associations of the Canadian Prairies[a]

Soil Association	Values for Native Prairie	60–70 Years Cultivation	Loss of C or P, %
Blain Lake			
Organic C, mg/kg	48,000 ± 10	33,000 ± 5	32
Organic P, mg/kg	645 ± 125	528 ± 54	18
Total P, mg/kg	823 ± 92	724 ± 53	12
Southerland			
Organic C, mg/kg	38,000 ± 7	24,000 ± 2	37
Organic P, mg/kg	492 ± 5	407 ± 30	17
Total P, mg/kg	756 ± 28	661 ± 31	12
Bradwell			
Organic C, mg/kg	32,000 ± 8	17,000 ± 2	46
Organic P, mg/kg	446 ± 46	315 ± 21	29
Total P, mg/kg	746 ± 101	527 ± 15	29

[a]Adapted from Tiessen et al.[19] Most soils were Typic Cryoborolls.

loss was accounted for by the organic fraction. It is of interest that losses of organic C (i.e., on a percentage basis) exceeded those of organic P.

Several studies have shown that losses of organic P by cultivation are less than for N, a result that has been attributed to greater uptake of N by plants and greater losses of N through leaching and denitrification in comparison to P.[18] Another factor to consider is that organic P is probably stabilized by a different mechanism than N, a point that was emphasized by Halstead and McKercher.[8]

Most of the P applied to soil as fertilizer, often as high as 90% or more, is not taken up by the crop but is retained in insoluble or fixed forms. Whereas a portion of the residual P can be used by subsequent crops, further additions of fertilizer are often required in order to maintain high crop yields. Repeated applications of P in amounts exceeding crop uptake will inevitably result in an accumulation of P in the soil. Accordingly, the P content of many culti-vated soils in industrial countries has been increasing over the years; to this extent, the soil has served as a "sink" for P. As one might expect, the extent of P accumulation will depend on both fertilizer application rate and years of application.

Results obtained by Barber[20] for the accumulation of P in soils of a rota-tion–fertility experiment where P was applied at several rates over a 25-year period are given in Table 9.4. In this case, accumulations of P occurred when P application rates exceeded 22 kg/ha/yr. Net losses of P at the lower rates can be accounted for by crop removal. A discussion of gains and losses of soil P over a 30-year period for a Typic Umprabult soil is given by Mc-Cullum.[21]

TABLE 9.4   Influence of Continuous P Fertilization over a 25-Year Period on Total P in Soil[a]

P Applied, kg/ha/year	Total P Applied kg/ha	Total Soil P mg/kg	Change in Soil P, mg/kg
0	0	400 ± 20	−10
11	275	455 ± 44	−4
22	550	492 ± 50	+5
40	1000	589 ± 57	+23
54	1350	632 ± 48	+32

[a]From Barber.[20]

Increases in soil P also occur through long-time applications of animal manures.[18,19] In the study of Haas et al.,[18] total P was increased by an average of 14% above the virgin soils where manure had been applied at an annual rate of 5,600 to 11,200 kg/ha, equivalent to about 6.2 to 12.3 kg P/ha, over a 30 to 48-year period.

In addition to changes in total P, long-time cropping of the soil, with or without addition of fertilizer P, leads to modifications in the distribution of P in the various pools (see Fig. 9.1). This work is discussed in the section below on Chemistry of Soil P.

PHOSPHORUS AS A PLANT NUTRIENT

Phosphorus is an essential constituent of all living organisms. Plants deficient in P are stunted in growth and maturity is delayed. The lower leaves are typically yellow and tend to wither and drop off. Leaf pigmentation in young leaf tissue is usually abnormally dark green, often shading to red and purple hues because of excess anthocyanin accumulation. Phosphorus is needed for favorable seed formation, root development, strength of straw in cereal crops, and crop maturity.

Considerable variation exists in the P requirements of plants, as suggested by the data given in Table 9.5. For the crops shown, total P removed by the crop amounts to from 19 to 54 kg/ha. At the higher value, a soil containing 500 mg P/kg will contain sufficient P to the plow depth for only a few decades of cropping without further P additions. However, much of this P is not in plant-available forms (discussed below). It should also be noted that substantial amounts of the P utilized by plants will be returned to the soil in crop residues. Unlike N, which can be returned to the soil by biological N_2 fixation (Chapter 4), P cannot be replenished except from external sources once it is removed in the harvest or by erosion.

Although most plants take up P throughout the entire growing season, about 50% of the seasonal total of P is taken up by the time plants have

TABLE 9.5   Approximate Amounts of P Removed from the Soil per Season by Specific Crops[a]

Crop	Yield kg/ha	Yield units/acre	P Removed, kg/ha
Grains			
Corn	12,544	200 bu	52
Grain sorghum	8,064	8,000 lb	54
Wheat	5,376	80 bu	34
Barley	5,376	100 bu	28
Oats	3,584	100 bu	22
Rice	7,280	145 bu	25
Forage crops			
Alfalfa	12,544	6 tons	35
Clovers	8,064	4 tons	20
Grasses (general)	8,064	4 tons	20
Oil crops			
Soybeans (beans only)	3,360	50 bu	25
Peanuts	3,360	3,000 lb	22
Fiber crops			
Sugarcane	67,200	30 tons	19
Sugar beets (roots and tops)	67,200	30 tons	28

[a]Adapted from Khasawneh et al.[5]

accumulated 25% of their seasonal total of dry matter. Plant species differ in their requirements for P, ability to extract P from the soil, and response to insoluble forms of P fertilizers (e.g., rock phosphate). Crops that are relatively efficient in using insoluble forms of P include alfalfa, buckwheat, millet, lupines, and sweet clover; inefficient crops include barley, cotton, corn, oats, potatoes, and wheat. For detailed information on the P nutrition of plants, the reader is referred to Khasawneh et al.[5]

Phosphorus exists in living organisms as the inorganic phosphate ion ($H_2PO_4^-$ or HPO_4^{2-}) and in combination with organic compounds, some of which are discussed in the section below on Organic Forms of Soil P. The organic P of plants exists as compounds in which phosphate is esterified with OH groups of sugars or alcohols or bound as a pyrophosphate bond to another phosphate group. Major P-containing compounds are the nucleic acids (see Chapter 5), phospholipids, phosphorylated polysaccharides, and phytin (Ca–Mg salt of inositol hexaphosphate). Vital compounds for metabolism include adenosine triphosphate (ATP), which functions in energy storage and transfer, and the phosphopyridine nucleotides (NAD^+, $NADP^+$), which serve as electron carriers.

Structures for ATP and two metabolic phosphates are shown below:

ATP   Glucose 6-phosphate   Fructose 1,6-diphosphate

ENVIRONMENTAL IMPACT OF SOIL P

Phosphorus has long been known to have an impact on the eutrophication of lakes, notably those that are oligotropic (i.e., those lakes where the concentration of P and/or other nutrients is very low and thus the growth of algae and other aquatic plants is limited). Alterations in the ecosystem, such as conversion of land from forest to agricultural use, can increase the input of P and thereby initiate eutrophication. For a discussion of the eutrophication process, the reader is referred to Chapter 6.

Because of the strong tendency of phosphate to be adsorbed on colloidal surfaces and to form insoluble complexes with di- and trivalent cations, losses of P from agricultural soils in subsurface and groundwater flow are generally small, amounting to from 0.1 to 1.2 kg P/ha per year. This behavior is in marked contrast to the high rate of N transfer (as NO_3^-) to subsurface waters (see Chapter 6). However, cumulative losses of soil P over a period of years, or over a broad area, can be appreciable. Leaching represents a major mechanism of P loss from forest lands.

Whereas short-term losses of P through leaching are usually of little significance from an agricultural standpoint, they may be of some importance when considered in terms of the P enrichment of natural waters, whereby the P contributes to eutrophication. The reader is referred to several reviews[11,12] for a discussion of the impact of soil and fertilizer P on the environment.

The concentration of dissolved inorganic P in drainage waters is generally low (Table 9.6) and depends on the concentration of inorganic phosphates in the soil solution, percolation rate, surface area exposed to percolating waters, and kind and amount of P-adsorbing material in the profile. Losses of P in surface runoff will be affected by slope, P fertilization history, cropping practice, type of soil, amount and intensity of rainfall, and (of some importance) tillage practices.[22,23]

In contrast to the low amounts of dissolved P in drainage waters from agricultural lands, usually on the order of 0.04 to 0.06 mg P/liter, rather high

TABLE 9.6 Concentrations of Dissolved Inorganic P in Subsurface Flow[a]

Location	Soil Texture	Drainage System	Crop	Dissolved Inorganic P, mg/L	
				Range	Mean
Ontario, Canada	Clay	Tile drains	Corn, oats	0.20–0.17	0.18
			Alfalfa, bluegrass	0.10–0.27	0.21
Snake Valley,	Calcareous loam	Irrigation return flow	Alfalfa, corn, root crops, pasture	0.01–0.23	0.10
Woburn, England	Sandy	Tile drains	Arable and grassland	0–0.30	0.08
South Central Michigan	Clay to sandy loams	Tile drains and ditches	Root crops	0.01–0.30	—
San Joaquin Valley California	Heavy silty clay	Irrigation return flow drains	Cotton, rice, alfalfa	0.05–0.23	0.08
Yakima Valley, Washington	Sandy loam	Surface return flow drains	—	0.07–0.30	0.16
		Subsurface return flow drains		0.03–0.46	0.18

[a]From Ryden et al.,[22] from which specific references can be obtained.

loads are found in surface waters from domestic wastes (4–9 mg P/liter). The contribution of domestic wastes to the P of surface waters of the United States (9–227 × 10^6 kg/yr) is of the same order of magnitude as for agricultural lands (5.5–545 × 10^6 kg P/yr). Unusually high amounts of P (> 100 kg/liter) have been found in surface runoff from cattle feedlots.[22,24]

Because of the low solubility and high sorption capacity of soil for inorganic forms of P (see next section), the main mechanism of P loss from most agricultural land is erosion. Excluding arid soils of the intermountain west, most surface soils of the United States contain from 200 to 900 mg P/kg, or about 450 to 2000 kg/ha to the depth of plowing. Thus, a metric ton of soil lost by erosion will carry with it from 0.22 to 1.88 kg of P.

CHEMISTRY OF SOIL P

The chemistry of P in soils is exceedingly complex, and full details are available elsewhere.[5,7,9] Only the overall aspects will be discussed herein.

The P compounds in soil can be placed into the following classes:

1. Soluble inorganic and organic compounds in the soil solution
2. Weakly adsorbed (labile) inorganic phosphate
3. Sparingly soluble phosphates:
   a. Of Ca in calcareous and alkaline soils of arid and semiarid regions;
   b. Of Fe and Al in acidic soils
4. Insoluble organic forms:
   a. Of the soil biomass;
   b. In undecayed plant and animal residues;
   c. As part of the soil organic matter (humus)
5. Phosphates strongly adsorbed and/or occluded by hydrous oxides of Fe and Al
6. Fixed phosphate of silicate minerals

Essentially all of the inorganic P in soil exists in the form of orthophosphates, or derivatives of phosphoric acid (H_3PO_4). The main compounds are phosphates of Ca, Al, and Fe with trace amounts of other cations. Various forms of apatite, such as fluorapatite, constitute the principal P minerals in calcareous and alkaline soils of arid and semiarid regions. Trace quantities of P may be present in the lattices of silicate minerals and as inclusions in minerals.

The principal water-soluble forms of P are $H_2PO_4^-$ and HPO_4^{2-}; the PO_4^{3-} ion is uncommon. Because the three forms are interconvertible (see section on Chemical Properties of Soil P), they will be referred to hereafter as P_i (i.e., soluble inorganic phosphate).

From the standpoint of plant nutrition, the P in soils can be considered in terms of "pools," as noted in Fig. 9.1. Only a small fraction of the P occurs in water-soluble forms at any one time. A portion of the insoluble P appears to be somewhat more available to plants than the bulk of the soil reserves. This fraction, designated as the "labile" pool, is believed to consist of easily mineralized organic P and phosphates weakly adsorbed to clay colloids. From 1 to 2% of the soil P occurs in microbial biomass.

The bulk of the soil P ($> 90\%$) occurs in insoluble or fixed forms, namely, as primary phosphate minerals, humus P, insoluble phosphates of Ca, Fe, and Al, and phosphate fixed by colloidal oxides and silicate minerals.

The ability of the soil to provide P to higher plants is determined by a variety of factors, including:

1. The quantities of P_i in the soil solution
2. The solubilities of Fe- and Al-phosphates and phosphate complexes with hydrous oxides and clay minerals in acid soils
3. The solubilities of Ca-phosphate and P-minerals in calcareous soils and the rates at which the minerals are solubilized
4. Amount and stage of decomposition of organic residues
5. Activities of microorganisms

As shown below, maximum P availability tends to occur at intermediate pH values of 6 to 7.

Chemical Fractionation Schemes

Knowledge of the forms of P in soils has come largely from chemical fractionations based on the ability of selective chemical reagents to solubilize discrete types of inorganic P compounds. The initial work was done by Chang and Jackson,[25] who developed a scheme in which the P present as Al-, Fe-, and Ca-phosphates was determined by sequential extraction with 0.1 N NH_4F, 0.1 N NaOH, and 0.5 N H_2SO_4; occluded P in Fe- and Al-phosphates was determined by extraction with dithionate–citrate and 0.5 N NH_4F, respectively.

Subsequent studies indicated that the various extractants were not as specific as first envisioned. Also, difficulties were encountered with calcareous soils and sediments due to readsorption of the phosphate from the initial NH_4F extract. The Chang–Jackson fractionation procedure was subsequently modified by others and applied to a wide variety of soils, including those subject to long-time cropping. The work is extensive, and the reader is referred to Khasawneh et al.[5] for detailed information.

Calcium phosphates constitute a series of compounds ranging from relatively soluble mono- and di-calcium phosphates (normally present in rather small amounts) to relatively insoluble hydroxyapatite and fluorapatite. Iron- and Al-phosphates also vary greatly in solubility, depending on the pH and amount of P present. In some soils, part of the Fe-and Al-phosphates may exist as the minerals dufrenite and wavellite, which are very insoluble except under neutral or alkaline conditions.

As one might expect, considerable variation exists in the kinds and amounts of any given form of P in soils. Thus, Fe- and Al-phosphates tend to accumulate in acid soils, whereas Ca-phosphates are predominant in neutral or alkaline soils. As noted below, a soil pH in the range 6 to 7 is best from the standpoint of P availability because this range is above the pH of maximum insolubility of Fe- and Al-phosphates but below the pH of maximum insolubility of Ca-phosphates.

A major disadvantage of the Chang–Jackson fractionation scheme, and modifications therefrom, is that some of the inorganic P, notably in alkaline extracts, is derived from the soil organic P. This factor has been taken into account in more recent schemes (see next section).

Fractionations Based on Biological Availability

Hedley et al.[26] introduced a sequential extraction procedure whereby soil P is separated into inorganic and organic fractions (pools) that vary in their availability to plants. Essentially, a progression of stronger reagents is used, with the P of each subsequent pool becoming increasingly less bioavailable and requiring harsher treatments to extract it. The method is described in detail by Kuo.[27]

A schematic diagram of the extraction procedure is given in Fig. 9.4. Biologically available inorganic P is removed first with an anion exchange resin, following which a mild extractant (0.5 M $NaHCO_3$) is used to remove "labile" inorganic and organic forms. At this step, inclusion of a chloroform treatment permits estimation of organic P originating from lysed microbial cells. Stable P forms (Fe- and Al-phosphates, apatite-type minerals, and highly insoluble organic and inorganic P compounds) are removed by stronger extractants.

The use of 0.5 M $NaHCO_3$ to determine "labile" organic P is based on the finding that this mild extractant, originally used to extract plant-available inorganic P, removes a small amount of organic P that is more readily mineralized than the bulk of the soil organic P. A portion of the biomass P can also be regarded as being readily available to plants because this fraction has a short residence time in soil.

The fractionation procedure of Headly et al.[26] has been used to:

1. Characterize the P in soils from long-term crop rotations, with and without applications of P fertilizers[28–31]
2. Assess the influence of soil management practices, organic residue additions, and texture on the forms and distribution of soil P[32–34]

Results of these and related studies have enhanced our knowledge of both the long-term and short-term dynamics of P in soil, and they have confirmed

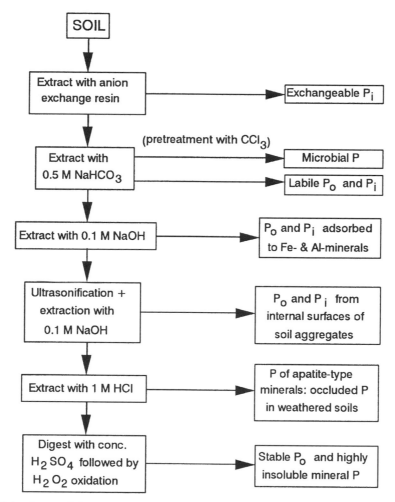

Fig. 9.4 Sequential extraction scheme for partition of soil P into pools based on availability to plants (see Fig. 9.1). Adapted from Hedley et al.[26]

the importance of associated transformations involving C when crop residues are returned to the soil. The subject of P dynamics is complex, and extensive coverage is beyond the scope of this chapter. For detailed information, the reader is referred to the papers cited above, as well as the reviews mentioned earlier.[5,6,12]

Phosphate Fixation

Numerous studies have shown that much of the P applied to soils in water-soluble forms (e.g., monocalcium phosphate, $Ca(H_2PO_4)_2$) does not remain as such for long but is converted to one of many insoluble or complex forms,

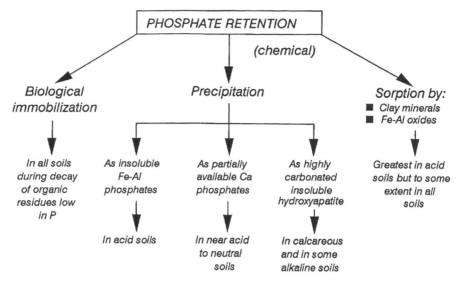

Fig. 9.5   Phosphate fixation reactions in soil. Adapted from Sauchelli.[3]

as illustrated by Fig. 9.5. Both biological and chemical processes are involved in fixation, with the latter being of greatest importance from the standpoint of retention of fertilizer P.

Figure 9.6 illustrates the importance of pH in governing fixation reactions and thereby the availability of phosphates to plants. For most soils, maximum availability would be expected in the slightly acid to neutral pH range. Phosphate fixation is known to be influenced by clay mineralogy and decreases in the following order: amorphous hydrous oxides > goethite = gibbsite > kaolin > montmorillonite. In highly acid soils, phosphate is readily precipi-

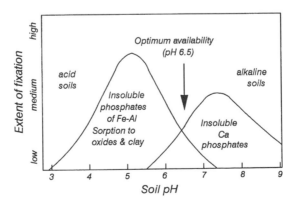

Fig. 9.6   Effect of pH on phosphate forms and extent of P fixation in soil. Adapted from Sauchelli.[3]

tated as highly insoluble Fe- or Al-phosphates or adsorbed to oxide surfaces. Both forms are poor sources of P for higher plants. This is particularly true for soils rich in Fe, such as Oxisols of the tropics and subtropics. Plants growing in such soils often show severe P deficiency symptoms, and relatively large amounts of fertilizer P must be applied in order to meet plant requirements.

In calcareous soils, the less soluble di- and tri-calcium phosphates $[Ca_2(HPO_4)_2$ and $Ca_3(PO_4)_2]$ are formed, with the latter being gradually converted to carbonate hydroxyapatite, $3[Ca_3(PO_4)_2] \cdot CaCO_3$. This highly insoluble compound is not a good source of P for plants, but availability may be enhanced in the presence of growing plants through the action of organic acids secreted from plant roots or synthesized by microorganisms in the rhizosphere, as noted below.

Adsorption of phosphate at hydrous oxide surfaces through a ligand binding mechanism (replacement of H_2O and/or OH) is one of the important processes affecting phosphate availability to plants.[35-38] The overall reaction for replacement of OH^- is as follows.

Labile P   Difficultly available P

Variations in the availability of "fixed" phosphate to plants have been attributed to the formation of two types of chemical bonds, a single coordinate linkage (labile P) and a chelate ring (difficulty available P). The concept of a "labile" pool of inorganic P is noted above.

Still another mechanism for the sorption of phosphate by soils and hydrous ferric oxide gels is through chemisorption at protonated ($—OH_2^+$) sites.[38] The reaction is as follows:

Because Fe- and Al-oxides can be positively charged at pH values below the point of zero charge, phosphate fixation by acid soils has sometimes been thought to be due to anion exchange (illustrated above).

Phosphate fixation at oxide surfaces is particularly important in highly weathered soils of the tropics and subtropics. The clay fraction of youthful soils formed on volcanic ash also fixes appreciable amounts of phosphate, which can be accounted for by their high content of allophane (chemical formula of $xSiO_2, Al_2O_3 \cdot yH_2O$). In a study of phosphate retention by Chilean allophanic soils, Borie and Zunino[39] observed an initial fast adsorption onto allophanic surfaces, following which the added phosphate was subjected to reactions leading to the formation of organic matter–P associations or complexes through Al bridges. The humus P complexes were found to comprise the major fraction of the total P in these soils and were viewed as a P sink in the overall P cycling process.

Attempts to increase the availability of P in highly weathered acidic soils through liming have generally been unsuccessful, a result that has been attributed to the formation of insoluble Ca-phosphates, thereby counteracting the effect of pH on the solubilization of fixed phosphate.[40]

Another mechanism that has been advanced to explain P fixation in soil is the combination of phosphate with clay minerals. Two reactions are possible, one being replacement of a –OH group from a structural Al atom and the second a linkage with Ca on the exchange complex to form a clay–Ca–PO_4 linkage. The phenomenon is greater for 1:1 type clay minerals than for the 2:1 type, which can be explained by their higher content of exposed Al–OH groups. Kafkafi et al.[41] identified three successive regions for adsorption of phosphate on kaolinite. The first two were associated with edge–face –OH groups and a third with occlusion of K-phosphate into the amorphous region of the clay surface.

The retention of phosphate by acid organic soils may be due to the formation of humate–Al-phosphate complexes,[42] as illustrated below. Some of the P in acid mineral soils may also occur in this form.

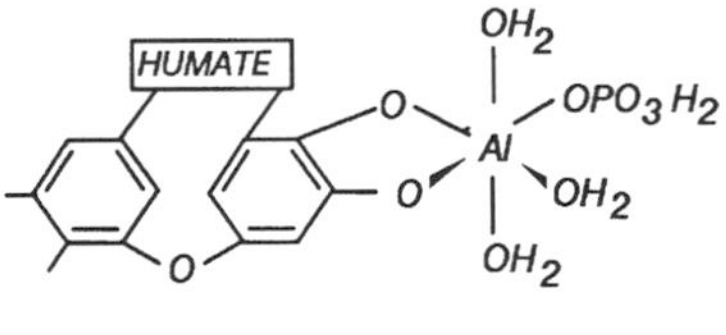

Water-Soluble P

Phosphorus is absorbed by plants largely as the negatively charged primary and secondary orthophosphate ions ($H_2PO_4^-$ and HPO_4^{2-}), which are present in the soil solution. Thus, the water-soluble pool is of particular interest because this P has a direct effect on plant growth, and hence yields. Plants have the ability to absorb certain soluble organic phosphates, but it is not known whether or not they serve as direct sources of P to plants.

Numerous attempts have been made to determine the minimum soil solution P concentration that will coincide with maximum crop yield. A range of from 0.01 to 0.3 mg P/liter in the soil solution has been recorded for a variety of crops.[43] Results obtained for corn grown on two dissimilar soils (Oxisol and Hydrandept) are shown in Fig. 9.7. For both soils, a value of 0.05 mg P/liter was required for 95% of maximum yield. Critical concentrations for other soils will vary from this value, depending on plant uptake conditions (climate, rooting volume, etc.).

Factors affecting levels of soluble P in soils include crop removal, chemical fixation reactions (conversion to insoluble forms), immobilization by microorganisms, and leaching.

The replenishment of P in the soil solution is of considerable importance from the standpoint of plant nutrition, for the reason that the quantity of inorganic P in the solution phase at any one time, usually on the order of 0.3 to 1 mg/liter, is insufficient to meet crop requirements over the whole of the growing season. At the higher concentration (1 mg/liter), the total amount of P in the soil solution (25% water on a dry-weight basis) would be about 0.13 kg/ha to a depth of 0.6 meters. Crop removal is usually between 4.5 and 22.4 kg/ha; thus, total soluble P at a given moment would be less than 5% of the crop requirement. Continuous renewal of P in the soil solution is therefore required to meet the needs of the growing plant. However, it should be pointed out that plants do not absorb P uniformly throughout the entire soil mass, and that the rate of replenishment of soil solution P must be greatest in small areas near the root tips, where the most active uptake of P occurs.

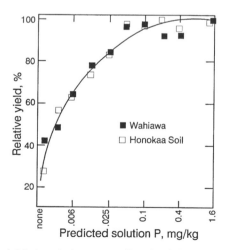

Fig. 9.7 Corn grain yields in relation to predicted soluble P in two soils with different mineralogical composition. The Wahiawa soil is a Oxisol, whereas the Honokoa soil is a Hydrandept. Adapted from Fox.[43]

Changes in inorganic P in the soil solution occur through mineralization (m) and/or immobilization (i) of organic P by microorganisms and by chemical equilibria between labile and relatively insoluble mineral forms of P, the postulated reactions being as follows:

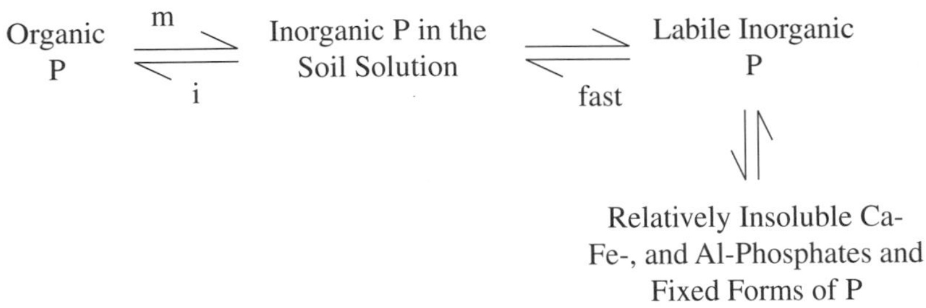

Organic P $\overset{m}{\underset{i}{\rightleftharpoons}}$ Inorganic P in the Soil Solution $\underset{\text{fast}}{\rightleftharpoons}$ Labile Inorganic P $\updownarrow$

Relatively Insoluble Ca-
Fe-, and Al-Phosphates and
Fixed Forms of P

A vital factor affecting plant uptake of soil P is the rate at which a suitable concentration of phosphate is maintained in the soil solution near the root surface. Movement of phosphate to plant roots by mass flow is believed to account for only a small portion of the phosphate consumed by plants. Accordingly, soil immediately adjacent to the root is a major source of P for the growing plant. This is the region where microorganisms are particularly active (the rhizosphere). A substantial portion of the soluble P may occur in organic forms (see Dalal[7]).

As noted above, the rate of conversion of insoluble phosphates to soluble forms in many soils is too slow to satisfy crop requirements, in which case fertilizer P must be applied in order to achieve optimum yields.

ORGANIC FORMS OF SOIL P

From 15 to 80% of the P in soils occurs in organic forms, the exact amount being dependent upon the nature of the soil and its composition.[7,8] The higher percentages are typical of peats and uncultivated forest soils, although much of the P in tropical soils and certain prairie grassland soils may occur in organic forms. In fertilized temperate-zone soils, however, the contribution of organic forms is likely to be rather small relative to inorganic P.

Determination of Organic P

Methods used for the determination of organic P can be divided into two main types, as follows:

Extraction methods: In this approach, organic and mineral forms of P are recovered from the soil by extraction with base and acid. Organic P is converted to orthophosphate, and the content of organic P is determined by difference.

Organic P = total P in alkaline extract − inorganic P in acid extract

Ignition method: Organic P is converted to inorganic P by ignition of the soil at elevated temperatures and is calculated as the difference between inorganic P in acid extracts of ignited and nonignited soil.

Organic P = inorganic P of ignited soil − inorganic P of untreated soil

Advantages and disadvantages exist for both approaches. In the extraction method, pretreatment of the soil with mineral acid (usually conc. HCl) is essential for removing polyvalent cations that render the P compounds insoluble. However, this treatment may cause some hydrolysis of organic P compounds. Other extractants (8-hydroxyquinoline and EDTA) have been used in attempts to eliminate this problem. The alkali used to solubilize organic matter (dilute NaOH) also causes some hydrolysis of organic P.

Of the two approaches, the ignition method is the easiest and is used often for survey types of investigations. Limitations of the method include alteration in the solubility of the native inorganic P and hydrolysis of organic phosphate during acid extraction of the nonignited soil. Methods for the determination of organic P are described by Kuo.[27]

Because organic P is always determined by the differences between total and inorganic P, accuracy is low when the soil is low in organic P, such as in the subsoil.

Organic P Content of Soils

Although organic P in soil is somewhat biologically stable, continuous turnover occurs through mineralization–immobilization (discussed below), and the equilibrium level will depend to a considerable extent on the nature of the soil and its environment. The situation is analogous to that of soil organic matter as a whole, as discussed in Chapter 2. As one might expect, the organic P content of the soil follows closely that for total organic matter.

Values for total organic P in soils are summarized in Table 9.7. In view of the diverse nature of soils, the widely different values are not surprising. Histosols, being rich in organic matter, usually contain high amounts of organic P. As one might expect, organic P decreases with depth in much the same way as organic C.[44] Depth distribution patterns for organic P and organic C in two typical soils from Iowa are shown in Fig. 9.8.

Factors influencing the proportion of P in the organic matter of mineral soils include P supply, parent material, climate, drainage, cultivation, pH, and depth of soil. The effect of each factor is not known with certainty, and contradictory results are the rule rather than the exception. Some workers have concluded that a low P content of the organic matter is a characteristic of P-deficient soils, but this hypothesis has not been confirmed. A slight effect has been noted for parent material, namely, the organic matter of soils derived from granite tends to be lower in P than the organic matter of soils derived

TABLE 9.7   Some Values for Organic P in Soil[a]

Location	Organic P mg/kg	Organic P % of Total P
Australia	40–900	—
Canada	80–710	9–54
Denmark	354	61
England	200–920	22–74
New Zealand	120–1360	30–77
Nigeria	160–1160	—
Scotland	200–920	22–74
United States	4–85	3–52

[a]From the review of Halstead and McKercher,[8] from which specific references can be obtained. The range shown for the U.S. soils must be regarded as a minimum.

from basalt or basic igneous materials. Also, the P content of organic matter may be higher in fine-textured soils than in coarse-textured ones. Baker[44a] found that organic P accumulated rapidly during the first 50 years of soil development, after which the rate declined until a steady-state was reached. As one might expect, levels of organic P can be modified by manure additions.

When soils are first placed under cultivation, the content of organic C (and N) usually declines (Chapter 2). This same pattern is also followed for organic P. At present, it is not known whether the mineralization rate of organic P is greater than, equal to, or less than the mineralization rate of organic C and N.

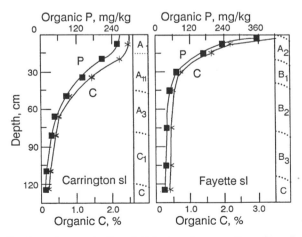

Fig. 9.8   Distribution of organic P and C in two Iowa soil profiles. Left = Mollisol; Right = Alfisol. From Pearson and Simonson.[44]

A combination of upward transport of P by plants and retention in the surface soil (i.e., as a result of plant litter decomposition and the low mobility of phosphates) may alter the vertical distribution of P in the soil profile. For soils subject to moderate weathering and leaching, the minimum in the concentration of total P is often found a short distance below the surface, which is due in part to enrichment of the surface soil with organic P at the expense of inorganic P.

The C/N/P/S Ratio

The amount of organic P in soils is roughly correlated to C and/or N, although C/organic P and N/organic P ratios are much more variable than C/N ratios.[45-49] The P content of soil organic matter varies from as little as 1.0% to well over 3.0%, which is reflected by the variable C/organic P ratios that have been reported for soils, as shown in Table 9.8 for the soils of several great soil groups in Canada.[45] One explanation for the extreme ratios is that N and S occur as structural components of humic and fulvic acids whereas P does not. It is noteworthy that essentially all of the organic P can be recovered from the soil by alkaline extractants and that most of the P resides in the fulvic acid fraction.

The percentage of the total organic P that occurs in association with living microorganisms (the soil biomass) is another factor that must be taken into account when comparing the C/organic P ratios of contrasting soil types. As noted below, the C/P ratio of the soil biomass is substantially lower than for the soil organic matter, and the percentage of the organic P as biomass P can vary over a broad range (> 3 to < 20%).

TABLE 9.8   Carbon/Organic P Ratios of Soils from Several Great Soil Groups in Canada[a]

Soil Order[b]	Number of Soils	C/Organic P Range	Mean
Chernozems	8	92–648	231
Solonetzs	5	106–177	149
Podzols	5	103–225	172
Brunizols	7	46–245	148
Regosols	5	106–417	209
Gleisols	8	62–274	124
All soils	38	46–648	172

[a]From John et al.[45]
[b]Nomenclature follows the 1938 Great Soil Groups System. Approximate equivalents in the Orders of Soil Taxonomy are: Chernozems, Mollisols; Solonetzs, Aridisols and Mollisols; Podzols, Spodosols; Brunizems, Mollisols; Regosols, Entisols and Inceptosols; Humic Gleisols, Mollisols, Inceptisols and Ultisols.

TABLE 9.9   Organic C, Total N, Organic P, and Organic S Relationships in Soil

Location	Number of Soils	C/N/P/S	Reference
Iowa	6	110:10:1.4:1.3	Neptune et al.[46]
Brazil	6	194:10:1.2:1.6	Neptune et al.[46]
Scotland[a]			
Calcareous	10	113:10:1.3:1.3	Williams et al.[49]
Noncalcareous	40	147:10:2.5:1.4	Williams et al.[49]
New Zealand	22	140:10:2.1:2.1	Walker and Adams[48]
India	9	144:10:1.9:1.8	Somani and Sarena[47]

[a]Values for S given as total S.

In view of the above, it is not surprising that considerable variation has also been observed in the C/N/P/S ratios of individual soils (Table 9.9). Notwithstanding, the mean for soils from different regions of the world is remarkably similar. As an average, the proportion of C/N/P/S in soil humus is 140:10:1.3:1.3.

Walker[4] concluded that the P content of the original parent material placed an upper limit on the amount of organic C and N that could accumulate in soils, postulating that, given sufficient time, the amount of N comes into balance with P (assuming that S is not limiting).

Specific Organic P Compounds

Little is known about the chemistry of organic P in soil. Principal forms are the inositol phosphates, phospholipids, and nucleic acids or their degradation products. Other organic P compounds present in trace quantities include sugar phosphates and phosphoproteins. The usual ranges for percentage of organic P in the various forms are as follows:

Inositol phosphates	10–50%
Phospholipids	1–5%
Nucleic acids	0.2–2.5%
Phosphoproteins	trace
Metabolic phosphates (ATP, etc.)	trace

Although somewhat over one-half of the organic P in select soils has been accounted for as inositol phosphates, recoveries well below this value are not uncommon (Table 9.10). As an average, no more than one-third of the organic P in soils can be identified in known compounds.

Inability to account for the bulk of the organic P in most soils may be due to:

TABLE 9.10   Distribution of the Forms of Organic P in Soils[a]

Source	Inositol Phosphates	Nucleic Acids	Phospholipids
	% of Organic P		
Australia	0.4–38	—	—
Britain	24–58	0.6–2.4	0.6–0.9
Canada	11–23	—	0.9–2.2
New Zealand	5–43	—	0.7–3.1
Nigeria	23–30	—	—
United States	3–52	0.2–1.8	—

[a]From Dalal,[7] from which specific references can be obtained.

1. Occurrence of polymeric phosphate-containing compounds, such as teichoic acids from bacterial cell walls and phosphorylated polysaccharides. The former consists of a polyol, ribitol, or glycerol phosphate in which adjacent polyol residues are linked by a phosphodiester bond.
2. Occurrence of stable complexes of inorganic phosphate with soil organic matter, which is recovered only after destruction of organic matter and is measured along with the organic P.
3. Formation of complexes between soil organic matter and phosphate esters (e.g., nucleotides and/or inositol phosphates), which modify the properties of the esters such that they are not estimated by conventional methods.

As shown above, recovery of soil organic P in the three principal forms follows the order: inositol phosphates $\gg$ phospholipids $>$ nucleic acids. The same types of organic P compounds are found in plants, but the order is reversed. The somewhat higher abundance of inositol phosphates in soils, as contrasted to plants, may be due to their tendency to form insoluble complexes with polyvalent cations, such as Fe and Al in acidic soils and Ca in calcareous soils. These insoluble complexes are undoubtedly less available to microorganisms and degradative enzymes than are the more soluble forms. Since the various forms of organic P differ in their ease of attack by microorganisms, losses of organic matter through intensive cultivation would be expected to be accompanied by a change in organic P composition, namely, a decrease in the percentage of the organic P in the more labile components (i.e., nucleic acids and phospholipids).

Inositol Phosphates   Inositol phosphates are esters of hexahydrohexahydroxy cyclohexane, commonly referred to as inositol (I). A variety of esters are possible, the most common being the hexaphosphate. In plants, mono-, di-, and tri-phosphates are sometimes found in rather large quantities. The hexaphosphate ester, or phytic acid (II), occurs in cereal grains as the mixed

Ca- and Mg-salt called phytin. For many years, the inositol hexaphosphates in soil were thought to be derived from the phytin of higher plants, but recent work indicates that they are probably of microbial origin.

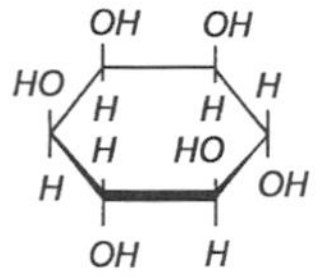

myo-Inositol

I

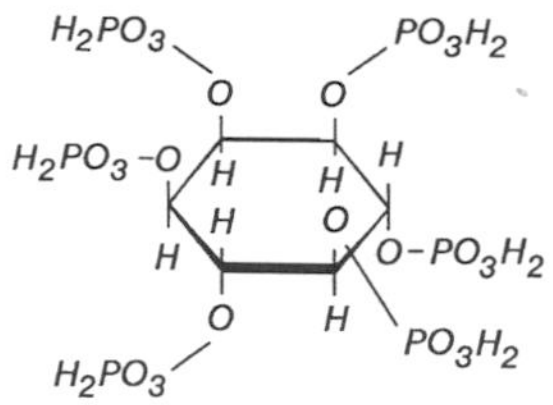

Phytic acid

II

A characteristic feature of inositol phosphates, particularly the hexaphosphate esters, is their high stabilities in acids and bases. Advantage is taken of this property for their extraction and preparation from soil. The steps involved have usually included the following:

1. Extraction with hot 0.5 N NaOH after pretreatment of the soil with 5% HCl
2. Oxidation of the extracted organic matter with alkaline hypobromite
3. Precipitation of inositol phosphates as their Fe (or Ca) salts
4. Alkaline hypobromite oxidation of the precipitate material
5. A second Fe (or Ca) precipitation of the inositol phosphates
6. Decomposition of the second Fe precipitate with NaOH solution and removal of the $Fe(OH)_3$.

The P recovered in the final solution is sometimes considered to be in the form of inositol phosphates, but other P compounds are usually present (e.g., inorganic phosphate). Most workers now use chromatographic techniques to separate the inositol phosphates from inorganic phosphate and from each other, in which case lower-phosphate esters of inositol have been found in addition to the hexaphosphate.

Caldwell and Black[50] analyzed 49 Iowa surface soils by anion-exchange chromatography and found that inositol hexaphosphate, the most abundant ester, accounted for 3 to 52% of the organic P, the average being only 17%. The percentage was higher for soils developed under forest vegetation than for soils developed under grasses.

From the structural formula of inositol, it can be demonstrated that nine positional stereoisomers are possible, depending on arrangements of H and OH groups. They include seven optically inactive forms and one pair of optically active isomers. The best-known form is myo-inositol (shown above), which is widely found in nature. This is the isomer from which phytin is constituted. Other naturally occurring isomers of more limited biological distribution are *D-chiro* and *L-chiro*-inositol (III and IV) and *scyllo*-inositol (V).

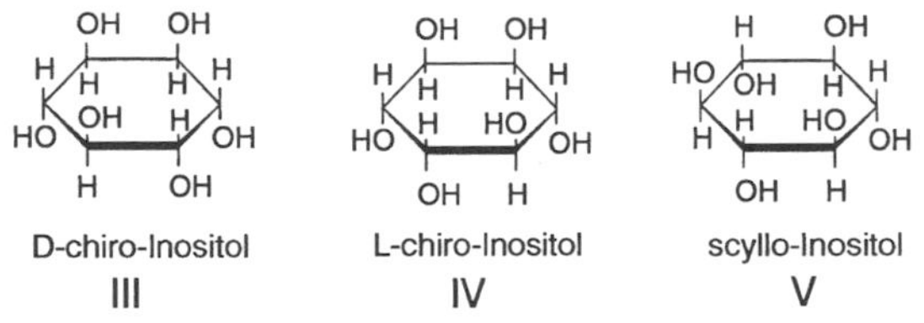

D-chiro-Inositol   L-chiro-Inositol   scyllo-Inositol
III   IV   V

The advent of chromatographic techniques has led not only to more precise values for inositol phosphates in soil, but to the isolation of a number of unusual isomers. In some early work, a hexaphosphate other than *myo*-inositol (I) was found to account for an average of 46% of the total inositol hexaphosphate material and to be synthesized by a variety of soil microorganisms.[50]

The unknown isomer was subsequently shown in studies by Cosgrove[51] to be the hexaphosphate of *scyllo*-inositol (V). Cosgrove also demonstrated the occurrence in soil of the hexaphosphate of *D-* and *L-chiro*-inositol (III and IV), as well as *neo*-inositol (structure not shown). Penta phosphates of *myo-*, *chiro-*, and *scyllo*-inositol were also found.

For additional information on inositol phosphate stereoisomers in soil, the reader is referred to the review of Cosgrove.[51] Results of the various studies show that considerable differences exist in the isomeric composition of inositol phosphates in soil and that some, if not all, of the soil inositol phosphates are synthesized *in situ* by microorganisms.

Nucleic Acids Nucleic acids are found in all living cells, and they are synthesized by soil microorganisms during degradation of plant and animal residues. Two types are known, ribonucleic acid (RNA) and deoxyribonucleic acid (DNA), each consisting of a chain of nucleotides (see Chapter 2). The few estimates that have been made thus far indicate that no more than 3% of the organic P in soil occurs as nucleic acids or their derivatives.[9]

Phospholipids The phospholipids represent a group of biologically important organic compounds that are insoluble in water but soluble in fat solvents, such as benzene, chloroform and ether. Included are the glycerophosphatides, such as phosphatidyl inositol, phosphatidyl choline or lecithin (VI), phosphatidyl serine (VII), and phosphatidyl ethanolamine (VIII), where RC=O is a long-chain fatty acyl group.

$$
\begin{array}{ccc}
\overset{\displaystyle O}{\overset{\displaystyle \|}{CH_2OCR}} & \overset{\displaystyle O}{\overset{\displaystyle \|}{CH_2OCR}} & \overset{\displaystyle O}{\overset{\displaystyle \|}{CH_2OCR}} \\
| & | & | \\
\overset{\displaystyle O}{\overset{\displaystyle \|}{CHOCR'}} & R'COOCH & R'COOCH \\
| & | & | \\
CH_2OP=O & H_2C-O-\overset{\displaystyle O}{\overset{\displaystyle \|}{P}}-CH_2CH\overset{+}{N}H_3 & H_2C-O-\overset{\displaystyle O}{\overset{\displaystyle \|}{P}}-OCH_2CH_2\overset{+}{N}H_3 \\
| & | \quad | & | \\
O & O^- \quad COO^- & O^- \\
\backslash CH_2CH_2\overset{+}{N}(CH_3)_3 & &
\end{array}
$$

Lecithin   Phosphatidyl serine   Phosphatidyl ethanolamine

VI   VII   VIII

The total quantity of phospholipids in soil is small, usually less than 5 mg P/kg, or from 0.6 to 3.0% of the soil organic P (see Table 9.10). The phospholipids in soil are undoubtedly of microbial origin.

Evidence for the presence of glycerophosphates in soil has come from the detection of glycerophosphate, choline, and ethanolamine in hydrolysates of extracted phospholipids. Phosphatidyl choline (see VI) appears to be the predominant soil phospholipid, followed by phosphatidyl ethanolamine (VIII).

APPLICATION OF ^{31}P–NMR SPECTROSCOPY

Nuclear magnetic resonance spectroscopy (i.e., ^{31}P–NMR) is a relatively new approach for characterizing soil P. Although often present in low concentrations in soil, ^{31}P is a sensitive nucleus, and the method has broad application for analysis of inorganic and organic forms of P in soil extracts. A major advantage of ^{31}P–NMR over other methods is that information can be obtained regarding organic P forms without resort to complex and time-consuming chromatographic procedures. For a discussion of NMR methodology, the reader is referred to Chapter 3.

The technique of ^{31}P–NMR has been used to obtain qualitative and quantitative estimates of the forms of P in alkaline extracts of soil;[52–57] some studies have been done by sequential extraction with several reagents, such as dilute HCl, aqueous acetylacetone, and base.[52,57]

Results of ^{31}P–NMR studies have shown that most of the organic P in soil extracts is present as orthophosphate monoesters, which includes the inositol phosphates. As noted above, these compounds constitute up to one-half of the soil organic P. Smaller amounts of the P occur as phosphate diesters (e.g., nucleic acids and phospholipids) and as phosphonates, in which the P is bound to C rather than O. In the work of Condron et al.,[52] choline phosphates were observed in some extracts. The relative proportions (%) of organic to inor-

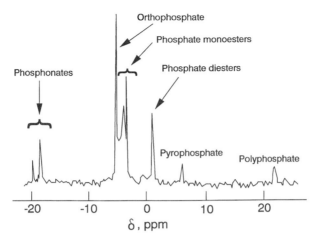

Phosphonates

Orthophosphate

Phosphate monoesters

Phosphate diesters

Pyrophosphate

Polyphosphate

-20   -10   0   10   20

δ, ppm

Fig. 9.9 ^{31}P–NMR spectrum of the alkaline extract from a New Zealand Typic Dystrochrept soil. Redrawn from Newman and Tate.[55]

ganic P in soil extracts, as determined by ^{31}P–NMR, have been found to compare favorably with those obtained by chemical analyses.[52]

Figure 9.9 shows the ^{31}P–NMR spectra of an alkali extract from a New Zealand Typic Dystrochrept soil, which illustrates the type of information that can be obtained about different forms of P.

PHOSPHORUS OF THE SOIL MICROBIAL BIOMASS

As a general rule, from 2 to 5% of the organic P of cultivated soils is present in the soil microbial biomass.[58–61] The P content of microbial cells has been reported to range from 1.5 to 2.5% for bacteria to as high as 4.8% for fungi.[58] These values are in accord with values estimated by Brookes et al.[60] for the P content of the soil biomass (1.4 to 4.7%). As one might expect, the amount of biomass P in soil will follow closely the content of biomass C (Fig. 9.10).

Since the microbial biomass is relatively "labile" when compared to other organic matter fractions, biomass P would be expected to be more available to plants than other organic P fractions. It should be noted that the P content of microbial cells is somewhat higher than that of plant tissue (0.1 to 0.5%).[13] Under P-limited conditions, microorganisms compete successfully with plants for available P, often with reduction in plant growth.

The following methods have been used to estimate biomass P in soils:

1. From cell mass based on microbial counts
2. From the inorganic P released during incubation of partially sterilized soil

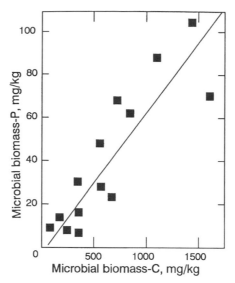

Fig. 9.10   Relationship between P and C of the soil microbial biomass. Adapted from Brookes et al.[60]

3. From ATP measurements

Calculations Based on Cell Mass

In their review on soil organic P, Halsted and McKercher[8] concluded that as much as 5 to 10% of the organic P in many soils could be associated with living microorganisms. Their estimate was based on a soil with 2 to 4% organic matter, a microbial biomass corresponding to 3.6 to 18.0 × 10^3 kg of fresh cell material per ha, and a N/organic P ratio of 10/1 for the microbial cells.

The Fumigation Method

In recent years, the quantity of P in the microbial biomass has been estimated from the amount of inorganic P that is produced during incubation of soil fumigated with chloroform ($CHCl_3$), the extractant being 0.5 M $NaHCO_3$. Biomass P is calculated from the following relationship:[58-61]

Biomass P

$$= \frac{\text{P extracted from fumigated soil} - \text{P extracted from nonfumigated soil}}{0.4}$$

The assumption is made that 40% of the P in the biomass is rendered extractable to 0.5 M $NaHCO_3$ (as inorganic P) by lysis and incubation. Also, a correction is required for phosphate fixation during extraction, as determined by adding a known amount of KH_2PO_4 during the extraction stage and measuring the recovery of added phosphate. As one might expect, measurements of biomass P in soils that contain large amounts of $NaHCO_3$-extractable phosphate will be subject to considerable error, for the reason that the percentage of the extracted P as biomass P will be very small in comparison to other P-forms.

As noted above, from 2 to 5% of the soil organic P will normally exist as biomass P. However, higher values are not uncommon, particularly for grassland soils, as shown by data obtained by Brookes et al.[60] and recorded in Table 9.11. In agreement with other work, C/P ratios for the biomass (11.7 to 35.9) are substantially lower than for the soil organic matter (i.e., compare these values also with the C/P ratios given earlier in Table 9.8). It appears evident, therefore, that the higher the percentage of the organic P as biomass P, the lower will be the C/organic ratio P of the soil.

Adenosine Triphosphate (ATP)

ATP is regarded as an indicator of live microbial biomass, and its measurement provides the basis for assays of the soil biomass, and hence biomass P.[62–65] The inherent instability of the high-energy phosphate bond precludes ATP from remaining intact outside of live cells. A survey of methods for extracting ATP from soils and sediments is given by Webster et al.[65]

TABLE 9.11   **Soil Biomass P of Some English Arable and Grassland Soils as a Percentage of Total Organic P**[a]

Agricultural History	Total Organic P, mg/kg	Biomass P		
		mg/kg	% Organic P	C/P
Arable soils				
1	180	6.0	3.3	26.3
2	210	5.3	2.5	35.9
3	242	7.0	2.9	14.1
4	810	27.5	3.4	17.9
Grassland soils				
1	330	72.3	21.9	22.5
2	352	61.7	17.5	13.7
3	190	15.0	7.9	20.0
4	210	12.0	5.7	12.3
5	500	24.8	5.0	25.6
6	200	48.6	24.3	11.7

[a]From Brookes et al.[60]

The basis for the determination of ATP in soil extracts is bioluminescence, i.e., light emission from the interaction of ATP with luciferin (LH_2), the enzyme luciferase (E), and atmospheric O_2, with the amount of light emitted being proportional to ATP content. The reactions are:

$$ATP + LH_2 + E \xrightarrow{Mg^{2+}} E\text{-}LH_2 \cdot AMP + P \cdot P$$

$$E\text{-}LH_2 \cdot AMP + O_2 \rightarrow E + \text{oxyluciferin} + AMP + \text{bioluminescence}$$

A linear relationship has been observed between ATP and biomass C, a typical result being shown in Fig. 9.11. The ATP content of the microbial biomass is on the order of 4.2 to 7.1 $\mu g/mg$ of biomass C.

The ATP content of the soil varies with season, storage of the soil, fertilizer P applications, and presence or absence of fresh plant residues. Hersman and Temple[66] concluded that ATP was the most satisfactory index of soil microbial activity in reclaimed coal strip mine spoils.

Phospholipid Analysis

The methods described above are laborious and time-consuming. A direct and simple method is desirable, one possibility being analysis for phospholipid P. In the method of Hill et al.,[67] phospholipids are extracted from the soil with chloroform, in which inorganic P compounds are insoluble and thus do not interfere with the analysis. For the soils examined, a linear relationship was found between phospholipid P and ATP content.

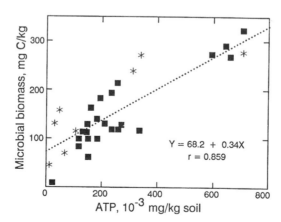

Fig. 9.11   Relationship between the microbial biomass (as biomass C) and ATP in soils. Values indicated by the symbol * are for tropical soils. Adapted from Verstraeten et al.[64]

SOIL TESTS FOR AVAILABLE P

Numerous soil tests have been developed over the years for measurement of plant-available P in soils. Some procedures vary only in minor detail; others represent extremes in extraction conditions. No single test will apply for all soils, and all methods have limitations that must be taken into account when making fertilizer P recommendations. A limitation common to most methods is that little, if any, information is provided on the rate at which the P in insoluble complexes is converted to plant-available (soluble) forms during the course of the growing season.

Dalal and Hallsworth[68] pointed out that a description of available soil P must include an intensity factor (I), a quantity factor (Q), and a capacity factor ($\Delta Q / \Delta I$), as well as rate and diffusion factors. As noted above, the immediate source of available P for plants is that contained in the soil solution—the intensity factor (I). The labile portion of the solid phase is the quantity factor (Q), which can be estimated by soil test extraction techniques. The capacity of the soil system to maintain P concentration in the solution phase as P is removed by plants ($\Delta Q / \Delta I$) is not directly determined by soil tests, nor is the rate of replenishment of soil solution P from solid phase forms.

The test procedures outlined below are those described in the 1996 edition of the Soil Science Society of America Book Series on *Methods of Soil Analysis.*[27] Additional information can be found in Fixen and Grove.[69]

Phosphorus Soluble in Water

The main objective of this method is to determine the P concentration in the soil solution that is required for optimum plant growth, the intensity factor (I). A water extract more closely approximates the P concentration in the soil solution than does the more drastic extractants discussed below, but such an extract is not easily obtained. To facilitate removal of clay through flocculation, a dilute salt solution (e.g., 0.01 M $CaCl_2$) is sometimes used. Under some soil and crop conditions, the method provides a satisfactory prediction of P fertilizer requirement. The approach would appear to be most useful for noncalcareous sandy soils, where conversion of fertilizer P to insoluble forms is minimal and recovery of the extract is not a problem.

Phosphorus Extractable with Dilute Acid-Fluoride (HCl · NH₄F)

This widely used extractant (0.025 N HCl:0.03 N NH_4F) is designed to remove easily acid-soluble P forms, largely Ca-phosphates and a portion of the Al- and Fe-phosphates. The NH_4F dissolves the latter through formation of fluoride complexes with Al and Fe. In general, the method has been found to be most successful for acidic soils ($<$ pH 6.5). Low estimates are obtained with calcareous soils due to neutralization of the acid by $CaCO_3$.

Phosphorus Soluble in NaHCO₃

Extraction of soil with 0.5 M $NaHCO_3$ (pH near 8.5) has been highly successful for predicting P availability in calcareous, alkaline, or neutral soils. This extractant decreases the concentration of Ca in solution by forming the insoluble $CaCO_3$, with the result that the concentration of phosphate in the solution is increased.

Extraction of soil with 0.5 M $NaHCO_3$ also leads to solubilization of a portion of the soil organic P, and, as mentioned earlier, the amount thus solubilized has been regarded as a quantitative measure of the potential contribution of soil organic P to plant uptake (i.e., labile organic P).[26]

Phosphorus Soluble in NH₄HCO₃-DTPA

This relatively new test was developed for simultaneous extraction of P, K, and micronutrient cations. The extractant solution is 1 M NH_4HCO_3:0.005 M diethylenetriaminepentaacetic acid (DTPA) at pH 7.6. The chelating agent (DTPA) is used for chelation and solubilization of micronutrients (i.e., Cu, Fe, Mn, and Zn). The reagent dissolves about half as much P as 0.5 M $NaHCO_3$.

Phosphorus Soluble in Dilute HCl:H₂SO₄

Soil tests involving mixed acids have been found useful for predicting P availability in soils which fix appreciable amounts of P, such as those of the Piedmont Region of North Carolina. The recommended extractant is 0.05 N HCl:0.025 N H_2SO_4.

Isotopic Dilution of ³²P

Phosphate in the soil solution exists in equilibrium with a portion of the solid-phase P. Through measuring dilution of applied ^{32}P-labeled phosphate in the aqueous phase, the combined amount of native ^{31}P in the two soil phases can be determined. The ^{31}P of the solid phase has sometimes been referred to as "labile" P.

When ^{32}P-labeled phosphate is added to a soil–water system, the following equilibrium reaction is established.

$$^{31}P_{solid} + {}^{32}P_{solution} \rightleftharpoons {}^{32}P_{solid} + {}^{31}P_{solution}$$

At equilibrium, the ratio of the two isotopes in a portion of the solid phase is equal to their ratio in solution. Therefore, $^{31}P_{solid}$ can be calculated from the relationship:

$$^{31}P_{solid} = \frac{^{32}P_{solid}}{^{32}P_{solution}} \times \, ^{31}P_{solution}$$

Experimental evidence indicates that much of the P taken up by plants is derived from the isotopically exchangeable P. Recent studies using this approach include those of Frossard et al.[70] and Morel et al.[71]

Other Approaches

Two additional methods for the estimation of plant-available P are extractions with anionic exchange resins and Fe–Al–oxide-impregnated filter paper strips. Both methods are believed to remove the more loosely bound (and available) inorganic phosphates (retained as anions on the resin or filter paper). Sharpley[72] found that, for a wide variety of soils, similar amounts of P were extracted by both methods.

General Observations Regarding Soil P Tests

Most of the methods commonly used for determining plant available P have been checked for their ability to predict P uptake by plants. Generally, it can be said that while no given test can be universally applied, one (and generally more) can be found for most soil types. Discretion must be exercised in method selection in that a given soil property (notably, pH) can negatively affect the performance of the test. In a comparison of five tests, Bates[73] reported that correlations between extractable P and P uptake by corn were highly variable and, for any given test, strongly affected by soil pH.

MICROBIAL TRANSFORMATIONS IN THE P CYCLE

Microorganisms can affect the P supply to higher plants in three different ways, namely:

1. By degradation of organic P compounds, with release of available inorganic phosphate
2. By immobilization of available phosphates into cellular material
3. By promotion of the solubilization of fixed or insoluble mineral forms of P, such as through the production of chelating agents

Mineralization and Immobilization

Turnover of P through mineralization–immobilization follows somewhat the same pattern as for N (Chapter 5) in that both processes occur simultaneously. Accordingly, the maintenance of soluble phosphate in the soil solution will depend to some extent on the magnitude of the two opposing processes.

$$\text{Organic P} \underset{\text{immobilization}}{\overset{\text{mineralization}}{\rightleftharpoons}} P_i \text{ (e.g., } H_2PO_4^-, HPO_4^{2-})$$

From an experimental standpoint, the turnover of organic P in soil is not as easily measured as that of organic N, one reason being that the end product of P mineralization (phosphate) can be removed from solution by adsorption to colloidal surfaces or by precipitation as insoluble phosphates of Ca, Fe, or Al. Furthermore, a decrease in soluble phosphates may result in some solubilization of insoluble forms of P through reestablishment of solubility equilibria, as discussed above.

Factors that affect the mineralization of organic P (and the immobilization of inorganic P) include the major influences on microbiol activity, such as temperature, moisture, aeration, soil pH, cultivation, presence of growing plants, and fertilizer P additions.[7] Seasonable changes in the inorganic P status of soils have been extensively documented,[6] and, for similar soil types having variable P contents, a positive correlation has often been observed between mineralized P and organic P content. Enwezor[74] observed a net immobilization of P for 21 of the 28 soils he examined. The amount immobilized was more closely related to the percentage of the total P in the organic form than to the total quantity of organic P. Although a poor correlation existed between immobilized P and the C/organic P ratio of the soil, there was a general

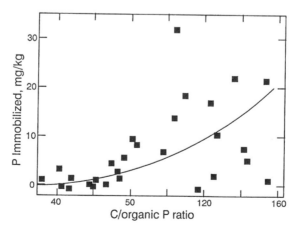

Fig. 9.12 Relationship between immobilized P and the C/organic P ratio of the soil. Adapted from Enwezor.[74]

tendency for P immobilization to be less at the lower C/organic P ratios than at the higher ratios, as can be seen by inspection of Fig. 9.12. Additions of fertilizer P can lead to increases in soil organic P through net immobilization, the effect being a lowering of the C/organic P ratio.[75]

The microbial reduction of phosphate to phosphine (PH_3) is possible under anaerobic conditions, but there is little evidence for significant evolution of this gas from waterlogged soils. Burford and Bremner[76] concluded that any phosphine formed in soils would be adsorbed by soil constituents and therefore would not escape to the atmosphere.

The P content of decaying organic residues plays a key role in regulating the quantity of soluble phosphates (P_i) in the soil at any one time. Net immobilization of P will occur when the C/organic P ratio is 300 or more; net mineralization will result when the ratio is 200 or less. In terms of the P content of crop residues, net immobilization is considered to occur when the decaying material contains less than 0.2 to 0.3% P; net release will occur at higher P contents. The relationship between the C/P ratio of crop residues and soil phosphate levels is as follows:

	C/P Ratio	
< 200	200–300	> 300
Net gain of P_i	Neither a gain nor loss of P_i	Net loss of P_i

As was the case for N (see Chapter 5), net immobilization of P during early stages of the degradation process will be followed by net mineralization of P as the C/P ratio of the decaying residue is lowered. The recycling of P in this manner is undoubtedly of tremendous importance in the P economy of many soils, particularly in those regions where soil P levels are low and where fertilizer P is in short supply or unavailable.

As noted earlier, a significant amount of the P consumed by plants, frequently on the order of one-third of the total, is contained in the nonharvested portion of the plant. Much of this organic P will become available to subsequent crops through mineralization; with mature crop residues, soil phosphate levels may temporarily decline due to net immobilization (as observed above). It should also be noted that availability of insoluble inorganic P may be enhanced during decay of plant residues.

Application of manures and other organic wastes to the soil invariably lead to changes in available P, but not in a predictable manner. Reddy et al.[77] found that manure amendments, with or without supplemental phosphate additions, initially led to decreases in available P. However, upon incubation, the amounts of available P increased to levels much higher than for the control soil (no-manure). This pattern can be attributed to an initial immobilization of available P, followed by net mineralization of biomass P, with release of inorganic P. In other work, Fossard et al.[70] found that isotopically exchangeable P either decreased or was unchanged after addition of sewage sludge to several soils.

Mineralization and immobilization of P may also be affected by the quantities of available N and S in the soil. The rate of release of P contained in crop residues is influenced by maturity, P content, application rate, time of contact of the crop residue with the soil, and kind of residue. It should be noted that most plant residues probably contain sufficient inorganic P to meet microbial needs during decay. However, a major factor affecting P availability is the extent to which the inorganic P initially present is converted to microbial organic P and how this is further transformed as decomposition proceeds.

Phosphatase Enzymes in Soil

The final stage in the conversion of organic P to inorganic phosphate in soils occurs through the action of phosphatase enzymes, both intracellular and those released into the soil solution following lysis of microbial cells.[78–79] Two different types exist in soils: acid phosphatases, which exhibit optimum activity in the pH range 4 to 6, and alkaline phosphatases, with pH optimum of 9 to 11. The subject of phosphatase activity in soil is reviewed by Tabatabai[80] and Halsted and McKercher.[8]

The general equation for the reaction catalyzed by phosphatases is:

$$R-O-\overset{\overset{\displaystyle O}{\|}}{\underset{\underset{\displaystyle O^-}{|}}{P}}-O^- + H_2O \xrightarrow{\text{phosphatase}} ROH + HPO_4^{2-}$$

Methods for the determination of phosphatase are outlined by Tabatabai.[80] The more common procedures are based on the colorimetric determination of the p-nitrophenol released by phosphatase when soil is incubated with buffered (pH 6.5 for acid phosphatase and pH 11 for alkaline phosphatase) sodium p-nitrophenyl phosphate solution. The results, which are expressed in such units as μg P/g soil/x hours, can then be correlated with other measured soil parameters or, in some cases, with soil management practices.[78,79]

For soils of pH < 7, phosphatase activity has been shown to be correlated with clay and organic C content.[78] Activity is also affected by such factors as pH, moisture, storage of the soil, P fertilization, and incubation of the soil with a C substrate. Attempts to isolate phosphatase enzymes from soil have apparently not been successful, as nothing has been reported in the literature.

Role of Microorganisms in the Solubilization of Insoluble Phosphates

As noted above, the availability of phosphate in soil is often limited by fixation reactions, which convert mono- and diphosphates (HPO_4^{2-}, $H_2PO_4^-$) to various insoluble salts. Insoluble Ca-phosphates predominate in calcareous soils, while Fe- and Al-phosphates are formed in acidic soils. Adsorption by clay minerals can affect phosphate availability under neutral or slightly acid conditions.

A discussion of the role of phosphate-solubilizing microorganisms in enhancing the availability of P in soils is given by Kucey et al.[81]

Many studies have shown that the availability of phosphate in soil is enhanced by additions of organic residues. Several independent, but not necessarily exclusive, reactions may be involved, including the following:[82]

1. Phosphorus tied up as insoluble Ca-, Fe-, and Al-phosphates may be released as soluble forms through the action of organic acids and other chelates that are produced during decomposition of crop residues and excreted from plant roots. The mechanisms involved are depicted in Fig. 9.13.

2. Humates produced during decomposition may compete with phosphate ions for adsorbing surfaces, thereby preventing fixation of phosphate.

3. Humates may form a protective surface over colloidal sesquioxides, with reduction in phosphate fixation.

4. Solubility of Ca- and Mg-phosphates may be increased through production of carbonic acid from CO_2 released during decay ($CO_2 + H_2O \rightarrow H^+ + HCO_3^-$).

5. Fresh organic matter may have a priming effect on the degradation of native humus, with mineralization of organic P.

6. Phosphohumate complexes may be formed.

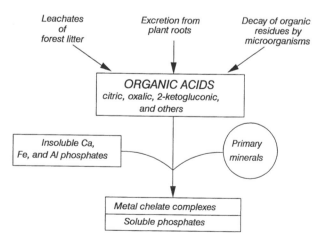

Fig. 9.13 Schematic diagram of the release of insoluble phosphates to soluble forms through the action of organic acids and other naturally occurring chelates. Adapted from Stevenson.[82]

At present, it is not possible to select any given mechanism from among these alternatives. Evidence that naturally occurring chelating agents enhance the availability of P to higher plants (item 1) is circumstantial, and some investigators have questioned whether or not organic acids and other chelating agents are produced (or persist) in sufficient quantity to appreciably influence phosphate solubility. The effectiveness of these compounds will undoubtedly be greatest in unfertilized soils low in natural fertility and where most of the P is tied up as insoluble Ca-, Fe- or Al-phosphates.

Another aspect that should be mentioned is that the slow but continued solubilization of inorganic phosphates through chelation, such as may occur during soil formation, would lead to conversion of mineral P to more available organic P forms, thereby enhancing the fertility of the soil.

The action of organic chelates in solubilizing phosphates and phosphate minerals (item 2) has been attributed to the formation of complexes with Ca, Fe, or Al, thereby releasing the phosphate in water-soluble forms. The reactions are:

$$CaX_2 \cdot 3Ca(PO_4)_2 + \text{ chelate} \rightarrow \text{ soluble } PO_4^{-2} + \text{Ca-chelate}$$
$$\text{where } X = OH \text{ or } F$$

$$Al(Fe) \cdot (H_2O)_3(OH)_2H_2PO_4 + \text{chelate} \rightarrow \text{soluble } PO_4^{2-} + \text{Al(Fe)-chelate}$$

Similar reactions may be involved in preventing the fixation of fertilizer-applied phosphate, as well as phosphate formed *in situ* by weathering of minerals or decay of organic matter. The reactions constitute a dynamic system, since microbial degradation of the chelating agent will result in the formation of soluble cations (Ca, Fe, or Al), which can then be precipitated as insoluble phosphate salts. Plant uptake of soluble P will facilitate the dissolution reaction as well. In a study of P forms in a chronosequence of soils in Hawaii, Crews et al.[14] found that Ca–phosphates predominated in the younger soils but that occluded P and organic P became the major components over time.

Numerous laboratory studies, reviewed elsewhere,[82] have shown that many organic acids are effective in releasing phosphates from insoluble mineral forms, as well as in reducing the precipitation of phosphate by Fe and Al. The most effective compounds are those capable of forming stable chelate complexes with metal ions, such as the di- and tricarboxylic hydroxy acids. The role of organic acids and other natural chelates in forming chelate complexes in soil is discussed in greater detail in Chapter 11.

The ability of soil microorganisms to solubilize insoluble phosphates and minerals of various types through the production of organic acids has been demonstrated in experiments carried out under laboratory conditions. This is normally done by first isolating the bacteria, actinomycetes, or fungi by using the agar-plate method for counting soil microorganisms. Individual isolates are selected and recultured, following which selected pure cultures are inoculated onto petri dishes containing scattered granules of insoluble mineral phosphates. A transparent halo is formed around individual microbial colonies when decomposition or solubilization occurs. In many instances, decomposition has been related to organic acid production, particularly of those compounds that form highly stable complexes with metal ions (e.g., citric acid).

Conditions that promote the activities of microorganisms would be expected to enhance the solubilization of insoluble phosphates through the production of chelating agents. Thus, it is not surprising that the addition of degradable organic materials to soil has been observed to increase phosphate uptake by plants, an effect that has been confirmed in experiments with ^{32}P-labeled products.

Gerretsen[83] is usually credited as the first to draw attention to the influence of rhizosphere microorganisms on phosphate uptake by plants. He grew a variety of crop plants in sterilized and nonsterilized soil, to which various insoluble phosphates were added, and observed greater phosphate accumulation by plants growing in nonsterilized soil. When bacteria isolated from root surfaces were put into sterile soil, enhanced phosphate uptake occurred. Normal rhizosphere roots were found to bring phosphate into solution far more effectively than uninfected roots.

Reasons given for the enhanced uptake of soil P in the rhizosphere include (1) stimulation of phosphate-solubilizing bacteria, (2) greater turnover of organic P, and (3) production of organic acids that dissolve insoluble phosphates through chelation of Ca, Fe, and Al. It should be noted that a high proportion of the microorganisms in the rhizosphere have been shown to produce organic acids when cultured on laboratory growth media (see Chapter 11).

As plant roots penetrate the soil, phosphate is absorbed at a high rate. The net result is that a zone of phosphate depletion is formed in the immediate vicinity of the root. This zone is only a few mm wide and coincides with the rhizosphere—the region near the root where microorganisms are particularly active. The net effect of the increased activity and solubilization of phosphate is that P is circulated at a higher rate, to the benefit of the plant.

Since plants contain a rich population of microorganisms around their roots, it is important to take into consideration the activities of these organisms when evaluating mechanisms for the uptake of soil P.

Role of Mycorrhizal Fungi in the Phosphorus Nutrition of Plants

A mycorrhizal association is a symbiotic relationship between the root systems of higher plants and a particular fungus. These associations are common in both crop plants and native grass, herbaceous, and woody species. In some instances, the plant either grows poorly or fails to survive unless the fungus is present. These fungi form a mantle of hyphae around the root (ectomycorrhizae), with penetration of hyphae either into surficial cortical cells of the root (ectoendomycorrhizal forms) or into deeper regions of the cortex, with formation of characteristic differentiated hyphae (arbuscules and vesicles in vesicular–arbuscular mycorrhizal (VAM) types): The hyphae can extend beyond the root surface and into the soil for substantial distances (to 70 mm), thereby increasing the soil volume from which the plant can obtain nutrients and water. Advantages of infected plants over uninfected plants include enhanced drought tolerance, greater uptake of nutrients (e.g,, N, P, and K), and improved resistance to root diseases.[84–86] The role of mycorrhizal fungi in enhancing the uptake of P by plants is particulary important in P-deficient soils. In such soils, mycorrhizal plants typically have two to three times the biomass of equal-age uninfected plants.[87,88]

Nonmycorrhizal roots have a limited ability to accumulate nutrients of low mobility, such as phosphate. For example, Jakobsen et al.[89] found that nonmycorrhizal roots of clover (*Trifolium subterraneum*) could not access ^{32}P–phosphate when the phosphate was placed 1 cm from the roots in a soil with a high phosphate adsorption capacity. In contrast, mycorrhizal roots accumulated phosphate placed as much as 7 cm from the root. The mycorrhizae facilitated uptake of P by the plant by "internalizing" the P within the hyphae and subsequently transferring the P through the hyphae to the root, thus circumventing the low mobility of phosphate in the soil. The hyphae also increased the number of contact points between soil and root, thereby increasing the volume of soil that was contacted. In Jacobsen's work, mycorrhizal roots had exploited a soil volume about 60 times larger than had nonmycorrhizal roots. Most evidence indicates that mycorrhizal fungi do not necessarily solubilize phosphate forms of low bioavailability, but rather, they act by increasing the soil–plant contact zone.

Extensive mycorrhizal development is exhibited in some forest systems (referred to as mycorrhizal mats because a substantial part of the soil volume is occupied by hyphae). In these cases, the chemistry of the soil solution, and not just the soil, is modified substantially. At first glance, the data given in Table 9.12[90] would appear to be in disagreement with information given earlier in this chapter in that Ca, Fe, and Al would be expected to react with the phosphate to form insoluble forms, thereby reducing the concentration of phosphate in the soil solution. However, in this case, the metal ions existed as *soluble chelate complexes* and insoluble phosphates were not formed (see Chapter 11); consequently, the cations (as chelate complexes) and phosphate coexisted in solution.

Reduced or delayed plant growth due to poor mycorrhizal colonization often occurs when uninfected nursery-grown plants are transplanted into the field,[91] when plants are introduced into soils that lack the appropriate fungal species, when soils are severely disturbed as a result of surface mining or landscape modification,[92,93] and, in some cases, after forest fires or clearcutting of woodlands.[94] In the absence of a suitable population of mycorrhizal species, inoculation is sometimes beneficial,[85,86] even with supplemental applications of phosphate.

TABLE 9.12   Soluble Phosphate, Ca, Fe, and Al Concentrations in Soil Solutions with or without Mycorrhizal Mats Formed by the Fungi *Hysterangium* or *Gautieria* with 42-Year-Old Douglas Fir (*Pseudotsuga menziesii*)[a]

	May Sample		August Sample	
Analyte	Mat	Nonmat	Mat	Nonmat
		mg/L		
PO_4^{3-}	52–62[a]	3–35	29–225	8–27
Ca^{2+}	710–2000	110–120	1700–2400	220–270
Fe (soluble)	39–320	6–16	98–117	36–72
Al^{3+}	353–8044	55–56	1094–2354	174–259

[a] Values for the mycorrhizal mats were significantly greater than for the non-mat solutions ($P \leq 0.05$) except for Fe in the soil solution for the mycorrhizal mat for *Hysterangium* at the August sampling date. From Griffiths et al.[90]

Inoculation of Soil with "Phosphobacterin"

Inoculation of soil with *Bacillus megatherium var. phosphaticum*, for which the term *phosphobacterin* is used, was a common practice in Russia for many years. The phosphobacterin was said to benefit 50 to 70% of the field crops to which it was applied. The organism, which converts organic P to available mineral forms, was reported to be particularly effective in Mollisols, which contain much organically bound P.

Early claims for beneficial effects through inoculation of soil with phosphobacterin were not substantiated by subsequent work (e.g., see Raf et al.[95]), and the practice cannot be recommended.

MODELING OF THE SOIL P CYCLE

Jones et al.[13] have designed a soil and plant P model (Fig. 9.14) based on pools of stable, active, and labile inorganic P; fresh organic and stable organic P; and grain, stover, and root P. The model simulates P uptake by plants in relation to soil chemical and physical properties, crop P requirements, tillage practice, fertilizer rate, soil temperature, and soil water content. Initial values for active inorganic P and stable organic P are calculated from chemical and/or adsorption properties. Estimates for initial labile P can be included, such as may be obtained from soil test data (i.e., P extracted with 0.5 M $NaHCO_3$). Promising results were obtained when the model was used to simulate long-time changes in soil organic P, soil test values, and yields of corn and wheat at several locations in the Great Plains of the United States.[13] Other P models have been devised by Cole et al.[96] and Smith et al.[97]

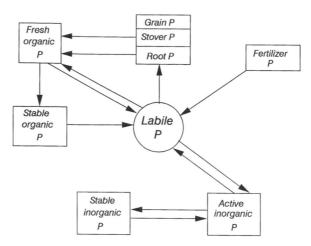

Fig. 9.14   Pools and flows of P in the soil–plant system. The pool of labile inorganic P supplies the plant, in which P is divided into root, stover, and grain. Organic P is partitioned into the fresh residue P, consisting of the P in microorganisms and undecayed residues, and the stable organic P. Adapted from Jones et al.[13]

SUMMARY

The phosphorus cycle is complex and involves the storage of P in living organisms, dead organic matter, and inorganic forms. Phosphorus is absorbed by plants largely as orthophosphate ($H_2PO_4^-$ or HPO_4^{2-}) which is present in the soil solution. However, replenishment of this P by solubilization of inorganic forms, or by mineralization of organic P, is of considerable importance from the standpoint of plant nutrition. A particularly useful concept is that the P in soil can be partitioned into "pools" based on availability of various organic and inorganic forms to plants.

Excellent progress has been made in determining the nature of organic P in soils, but much of the organic P has yet to be identified in known compounds. As a general rule, somewhat less than one-half of the organic P can be accounted for. Approximate recoveries of organic P are: inositol phosphates, 2 to 50%; phospholipids, 1 to 5%; nucleic acids, 0.2 to 2.5%; phosphoproteins, trace; metabolic phosphates, trace.

Fixation reactions of various types are important from the standpoint of the efficiency of fertilizer P use by plants. In acid soils, phosphate is precipitated as the highly insoluble Fe- or Al-phosphates or adsorbed to oxide surfaces. A major factor affecting P availability in calcareous soils is the formation of insoluble Ca-phosphates. Mycorrhizal fungi often play an important role in enhancing the availability of phosphate to higher plants.

As much as two-thirds or more of the total P in soils occur in organic forms, with the higher percentages being typical of peats, uncultivated forest soils, and tropical soils. An active P cycle exists in soil, in which turnover between organic and inorganic forms (mineralization–immobilization) plays a major role. Addition of degradable organic matter to soil has been shown in laboratory and greenhouse studies to modify the availability of insoluble Ca-, Fe-, and Al-phosphates to plants, but the importance of the process under conditions existing in the field is unknown.

REFERENCES

1. J. W. B. Stewart, "The Importance of P Cycling and Organic P in Soils," in *Better Crops with Plant Food: Winter Issue,* American Potash–Phosphate Institute, Atlanta, 1980–1981, pp. 16–19.
2. J. E. Richey, "The Phosphorus Cycle," in B. Bolin and R. B. Cook, Eds., *The Major Biogeochemical Cycles and their Interactions,* Wiley, New York, 1983, pp. 51–56.
3. V. Sauchelli, *Manual on Phosphates in Agriculture,* Davidson Chemical Corp., Baltimore, 1951.
4. T. W. Walker, "The Significance of Phosphorus in Pedogenesis," in *Experimental Pedology,* Easter School in Agricultural Sciences, University of Nottingham, England, 1964, pp. 295–316.
5. F. E. Khasawneh, E. C. Sample, and E. J. Kamprath, Eds., *The Role of Phosphorus in Agriculture,* American Society of Agronomy, Madison, Wisconsin, 1980.
6. J. Magid, H. Tiessen, and L. M. Condron, "Dynamics of Organic Phosphorus in Soils under Natural and Agricultural Ecosystems," in A. Piccolo, Ed., *Humic Substances in Terrestrial Ecosystems,* Elsevier, Amsterdam, 1996, pp. 429–466.
7. R. C. Dalal, *Adv. Agron.,* **29,** 83 (1977).
8. R. L. Halstead and R. B. McKercher, "Biochemistry and Cycling of Phosphorus," in E. A. Paul and A. D. McLaren, Eds., *Soil Biochemistry,* Vol. 4, Marcel Deckker, New York, 1975, pp. 31–63.
9. S. K. Sanyal and S. K. De Datta, *Adv. Soil Sci.,* **16,** 1 (1991).
10. J. W. B. Stewart and A. N. Sharpley, "Controls on Dynamics of Soil and Fertilizer Phosphorus and Sulfur," in R. F. Follett, J. W. B. Stewart, and C. V. Cole, Eds., *Soil Fertility and Organic Matter as Critical Components of Production Systems,* American Society of Agronomy, Madison, Wisconsin, 1987 pp. 101–121.
11. A. N. Sharpley and R. G. Menzel, *Adv. Agron.,* **41,** 297 (1987).
12. G. M. Pierzynski, J. T. Sims, and G. F. Vance, *Soils and Environmental Quality,* Lewis, Boca Raton, Florida, 1994, Chapter 5, pp. 103–141.
13. C. A. Jones, C. V. Cole, A. N. Sharpley, and J. R. Williams, *Soil Sci. Soc. Amer. J.,* **48,** 800, 805, 810 (1984).
14. T. E. Crews, K. Kitayama, J. H. Fownes, R. H. Riley, D. A. Herbert, D. Mueller-Dombois, and P. O. M. Vitousek, *Ecology,* **76,** 1407 (1995).
15. A. Finck, *Fertilizers and Fertilization,* Verlag Chemie, Deerfield Beach, Florida, 1982.

16. R. W. Simonson, "Loss of Nutrient Elements During Soil Formation," in O. P. Engelstad, Ed., *Nutrient Mobility in Soils: Accumulations and Losses,* Soil Science Society of America, Madison, Wisconsin, 1970, pp. 21–45.

17. A. Oberson, J. C. Fardeau, J. M. Besson, and H. Sticher, *Biol. Fert. Soils,* **16,** 111 (1993).

18. H. J. Haas, D. L. Grunes, and G. A. Reichman, *Soil Sci. Soc. Amer. Proc.,* **25,** 214 (1961).

19. H. Tressen, J. W. B. Stewart, and J. R. Bettany, *Agron. J.,* **74,** 831 (1982).

20. S. A. Barber, *Commun. Soil Sci. Plant Anal.,* **10,** 1459 (1979).

21. R. E. McCullum, *Agron. J.,* **83,** 77 (1991).

22. J. C. Ryden, J. K. Syers, and R. F. Harris, *Adv. Agron.,* **25,** 1 (1973).

23. B. J. Andraski, D. H. Mueller, and T. C. Daniel, *Soil Sci. Soc. Amer. J.* **49,** 1523 (1983).

24. R. Thomas and J. P. Law, "Properties of Waste Waters," in L. F. Elliott and F. J. Stevenson, Eds., *Soils for Management of Organic Wastes and Waste Waters,* American Society of Agronomy, Madison, Wisconsin, 1977, pp. 47–72.

25. S. C. Chang and M. L. Jackson, *Soil Sci.,* **84,** 133 (1957).

26. M. J. Hedley, J. W. B. Stewart, and B. S. Chauhan, *Soil Sci. Soc. Amer. J.,* **46,** 970 (1982).

27. S. Kuo, "Phosphorus," in D. L Sparks, Ed., *Methods of Soil Analysis, Part 3, Chemical Methods,* Soil Science Society of America, Madison, Wisconsin, 1996, pp. 869–919.

28. M. A. Beck and P. A. Sanchez, *Soil Sci.,* **34,** 1424 (1994).

29. R. H. McKenzie, J. W. B. Stewart, J. F. Dormaar, and G. B. Schaalje, *Can. J. Soil Sci.,* **72,** 569, 581 (1992).

30. J. E. Richards, T. E. Bates, and S. C. Sheppard, *Can. J. Soil Sci.,* **75,** 311 (1995).

31. J. P. Schmidt, S. W. Buol, and E. J. Kamprath, *Soil Sci. Soc. Amer. J.,* **60,** 1168 (1996).

32. I. P. O'Halloran, J. W. B. Stewart, and R. G. Kachanoski, *Can. J. Soil Sci.,* **67,** 147 (1987).

33. S. A. Huffman, C. V. Cole, and N. A. Scott, *Soil Sci. Soc. Amer. J.,* **60,** 1095 (1996).

34. A. N. Sharpley, *Soil Sci. Soc. Amer. J.,* **60,** 1459 (1996).

35. J. H. Kyle, A. M. Posner, and J. P. Quirk, *J. Soil Sci.,* **26,** 32 (1975).

36. R. L. Parfitt, R. J. Atkinson, and R. S. C. Smart, *Soil Sci. Soc. Amer. Proc.,* **39,** 837 (1975).

37. R. L. Parfitt, A. R. Fraser, J. D. Russell, and V. C. Farmer, *J. Soil Sci.,* **28,** 40 (1977).

38. J. C. Ryden, J. R. McLaughlin, and J. K. Syers, *J. Soil Sci.,* **28,** 62, 72, 585 (1977).

39. F. Borie and H. Zunino, *Soil Biol. Biochem.,* **15,** 599 (1983).

40. W. A. Stoop, *Geoderma,* **31,** 57 (1983).

41. U. Kafkafi, A. M. Posner, and J. P. Quirk, *Soil Sci. Soc. Amer. Proc.,* **31,** 348 (1967).

42. G. F. Vance, F. J. Stevenson, and F. J. Sikora, "Naturally Occurring Aluminum–Organic Complexes," in G. Sposito, Ed., *The Environmental Chemistry of Aluminum,* 2nd ed., CRC Press, Boca Raton, Florida, 1996, pp. 169–220.

43. R. L. Fox, "External Phosphorus Requirements of Crops," in R. H. Dowdy et al., Eds., *Chemistry in the Soil Environment,* American Society of Agronomy, Madison, Wisconsin, 1981, pp. 223–239.

44. R. W. Pearson and R. W. Simonson, *Soil Sci. Soc. Amer. Proc.,* **4,** 162 (1939).

44a. R. T. Baker, *J. Soil Sci.* **27,** 504 (1976).

45. M. K. John, P. N. Sprout, and C. C. Kelley, *Can. J. Soil Sci.,* **45,** 87 (1964).

46. A. M. L. Neptune, M. A. Tabatabai, and J. J. Hanway, *Soil Sci. Soc. Amer. Proc.,* **39,** 51 (1975).

47. L. L. Somani and S. W. Sarena, *Anales de Edafologia y Agrobiologia,* **37,** 809 (1978).

48. T. W. Walker and A. F. R. Adams, *Soil Sci.,* **85,** 307 (1958).

49. C. H. Williams, E. G. Williams, and N. M. Scott, *J. Soil Sci.,* **11,** 334 (1960).

50. A. G. Caldwell and C. A. Black, *Soil Sci. Soc. Amer. Proc.,* **22,** 290, 293, 296 (1958).

51. D. J. Cosgrove, *Inositol Phosphates,* Elsevier, New York, 1980.

52. L. M. Condron, K. M. Goh, and R. H. Newman, *J. Soil Sci.,* **36,** 199 (1985).

53. L. M. Condron, E. Frossard, H. Tiessen, R. H. Newman, and J. W. B. Stewart, *J. Soil Sci.,* **41,** 41 (1990).

54. K. R. Tate and R. H. Newman, *Soil Biol. Biochem.,* **14,** 191 (1982).

55. R. H. Newman and K. R. Tate, *Commun. Soil Sci. Plant Anal.,* **19,** 47 (1980).

56. G. E. Hawkes, D. S. Powlson, E. W. Randall, and K. R. Tate, *J. Soil Sci.,* **35,** 35 (1984).

57. F. Gil-Sotress, W. Zech, and H. G. Alt., *Soil Biol. Biochem.,* **22,** 97 (1990).

58. J. P. E. Anderson and K. H. Domsch, *Soil Sci.,* **130,** 211 (1980).

59. P. C. Brookes, D. S. Powlson, and D. S. Jenkinson, *Soil Biol. Biochem.,* **14,** 319 (1982).

60. P. C. Brookes, D. S. Powlson, and D. S. Jenkinson, *Soil Biol. Biochem.,* **16,** 169 (1984).

61. M. J. Hedley and J. W. B. Stewart, *Soil Biol. Biochem.,* **14,** 377 (1982).

62. G. P. Sparling and F. Eiland, *Soil Biol. Biochem.,* **4,** 227 (1983).

63. K. R. Tate and D. S. Jenkinson, *Soil Biol. Biochem.,* **14,** 331 (1982).

64. L. M. J. Verstraeten, K. De Coninck, and K. Valassak, *Soil Biol. Biochem.,* **15,** 391, 397 (1983).

65. J. J. Webster, G. J. Hampton, and F. R. Leach, *Soil Biol. Biochem.,* **16,** 335 (1984).

66. L. E. Hersman and K. L. Temple, *Soil Sci.,* **127,** 70 (1979).

67. T. C. J. Hill, E. F. McPherson, J. A. Harris, and P. Birch, *Soil Biol. Biochem.,* **25,** 1779 (1993).

68. R. C. Dalal and E. G. Hallsworth, *Soil Sci. Soc. Amer. J.,* **40,** 541 (1976).

69. P. E. Fixen and J. H. Grove, "Testing Soils for Phosphorus," in R. L. Westerman, Ed., *Soil Testing and Plant Analysis,* American Society of Agronomy, Madison, Wisconsin, 1990, pp. 141–180.

70. E. Frossard, D. Lopez-Hernandez, and M. Brossard, *Soil Biol. Biochem.,* **28,** 857 (1996).

71. C. Morel, H. R. Tiessen, J. O. Moir, and J. W. B. Stewart, *Soil Sci. Soc. Amer. J.,* **58,** 1439 (1994).

72. A. N. Sharpley, *Soil Sci. Soc. Amer. J.,* **55,** 1038 (1991).

73. T. E. Bates, *Commun. Soil Sci. Plant Anal.,* **21,** 1009 (1990).

74. W. O. Enwezor, *Soil Sci.,* **103,** 62 (1967).

75. A. van Diest, *Plant Soil,* **29,** 241, 248 (1968).

76. J. R. Burford and J. M. Bremner, *Soil Biol. Biochem.,* **4,** 489 (1972).

77. K. S. Reddy, A. S. Rao, and P. N. Takkar, *Biol. Fert. Soils,* **22,** 279 (1996).

78. A. F. Harrison, *Soil Biol. Biochem.,* **15,** 93 (1983).

79. R. E. Malcolm, *Soil Biol. Biochem.,* **15,** 403 (1983).

80. M. A. Tabatabai, "Soil Enzymes," in A. L. Page, R. H. Miller, and D. R. Keeney, *Methods of Soil Analysis: Part 2, 2nd ed.,* American Society of Agronomy, Madison, Wisconsin, 1982, pp. 903–947.

81. R. M. N. Kucey, H. H. Janzen, and M. E. Leggett, *Adv. Agron.,* **42,** 199 (1989).

82. F. J. Stevenson, *Humus Chemistry: Genesis, Composition, Reactions,* 2nd ed. Wiley, New York, 1986.

83. F. C. Gerretsen, *Plant Soil,* **1,** 51 (1948).

84. W. L. Pritchett and R. F. Fisher, *Properties and Management of Forest Soils,* Chapter 10, Wiley, New York, 1987, pp. 165–178.

85. G. J. Bethlenfalvay and R. G. Linderman, Eds., *Mycorrhizae in Sustainable Agriculture,* American Society of Agronomy, Madison, Wisconsin, 1992.

86. G. R. Safir, *Ecophysiology of VA Mycorrhizal Plants,* CRC Press, Boca Raton, Florida, 1987.

87. S. Al-Nahidh and F. E. Sanders, "Comparison Between Vesicular–Arbuscular Mycorrhizal Fungi with Respect to the Development of Infection and Consequent Effects on Plant Growth," in G. F. Pegg and P. G. Ayres, Eds., *Fungal Infection of Plants,* Cambridge University Press, Cambridge, 1987, pp. 239–252.

88. J. M. Barea, R. Azcón, and C. Azcón-Aguilar, *Adv. Plant Path.* **9,** 167 (1993).

89. I. Jakobsen, L. K. Abbott, and A. D. Robson, *New Phytol.,* **120,** 509 (1992).

90. R. P. Griffiths, J. E. Baham, and B. A. Caldwell, *Soil Biol. Biochem.,* **26,** 331 (1994).

91. D. H. Marx and C. E. Cordell, "Specific Ectomycorrhizae Improve Reforestation and Reclamation in the Eastern United States," in M. Lalonde and Y. Piche, Eds., *Canadian Workshop on Mycorrhizae in Forestry,* Université Laval, Sainte Foy, Québec, 1988, pp. 75–86.

92. D. A. Jasper, L. K. Abbott, and A. D. Robson, *Australian J. Botany,* **37,** 33 (1989).

93. R. M. Miller and J. D. Jastrow, "The Application of VA Mycorrhizae to Ecosystem Restoration and Reclamation," in M. F. Allen, Eds., *Mycorrhizal Functioning,* Chapman & Hall, New York, 1992, pp. 438–467.

94. M. P. Amaranthus, "Mycorrhizas, Forest Disturbance and Regeneration in the Pacific Northwestern United States," in D. J. Read, D. H. Lewis, A. H. Fitter, and

I. J. Alexander, Eds., *Mycorrhizas in Ecosystems,* CAB International, Wallingford, England, 1992, pp. 202–207.

95. J. Raf, D. J. Bagyaraj, and A. Manjunath, *Soil Biol. Biochem.,* **13,** 105 (1981).

96. C. V. Cole, G. S. Innis, and J. W. B. Stewart, *Ecol.,* **58,** 1 (1977).

97. O. L. Smith, *Soil Biol. Biochem.,* **11,** 585 (1979).

10

THE SULFUR CYCLE

Sulfur, the 10th most abundant element in the universe (about 5th by weight) occurs naturally as gypsum ($CaSO_4 \cdot 2H_2O$); as pyrite (FeS_2) in shales, coal, limestone, and sandstone; as elemental S^0 in bituminous shales and various sedimentary traps, such as salt domes; and as part of soil and marine humus. Sulfur is a constituent of all living organisms, where it is a constituent of proteins and some polysaccharides. Sulfur is also an integral component of several vitamins and hormones.

Despite its essentiality, S has been described as the neglected plant nutrient. In comparison to N, only sporadic attention has been given to this nutrient because few crops give positive yield responses to applied S, and those in only a few geographical areas. Sulfur deficiencies are becoming increasingly more common and are being seen in a wider variety of crops, a result that has been attributed to increased use of high-analysis S-free fertilizers, a reduction in the amount of S used as a pesticide, and higher crop yields, which means that requirements for all essential plant nutrients will be greater. Another contributing factor is a reduction in the amount of S that reaches the soil in precipitation or dry deposition from the atmosphere. This reduction is the result of using effective emission-control systems for reducing the sulfur dioxide (SO_2) content of the atmosphere through burning of fossil fuels. Due to the considerations noted above, the size of S-deficient areas may increase and new deficient areas may develop, particularly in rural areas that are distant from major industrial areas and electricity-generating plants.

On a localized basis, S is a serious environmental hazard, particularly at abandoned mine sites, where microbial and chemical reactions convert pyrite (FeS_2) to sulfuric acid (H_2SO_4) in sufficient quantities to decrease soil and aquatic pH to levels that are too low to support most lifeforms. In minute amounts, some volatile S-containing compounds are extremely malodorous, and they are a major nuisance that has required establishment of constraints on activities that generate them in large quantities.

A pictorial representation of the S cycle is shown in Fig. 10.1.[1] During weathering, the S of primary minerals is converted to sulfate (SO_4^{2-}), which, in turn, is used by plants and microorganisms and converted into organic forms, such as the cysteine and methionine of proteins and sulfates of polysaccharides. When plant and animal residues are returned to the soil and subjected to microbial decay, part of the organic S reappears as H_2SO_4; part is incorporated into microbial biomass, and hence into humus. Sulfur is added to soil in fertilizers, in pesticides of various types, in irrigation water, in precipitation, and through absorption of S gases (e.g., SO_2) from the atmosphere. Losses of soil S result from runoff and leaching. Under certain circumstances, soil and anaerobic environments such as sediments and wetlands constitute a source of hydrogen sulfide (H_2S) and other S gases to the atmosphere.

The importance of organic matter in providing S to growing plants will, of course, depend on the adequacy of external additions in meeting crop requirements. As far as atmospheric S is concerned, rainfall deposition ranges from less than 5 kg/ha/yr in rural areas remote from industrial activity to as much as 200 kg/ha/yr in areas of concentrated industrial activity.

In this chapter, a brief examination is given of pertinent facts regarding the S cycle in soil. Additional data and documentation can be found in numerous monographs and reviews covering such topics as biogeochemistry and S cycling,[2-6] plant nutrition,[7-9] forms and reactions in soil,[10-15] microbial ox-

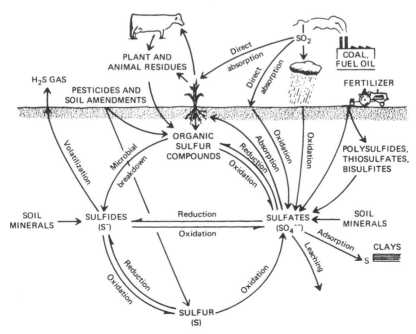

Fig. 10.1 Pictorial representation of the S cycle in soils. From Brady,[1] reprinted with permission of Macmillan Publishing Co.

idations and reductions,[16-17] and biogenic S emissions.[18] Special attention has been given to the dynamics of S cycling in forest ecosystems.[19]

BIOGEOCHEMISTRY OF THE GLOBAL S CYCLE

Major reservoirs of S in the earth (Table 10.1) are the lithosphere (24.3 × 10^{18} kg) and the hydrosphere (1.3 × 10^{18} kg). Moderate amounts of S occur in soils (2.6 × 10^{14} kg); lesser amounts are found in the atmosphere (4.8 × 10^9 kg) and land plants (7.6 × 10^{12} kg). Fluxes of S among some of the more important components in a typical agricultural soil are as follows:[15]

	kg/ha/yr
Plant uptake	10–50
Removal in harvested portion	5–20
Removal in animal products	0–5
Atmospheric S inputs	2–20
Leaching	0–50
Weathering of S-bearing minerals	0–5
Volatile S emissions	unknown

Atmospheric inputs of S to the soil vary greatly from location to location. In remote continental areas, the annual deposition of S is less than 5 kg/ha; in other areas, much higher amounts will be found. In western and central Europe, for example, annual deposition has been estimated to range from 10 to 100 kg/ha.[4] Even higher amounts are added to soils near industrial plants where fossil fuels are used. Sulfur is lost from the soil (as well as plants) as H_2S and volatile organic compounds of various types.[5]

TABLE 10.1   Major S Reserves in the Earth[a]

Reservoir	Amount of S, kg
Atmosphere	4.8 × 10^9
Lithosphere	24.3 × 10^{18}
Pedosphere	
Soil	2.6 × 10^{14}
Soil organic matter	0.1 × 10^{14}
Land plants	7.6 × 10^{11}
Hydrosphere	
Oceans	1.3 × 10^{18}
Fresh water	3.0 × 10^{12}
Marine organisms	2.4 × 10^{11}

[a]The lithosphere refers to igneous, metamorphic, and sedimentary rocks of the earth's crust.

Principal geologic sources of S for industrial uses include elemental S^0 deposits associated with salt domes and sedimentary rocks. Total annual production of elemental S^0 in the United States exceeds 5×10^9 kg, most of which is converted to H_2SO_4 for use in agricultural, chemical, and related industries.[20] Known reserves of mineable S in the United States exceed 200×10^9 kg.

Chemical Properties of S

Sulfur, with an atomic weight of 32.064, exists in several oxidation states, the lowest being -2 in sulfide (gaseous H_2S; ferrous sulfide, FeS) and the highest being $+6$ in sulfate (SO_4^{2-}). Important inorganic S species in the environment are given in Table 10.2. Common oxidation states in soil and biological materials are -2 for sulfide ion and most organic S, and $+6$ for sulfate and sulfate-containing biochemicals. Oxidation and reduction of S occurs rather easily under proper conditions, thereby accounting for the great diversity of reactions that S undergoes in the environment.

Sulfur exists in four stable isotopes—^{32}S, ^{33}S, ^{34}S, and ^{36}S—with natural abundances of about 95.1%, 0.74%, 4.2%, and 0.017%, respectively. The $^{34}S/^{32}S$ ratio of naturally occurring S compounds is highly variable ($\pm 4.5\%$ from a mean value),[20] in part because biological and chemical reactions display rather large isotope discrimination effects for the various isotopes. Ester sulfate forms of S in soils were found by Schoenau and Bettany[21] to be enriched in ^{34}S, as compared to C-bonded forms. Sedimentary SO_4^{2-} is normally enriched with ^{34}S as compared to other deposits.[20]

Radioactive isotopes are ^{31}S, ^{35}S, and ^{37}S. The use of ^{35}S to follow S transformations in soil is discussed below.

TABLE 10.2 Important Inorganic S Species and Their Oxidation State in the Environment

Sulfides, -2		Thiosulfates, -2 and $+6$	(avg. $+2$)
Sulfide ion	S^{2-}	Thiosulfate ion	$+2S_2O_3^{2-}$
Bisulfide ion	HS^-	Bithiosulfate ion	$HS_2O_3^-$
Hydrogen sulfide (gas)	H_2S	Thiosulfuric acid	$H_2S_2O_3$
Carbonyl sulfide (gas)	COS		
Polysulfides, -1		Sulfites, $+4$	
Disulfide	S_2^{2-}	Sulfite ion	SO_3^{2-}
Pyrite	FeS_2	Hydrogen sulfite	HSO_3^-
		Sulfur dioxide (gas)	SO_2
Elemental, 0			
Sulfur	S^0	Sulfates, $+6$	
		Sulfate ion	SO_4^{2-}
		Bisulfate ion	HSO_4^-
		Sulfuric acid	H_2SO_4

Origin of S in Soils

The principal original source of S in soils is the pyrite (FeS_2) of igneous rocks. During weathering and soil formation, the S of pyrite undergoes oxidation to the SO_4^{2-} form, which is ultimately assimilated by plants and microorganisms and incorporated into the soil organic matter. In some soils, part of the S is retained as gypsum ($CaSO_4 \cdot 2H_2O$) and epsomite ($MgSO_4 \cdot 7H_2O$), or leached. In dry regions, where rainfall is insufficient to leach SO_4^{2-} from the soil profile, gypsum often accumulates in a horizon below a zone of $CaCO_3$ accumulation.

Sulfur Content of Soils

Soils vary greatly in their content of S, being lowest in those developed from sands (~20 mg S/kg) and highest in those developed in tidal areas, where sulfides have accumulated (~35,000 mg S/kg). The normal range in agricultural soils of humid and semihumid regions is 100 to 500 mg/kg, or 0.01 to 0.05% S (224 to 1120 kg/ha within the plow layer). For the soils of these regions, most of the S is tied up in organic forms, and the amounts present will follow closely that for organic C and will be related to those factors affecting organic matter content, as discussed in Chapter 2. Tropical soils generally contain low amounts of S, which can be explained by their low organic matter content.

The long-term effect of cropping on S levels (without additions of agricultural chemicals) is a decrease in the total amount present, due to mineralization of organic matter and subsequent plant removal and leaching. As noted below, the amount of S consumed by plants is highly variable but is generally on the order of 10 to 30 kg/ha. However, a portion of the S in normally returned to the soil as crop residues. In grazed pastures, S removal in animal products (meat, wool, etc.) is rather low (3–5 kg/ha), the remainder being returned as residues or excreta.[22] In forest ecosystems, much of the S taken up is returned to the soil via litter and turnover is about 3 to 7 kg/ha/yr (see Howarth et al.[3]).

As noted in Chapter 2, losses of organic matter through cultivation have ranged from as little as 15% in well-managed soils to well over 50% in soils subject to continuous and intensive cropping. Losses of organic S have undoubtedly been of the same order of magnitude. At steady-state levels of organic matter (usually attained after several decades of cropping), losses of soil S through crop removal and leaching must equal gains of S through atmospheric deposition, return in crop residues, or additions in S-containing fertilizers and pesticides. Studies using lysimeters installed at the Rothamsted Experimental Station in England over a century ago have shown that the amount of S deposited from the atmosphere has been about equal to the amount of SO_4^{2-} lost in drainage waters.[23] In this case, S must be added from some external source to avoid S-depletion by removal of harvested crops.

Agricultural practices that result in increases in the organic matter content of the soil, such as long-time applications of organic residues, can increase S content. Larson et al.[24] found that the S (and C) content of the soil increased in proportion to the amount of plant residues added; a 15% increase in S content was obtained by the annual application of 16 tons of plant residues per hectare over an 11-year period.

Sulfur in the form of $CaSO_4$ is a constituent of phosphate fertilizers. For this reason, S deficiencies seldom occur on soils well fertilized with P. Several "controlled-release" fertilizers contain elemental S^0 as a coating, the best-known example being S-coated urea. A coating of about 20% elemental S^0 is normally used, thereby providing a source of SO_4^{2-} to plants through microbial oxidation (discussed below).

ENVIRONMENTAL ASPECTS OF THE S CYCLE

Environmental concern regarding the S cycle centers on:

1. Inputs of SO_2 into the atmosphere
2. Leaching of SO_4^{2-}
3. Acid sulfate soils
4. Formation of H_2SO_4 from pyrite in mine spoils

Atmospheric Sulfur Dioxide

There has been much interest in the flux of S gases between soil and the atmosphere.[5,18,25,26] These constituents have the potential for adversely affecting climate and the environment through their aerosol-forming properties.

Some items of importance regarding emissions of S gases into the atmosphere are as follows:

1. Sulfur dioxide (SO_2) and other miscellaneous S compounds are emitted into the atmosphere from power and industrial plants that burn fossil fuels, as well as from natural sources such as volcanic activity. Atmospheric SO_2 may be absorbed directly from the air or washed into the soil in rain water. The gas readily combines with water to form *sulfurous* acid ($SO_2 + H_2O \rightarrow H_2SO_3$), but this acid is soon oxidized to H_2SO_4 by chemical reactions (in air) or by S-oxidizing microorganisms in soil.
2. Anthropogenic input of S into the atmosphere (e.g, by combustion of fossil fuels) is about one-half as much as from natural sources.[6a] In highly industrial regions (e.g., Europe, parts of eastern Asia, and the eastern portion of North America), emissions of S gases into the atmosphere have probably increased by at least one order of magnitude over the past century due to burning of fossil fuels. The net effect of

this increase is that, because of fallout of atmospheric S, the S content of many lakes has nearly doubled. Point sources can emit incredible quantities of SO_2 and SO_4^{2-}; one smelter in Canada, for example, has been reported to emit about 1.1 million metric tons of SO_2 annually.[27]

3. Biologically produced S compounds represent a significant contribution to S entering the atmosphere each year. Most of the biologically produced H_2S arises from coastal areas or land surfaces, and this H_2S accounts for about 2×10^9 kg/yr of the estimated 106×10^9 kg/yr of S believed to be emitted into the atmosphere by biological processes. An alternative biological source of atmospheric S is dimethyl sulfide (CH_3—S—S—CH_3).[25] In considering the soils contribution to S gases in the atmosphere, one must keep in mind that most soils are able to adsorb S gases of various types.

Leaching of Sulfate

The sulfate ion (SO_4^{2-}), being negatively charged, is subject to leaching. The extent to which leaching occurs depends on rainfall, SO_4^{2-} retention capacity of the soil, drainage characteristics, and immobilization by microorganisms during decay of plant residues. As one might expect, leaching is greatest in coarse-textured soils under high rainfall. Due to leaching, SO_4^{2-} seldom accumulates in soils of humid and semihumid regions. Leaching of organic S may also be of importance under certain circumstances.[3,28]

Acid Sulfate Soils

Drainage of tidal marshes, as well as exposure of acid-forming underlying clays of coal-bed outcrops, causes H_2SO_4 to form from pyrite (FeS_2), which results in extremely acid soils (pH's as low as 3.5 are not uncommon). Thus, when wetland areas (e.g., marshes and swamps) are drained and reclaimed for agricultural use, serious problems arise due to the exceptional acidity that results when pyrite is oxidized to H_2SO_4. Such soils (sometimes known as "catclays") are called "acid sulfate" soils, referring to soils that have been drained, contain free and adsorbed SO_4^{2-}, show pale yellow mottles of jarosite [$KFe_3(SO_4)_2(OH)_6$], and usually have a pH below 4 in water.[29] A review of pyrite oxidation chemistry is by Evangelou.[30] Oxidation is partly chemical (chem) and partly biological (bio), as seen in the following sequence:

$$2FeS_2 + 2H_2O + 7O_2 \xrightarrow{\text{chem or bio}} 2Fe^{2+} + 4H^+ + 4SO_4^{2-}$$

$$2Fe^{2+} + O_2 + 4H^+ \xrightarrow{\text{chem or bio}} 2Fe^{3+} + 2H_2O$$

$$2Fe^{3+} + FeS_2 \xrightarrow{\text{chem}} 3Fe^{2+} + 2S^0$$

$$2S^0 + 3O_2 + 2H_2O \xrightarrow{\text{bio}} 4H^+ + 2SO_4^{2-}$$

The overall reaction is:

$$3FeS_2 + 11O_2 + 2H_2O \rightarrow 3Fe^{3+} + 4H^+ + 6SO_4^{2-}$$

Sulfates from sea water are the original source of sulfides in acid sulfate soils of deltaic and coastal regions. In inland swamps, the sulfides are derived from the FeS_2 of surrounding rocks through oxidation and movement of soluble SO_4^{2-} salts to the lower-lying areas in drainage waters. In each case, bacterial reduction of SO_4^{2-} (discussed below) produces the sulfides from which SO_4^{2-} is regenerated.

Coastal deposits of sulfide-bearing muds occur over broad areas, particularly in the tropics, a typical example being Malaysia, where deposits up to 60 km (40 miles) wide and 105 m (450 ft) deep are known. Documentation of the occurrence of acid sulfate soils is available.[29]

Acid Mine Drainage and Mine Spoil Leachate

Many common ores of metals such as Cu, Zn, and Pb are associated with pyritic materials. In addition, high-S coals of the U.S. Midwest also contain substantial amounts of pyrite and organic S. When the valuable ores or coal are removed, the historic practice has been to leave the mine spoils on the earth's surface. Upon exposure to air and water, the reactions described above occur, and drainage waters from these sites often have pH's of 3 or less. At this pH most fish, invertebrates, plants, and microorganisms cannot grow or survive, and, as a result, surface waters into which the acidic water flows become nearly sterile.[31] The high content of soluble Fe also contributes to toxicity until diluted to the extent that Fe is precipitated (as $Fe(OH)_3$). At one time, the amount of H_2SO_4 entering the Ohio River and its tributaries in leachate waters from coal-mining areas was estimated to be 5×10^9 kg of H_2SO_4 per year. Based on the chemistry and biochemistry of the acid-forming process, a simple solution to enviromental problems arising from exposure of new spoils was to cover the fresh spoil material with about 1 meter of soil or pyritic-free material. When buried to this depth, little O_2 reaches the spoil material, with the result that the O_2-requiring reactions (shown above) do not occur and soil and water acidification is kept to a minimum.

SULFUR IN PLANT NUTRITION

Plants contain about the same amounts of S as of P, the usual range being 0.2 to 0.5% on a dry-weight basis. Among the organic S compounds in plants are vitamins and metabolic cofactors like glutathione, thiamine, vitamin B, biotin, ferredoxin, coenzyme A, and amino acids (cysteine, cystine, and methionine). The last group accounts for the bulk of the S in most plants (~90%). Small but variable amounts of the S occur in inorganic forms. The SO_4^{2-} ion is the main form of S taken up by the plant.

Sulfur has numerous functions in plants, many of which are related to enzyme-mediated transformations. Ferredoxin, an Fe–S protein, is the first stable redox compound of the photosynthetic electron chain and it is an important reactant in N_2 fixation. The disulfide (—S—S—) bond serves as a linkage between polypeptide chains in proteins; the sulfhydryl (—S—H) group provides sites for metal cation binding and for attachment of prosthetic groups to enzymes. The methyl group (—CH₃) of methionine is involved in the biosynthesis of lignin, pectin, chlorophyll, and the flavonoids.

Plants suffering from lack of S show deficiency symptoms characteristic of reduced photosynthetic activity. Growth is retarded and maturity is often delayed. In most plants, the younger leaves are light green in color, which is sometimes confused with rather similar symptoms of N deficiency (see Chapter 6). Sulfur deficiencies are widely found in leguminous crops.

Reviews of S metabolism in plants include those of Anderson[7] and Thompson et al.[8] Sulfate taken up by the plant must first be reduced before incorporation into organic forms. The first step of S incorporation is a reaction between SO_4^{2-} and ATP (adenosine 5'-triphosphate) to form APS (adenosine 5'-phosphosulfate):

$$ATP + SO_4^{2-} + 2H^+ \longrightarrow Adenosine -O-\overset{\overset{O}{\|}}{\underset{\underset{OH}{|}}{P}}-O-\overset{\overset{O}{\|}}{\underset{\underset{O}{\|}}{S}}-OH \; + \; HO-\overset{\overset{O}{\|}}{\underset{\underset{OH}{|}}{P}}-O-\overset{\overset{O}{\|}}{\underset{\underset{OH}{|}}{P}}-OH$$

$$APS$$

Further transformations occur according to the sequence shown in Fig. 10.2, with formation of cysteine, methionine, S-containing proteins, and ester sulfates; additional S compounds include vitamins (biotin, thiamine), glutathione, lipoic acid, and coenzyme A.[7–9]

The total amount of S consumed by plants is highly variable, as illustrated in Table 10.3. Cruciferous crops (e.g., cabbage and turnips) and onions require considerably more S than most crops because they contain unusually large quantities of mercaptans and thioglucosides. These compounds, or metabolites from them, account for the characteristic odors of cooked cabbage and cut onions. For the common field crops, total S taken up by the plant (grain plus straw or stover) amounts to 10 to 30 kg/ha. At a medium value, a typical agricultural soil (224–1,120 kg S/ha to plow depth) will contain sufficient S for only a few decades of cropping without any further S additions. However, as noted below, the S in many soils is tied up in organic forms and is not directly available to plants. It should also be noted that part of the plant S is recycled when crop residues are returned to the soil following harvest.

The amount of SO_4^{2-} used by the plant depends on the concentration of SO_4^{2-} at the root surface, which in turn depends on the rate at which SO_4^{2-} is replenished through mass flow and diffusion. In most soils, the reserve of SO_4^{2-} in the solution phase is inadequate for optimum crop yields, and S deficiencies would develop within a short time if not for additions in rainfall and/or mineralization of organic matter.

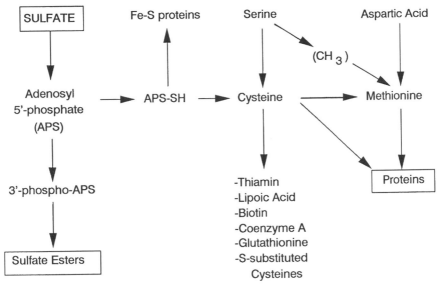

Fig. 10.2   Pathways for the transformation of SO_4^{2-} to amino acids and other organic compounds in plants. Adapted from Thompson et al.[8]

INORGANIC FORMS OF SOIL S

Sulfur occurs in soils in both organic and inorganic forms. The amounts of S in the two forms vary widely, depending on the nature of the soil (pH, drainage status, organic matter content, mineralogical composition) and depth in the profile.[32,33]

Inorganic S occurs in soil largely as SO_4^{2-}, although compounds of lower

TABLE 10.3   Amounts of S Removed Annually from the Soil in Specific Crops[a]

Crop	S Removed, kg/ha
Corn	9–11
Wheat	10–13
Potatoes	8–11
Grasses	9–11
Cotton	13–17
Clovers	17–25
Alfalfa hay	22–27
Sugar beet	21–31
Onions	20–22
Cabbage	21–43
Turnip	28–39

[a]From data recorded by Tabatabai.[14]

oxidation states are often found, such as sulfide (e.g., FeS), sulfite, thiosulfate, and elemental S^0. Under waterlogged conditions, inorganic S occurs in reduced forms, such as H_2S, FeS, and FeS_2. Pyrite is often the main inorganic form of S in wetland and submerged soils. In some instances, elemental S^0 can be formed. A major fraction of the S in calcareous and saline soils occurs as gypsum ($CaSO_4 \cdot 2H_2O$).

Unlike phosphate, SO_4^{2-} is subject to leaching; thus, in highly leached soils, inorganic forms of S are continuously being removed and only the S in organic forms remains. Due to a combination of leaching and plant uptake, seasonable fluctuations are often observed in the amounts of water-soluble SO_4^{2-} in the surface soil and surface waters;[34] other factors that may affect SO_4^{2-} levels include additions in fertilizers, as well as in rain and irrigation waters.

Some soils have the capacity to retain SO_4^{2-} in an adsorbed form.[35] Sorption is restricted to acidic soils and is due primarily to anion exchange by positive charges on Fe- and Al-oxides and clay minerals. Some typical features of SO_4^{2-} adsorption are:

1. More SO_4^{2-} is adsorbed by the lower soil horizons than by the surface layer.
2. Adsorption increases as soil pH decreases.
3. Kaolinite adsorbs more SO_4^{2-} than montmorillonite; soils rich in Fe- and Al-oxides adsorb even more.

Organic matter exerts a negative effect on SO_4^{2-} adsorption,[35] which may be the reason for the low sorption capacities of surface soils. Accumulations of adsorbed SO_4^{2-} have been observed in the B horizons of forest soils due to inputs from atmospheric deposition.[36] The reactions involved in SO_4^{2-} adsorption are similar to those for phosphate adsorption (see Chapter 9).

ORGANIC FORMS OF SOIL S

As noted above, essentially all of the S in soils of humid and semihumid regions occurs in organic forms.[37-39] For these soils, the absolute amount of both total and organic S is influenced to a large extent by factors that affect the organic matter content of the soil (see Chapter 2).

Table 10.4 summarizes data obtained by Tabatabai and Bremner[38] for total and organic S in 37 Iowa surface soils (0–15 cm depth). Organic S accounted for 95 to 99% of the total S and was significantly correlated with organic C and total N. Both the amount of total S (55 to 618 mg/kg) and the percentages of organic S are similar to values generally reported for agricultural soils (see Freney[13]). In an early study, Evans and Rost[33] found that from 46 to 73% of the S in some Minnesota soils was organic, being highest in some Spodosols

**TABLE 10.4   Organic and Inorganic S in 37 Iowa
Surface Soils**[a]

Form of S	Range, mg/kg	Average Values	
		mg/kg	as % of S
Organic	55–604	283	97
Inorganic sulfate	1–26	8	3
Totals	55–618	292	100

[a]From Tabatabai and Bremner.[38]

and lowest in some prairie soils. Very high percentages of the S in the sub-surface soils examined by Tabatabai and Bremner[38,39] occurred in organic forms (84–99%).

Determination of Organic S

Organic S is usually taken as the difference between total soil S, as determined by dry- or wet-ashing techniques, and inorganic S (SO_4^{2-} and sulfide) extracted from the untreated soil by reagents such as dilute HCl or $NaHCO_3$ (i.e., organic S = total S–inorganic S). There is some uncertainty in these procedures because extractants used to remove inorganic S compounds undoubtedly remove small amounts of organic S as well.

In lieu of extraction of inorganic forms, SO_4^{2-} can be analyzed in ignited soil with and without addition of $NaHCO_3$, in which case organic S is calculated from the difference between the two values.[40,41] A direct method based on ignition of acid-leached soil has also been described. The SO_4^{2-} content of digests or extracts is commonly determined by the Johnson and Nishita[42] reduction–distillation method, which involves these steps:

1. Conversion of SO_4^{2-} to H_2S by reduction with a reducing mixture containing HI
2. Recovery of H_2S by distillation
3. Colorimetric determination of the H_2S by the methylene blue method

Other methods are available for total and organic S, as described in the reviews of Tabatabai[40] and Blanchar.[41]

The C/S Ratio

In Chapter 9 it was noted that, as an average, the proportion of C, N, P, and S in soil humus is on the order of 140:10:1.3:1.3, which translates into an average C/S ratio of about 110 (mean N/S ratio of about 8). However, C/S and N/S ratios of the soil deviate considerably from these values, as illustrated by the data given in Table 10.5.[43-45] In general, variations in the C/S ratio are greater than for C/N.

Differences in the C/S ratios between soil groups have been attributed to such factors as parent material and type of vegetative cover. Bettany et al.[43] found that the C/S ratios of some grassland soils of Saskatchewan, Canada, were lower than some comparable forest soils of the region; mean C/S ratios for the two groups were 58 and 130, respectively.

Chemical Fractionation of Organic S

Much of the information regarding soil organic S has come from studies on the reactivity of organic S to reduction with hydriodic acid (HI).[39,40,43,46-50] A flow diagram is shown in Fig. 10.3. The soil is first leached with a phosphate

TABLE 10.5   Ratios of C/S and N/S in Soils from Different Regions of the Earth

Soils and Locations[a]	Ratios	
	C/S	N/S
Australia		
Agricultural (24)	115	7.7
Grassland	94	8.3
Brazil		
Agricultural (6)	121	6.3
Canada (Saskatchewan)		
Haploborolls (28)	62	6.7
Transitional (14)	96	7.7
Typic Cryoboralf (12)	132	11.1
New Zealand		
Grassland	250	8.3
Pasture, 25	92	8.3
Scotland		
Grassland, acid (40)	105	7.1
Calcareous (10)	89	7.9
United States (Iowa)		
Cultivated (6)	85	7.7

[a]For specific references see Freney.[13] Results for Canada, Iowa, and Brazil are more recent.[56,58] Figures in parentheses signify number of samples. High correlation coefficients (>0.85) were observed for both C/S and N/S ratios.

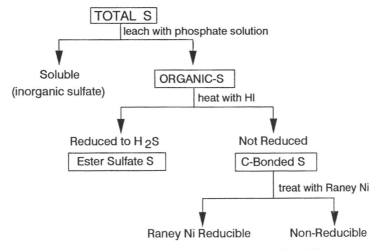

Fig. 10.3   Scheme for the fractionation of soil S.

solution to remove inorganic SO_4^{2-}, following which the leached soil is treated with HI. The S recovered as H_2S is believed to occur as ester sulfates; that which is not reducible is C-bonded S. The C-bonded S is further separated into two fractions based on its ability to be reduced (or not) to H_2S with Raney Ni.

There are two main categories of organic S:

1. *Ester sulfate S:* This S is readily reduced to H_2S by HI and occurs in compounds containing the C—O—S linkage, such as phenolic sulfates and sulfonated polysaccharides.[10,13] A typical biochemical compound containing S in the ester linkage is choline sulfate $[(CH_3)_3N^+—CH_2—CH_2O—SO_3H]$. The S that is reducible with HI occurs in both low- and high-molecular weight compounds of soil organic matter, but mostly in the latter.[47] Lowe and deLong[48] found that sulfonated polysaccharides accounted for <2% of the S in some Canadian soils.

2. *Carbon—bonded S:* This S is directly attached to C through the S—C linkage and is taken as the difference between total organic S and ester sulfate S. Included in this group are the S-containing amino acids, cysteine $(HOOC—CH(NH_2)—CH_2—SH)$ and methionine $(HOOC—CH(NH_2)—CH_2—CH_2—S—CH_3)$. Additional biochemicals containing C-bonded S are the vitamins biotin and thiamine.

$$O$$
$$\|$$
$$C$$

HN   NH

| |

HC —— CH

| |

H_2C CH – $(CH_2)_4$ – COOH

S

Biotin

H_2C–C   N NH_2 S   C – CH_2 – CH_2 OH

HC

H_2C–C $\|$ N   C $\|$ C   HC $\|$ N — C – CH_3

C   C
H H_2

Thiamine

As noted in Fig. 10.3, not all of the C-bonded S is reducible to H_2S by Raney Ni. This is not surprising, as it is well known that Raney Ni does not reduce the C-bonded S of some organic compounds, typical examples being cysteic acid and methionine sulfone. Furthermore, the S combined with humic and fulvic acids may not be reduced. Iron and other soil constituents can lead to low results using the Raney Ni method.[46]

For a range of Australian soils, Freney et al.[13,46] accounted for an average of 56% (range of 45–93%) of the C-bonded S by Raney Ni reduction. The average distribution of total S in 24 Australian soils was: ester sulfate S, 52%; C-bonded S, 41%; inorganic S, 7%. The average percent as ester sulfate S (52%) is similar to that reported in Bettany et al.[43] for the Ap layers of 54 Saskatchewan soils (36–50%) and by Tabatabai and Bremner[39] for 37 Iowa surface soils (50%). The fraction of the soil S as ester sulfates appears to be higher in grassland soils than in forest soils; for the latter, values as low as 18% have been observed.[37] Bettany et al.[51] found that total S increased as temperature decreased and moisture content increased in Canadian soils, but the ester sulfate S content did not follow the same pattern.

The percentage of the total S as ester sulfates increases with depth in the soil profile, whereas the percentage as C-bonded S decreases.[38,52] Results obtained by Tabatabai and Bremner[38] are shown in Fig. 10.4.

McLaren and Swift[50] found that a high proportion (75%) of the S lost when soils are subject to long-time cultivation consisted of C-bonded forms. McLachlan and DeMarco[49] also found that, on cropping, more S was withdrawn from C-bonded forms than from sulfate esters. In related work, Ghani et al.[53] found that C-bonded forms of S represented the major source of *mineralizable* S in soils; David et al.[37] observed a decrease in C-bonded S and an increase in ester sulfate S during incubation of some forest soils. Total S and HI-reducible S have been found to be distributed differently among various soil size fractions (Table 10.6).[13,54]

Sulfur-Containing Amino Acids

The proportion of the soil S in amino acids is not known with certainty; published data for amino acids in soil hydrolysates indicate that 11 to 16%

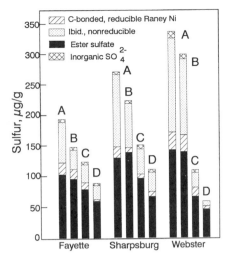

Fig. 10.4 Amounts of different forms of S in three Iowa soils. A = 0–15 cm, B = 15–30 cm, C = 30–60 cm, D = 60–90 cm, E = 90–120 cm. Adapted from Tabatabai and Bremner.[38]

of the soil S occurs in this form.[12] This estimate may be low, because cystine and methionine undergo extensive destruction during acid hydrolysis and thus are not recovered quantitatively. In an attempt to avoid this problem, Freney et al.[55] oxidized cystine and methionine to more stable forms (cysteic acid and methionine sulfone) prior to acid hydrolysis. For two soils, an average of 26% of the soil S, equivalent to 46% of the C-bonded S, was estimated to be present as amino acids (Table 10.7). By using a somewhat similar procedure, Scott et al.[56] found that from 11 to 17% of the total organic S (19–31% of the C-bonded S) in some Scottish soils occurred as amino acid S. Other results have shown a range of 27 to 125 mg/kg amino acid S/g of soil.

TABLE 10.6 Distribution of S and HI-Reducible S (Ester Sulfate S) in the Size Fractions of a Canadian Mollisol[a]

Fraction	Organic C g/kg	Total S g/kg	HI Reducible S % of S	C/Total S
Sand	1.26	0.011	27	115
Coarse silt	7.98	0.073	25	109
Fine silt	5.24	0.042	50	125
Coarse clay	12.03	0.140	63	86
Fine clay	5.45	0.151	77	36
Whole soil	33.20	0.418	62	79

[a]From Anderson et al.[54] as recorded by Freney.[13]

TABLE 10.7   Distribution of Amino Acid S in Two Soils[a]

Sulfur Fraction	Virgin Soil mg/kg	Improved Pasture mg/kg
Total S	145	266
Hydriodic acid-reducible	78	94
C-bonded	67	172
Cysteine and Cystine[b]	19	48
Methionine[c]	11	33
Total	30	81
Amino acid S		
As % of total S	20.7	30.5
As % of C-bonded S	44.8	47.0

[a]From Freney et al.[55] The soil was a yellow "Podzolic." The pasture soil had been improved through growing of legumes and by regular fertilization with superphosphate over a 46-year period.
[b]Determined as cysteic acid.
[c]Determined as methionine sulfone.

Lipid S

The presence of sulfolipids in soil has been demonstrated.[57–59] Chae and Tabatabai[59] found that the amount of lipid S in 10 Iowa surface soils ranged from 0.87 to 2.63 mg/kg and accounted for only 0.29 to 0.45% (average = 0.37%) of the total soil S. From 20 to 40% of the sulfolipid S was HI-reducible (i.e., in ester-type compounds). For 27 British Columbia (Canada) soils, from 0.5 to 3.5% of the total S has been accounted for as lipid S.[57] In other studies, column chromatography has been used to fractionate soil lipid S into three classes: polar, glycolipids, and less polar lipids.[58] Glycolipids were the dominant form (mean of 64%), and polar lipids were the least abundant (mean of 4%).

Complex Forms of Organic S

Extractions and fractionations of soil organic S based on solubility characteristics (see Chapter 2) indicate that some of the soil S is incorporated into humic substances.[51,60] By extraction with a 0.15 M NaOH:0.1 M $Na_4P_2O_7$ solution, Bettany et al.[51] recovered over 60% of the S from a series of soils along an environmental gradient in Canada. The distribution of S in the three main humic fractions was: humic acid = 34%; fulvic acid = 39%; humin = 26%.

Milder extractants have also been used,[47,61] including dilute Na_2CO_3 and aqueous acetylacetone.[61] In the latter case, over 80% of the soil organic S was extracted with 0.2 M aqueous acetylacetone at pH 8 in combination with an ultrasonic treatment; most of the extracted S occurred in components with

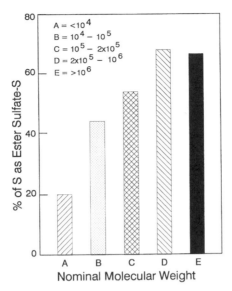

Fig. 10.5 Distribution of ester sulfate S in different molecular weight fractions of soil organic matter. Adapted from Kerr et al.[61]

molecular weights >200,000 Daltons, as estimated by gel permeation chromatography. As shown in Fig. 10.5, the percentage of the S as ester sulfate S increased with an increase in molecular weight of the organic matter.

Part of the S of humic substances may occur as complexes resulting from the reaction of thiol compounds with quinones and reducing sugars. The mechanisms involved are similar to those discussed in Chapter 1 for the formation of humic substances. For example, cysteine, thiourea, and glutathione react with quinones (formed from polyphenols derived from lignin or synthesized by microorganisms) to form brown-colored pigments containing S.

The reaction between cysteine and p-benzoquinone is illustrated in Fig. 10.6. The first product is 1,4-benzoquinone-2-cysteine, which subsequently undergoes an inner condensation between the quinone C=O group and the free NH_2 group to form a cyclic product. Part of the unknown C-bonded S in soils, notably the fraction not reduced to H_2S with Raney Ni, may be in this form.

The resistance of soil organic S to attack by microorganisms may be partly caused by the existence of such C–S linkages in high-molecular-weight humic components.

Soluble Forms

As a result of microbial activity, numerous S-containing biochemicals would be expected to be produced in soil. However, these same compounds are very

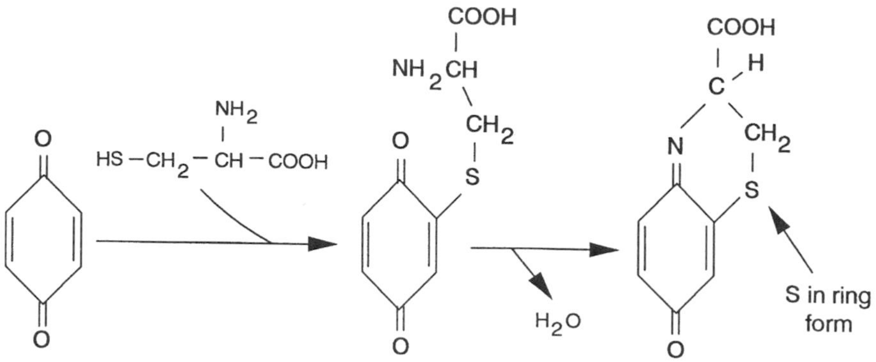

Fig. 10.6   Reaction between cysteine and *p*-quinone to form complex structures containing C–S linkages.

susceptible to degradation, and the amounts found in the soil solution at any one time will represent a balance between synthesis and destruction by microorganisms. In accordance with this concept, only trace quantities of S-containing amino acids (cysteine, cystine, methionine, methionine sulfoxide, and cysteic acid) have been recovered from soil by extraction with water or such solvents as neutral NH_4OAc (see Chapter 6). The concentration of S-containing organics may be higher in rhizosphere soil than nonrhizosphere soil, since they are found in plant root exudates.

Zhao and McGrath[62] recovered from 1 to 2% of the soil organic S by extraction with 0.016 M potassium phosphate solution; the extracted organic S was found to be of some importance in S cycling through mineralization–immobilization.

Physical Fractionation

Ultrasonic dispersion of the soil, followed by particle size separations (see Chapter 2), has been used for the physical fractionation of organic S.[54] In this work, consistent differences were observed for both the distribution and the composition of S in the various size fractions (Table 10.6). Most of the S (70%) was recovered in the clay fractions, and most of the organic S occurred in HI-reducible forms. Carbon/S ratios were also lowest in the clay fractions. The bulk of the S in the coarse fractions occurred in C-bonded forms.

DYNAMICS OF ORGANIC S TRANSFORMATIONS

The conversion of S in soil organic matter and organic residues to plant-available forms is strictly a microbiological process. When the soil is well aerated, organic S is mineralized and subsequently oxidized (if necessary) to SO_4^{2-}, which is the form taken up by most plants. Concurrently, SO_4^{2-} is assimilated by microorganisms and incorporated into microbial biomass. Increases in the organic S content of the soil will occur only when conditions are suitable for the accumulation of organic matter, such as by frequent additions of organic residues.[24] Williams and Donald[63] found that essentially all of the fertilizer S added in superphosphate to soils over a 15- to 25-year period could be recovered in the organic matter.

The turnover of S through mineralization–immobilization follows somewhat the same pattern seen for N (Chapter 6) and P (Chapter 9) in that both processes occur *simultaneously*. Accordingly, the quantity of plant-available SO_4^{2-} in the soil solution at any one time represents the difference between the magnitude of the two opposing processes.

$$\text{Organic S} \underset{\text{immobilization}}{\overset{\text{mineralization}}{\rightleftharpoons}} SO_4^{2-}$$

Hence, data obtained for *net mineral S accumulation* (i.e., SO_4^{2-}) during laboratory incubations do not provide accurate estimates for *actual* mineralization rates of organic S, although use of *net* mineralization is frequently used as the equivalent to *actual* mineralization. Numerous studies, for example, have shown that the addition of C-containing substrates (e.g., glucose, cellulose, crop residues) to the soil leads to *net* loss of both applied and native soil SO_4^{2-} through immobilization.

Factors Affecting the Turnover of Soil Organic S

Turnover of soil organic S through mineralization–immobilization will be influenced by those factors that affect the activities of microorganisms, such as addition of crop residues,[64,65] soil properties and liming,[66] temperature and moisture,[67–69] nutrient additions,[70] and cultural practices.[71,72] Ghani et al.[73] found that as much as one-half of newly immobilized S could be recovered as SO_4^{2-} by subsequent incubation under laboratory conditions. The subject of S turnover through mineralization–immobilization is covered in review articles.[13,15]

As noted in the previous section, results of studies on changes in mineral S levels over time have generally been expressed in terms of "mineralization" of organic S. This practice will be followed in the discussion that follows, even though concurrent immobilization has invariably occurred as well.

General observations concerning the release of nutrients from soil organic matter were outlined in Chapter 2. Those aspects that pertain to S can be summarized as follows:[13–15]

1. The amount of S mineralized from unamended soils does not appear to be directly related to soil type, total amount of C or S, the C/S ratio, soil pH, or mineralizable N.

2. Sulfate content follows a number of patterns:
   a. Initial immobilization followed by net mineralization in later stages;
   b. A steady, linear release of SO_4^{2-} with time;
   c. An initial rapid release followed by a slower release; and
   d. A rate of release that decreases with time.
3. Mineralization of S in the presence of plants is greater than in fallow soil, a result that may be due to greater proliferation of microorganisms in the rhizosphere.
4. Mineralization of S is affected by those factors that influence the growth of microorganisms, notably temperature, moisture, and pH:
   a. *Temperature:* Mineralization is markedly suppressed at 10°C but increases with increasing temperature from 20 to 40°C and decreases thereafter.
   b. *Moisture:* Mineralization is considerably retarded at low ($< 15\%$ of capacity) and high ($> 80\%$ of capacity) moisture levels. Optimum moisture content for mineralization is 60% of the maximum water-holding capacity.
   c. *pH:* Mineralization is usually most rapid in the pH range of 5.5 to about 7.5 and is lesser at values outside this range.
5. More SO_4^{2-} is released when soils are dried and remoistened prior to incubation than when they are incubated without prior drying.
6. Newly formed soil organic S is more readily mineralized than the indigenous organic S, as observed by Ghani et al.[73]

Reasons for differences in the mineralization rate of organic S in soils are not fully understood. Errors involved in estimating mineralization in the presence of growing plants include atmospheric contamination by SO_2, incomplete recovery of roots from soil, additions of S during watering, and failure to account for all of the S in plant material due to inadequacies of analytical methods.

Some, but not all, workers have shown a direct relationship between the amount of SO_4^{2-} released from soil during incubation and the S content of the soil. A plausible explanation for the divergent results is that the S content of recently added plant residues regulates the amount of SO_4^{2-} produced, rather than total organic S in the soil.

It has long been known that the C/S ratio of organic residues provides a rough guide to the amount of SO_4^{2-} that accumulates during decay. As shown below, when the C/S ratio of added plant residues is below 200, there is a net gain of SO_4^{2-}; when the C/S ratio exceeds 400, there is a net loss. For C/S ratios between 200 and 400, there is neither a gain nor loss of SO_4^{2-}. Carbon/S ratios of 200 and 400 in plant residues correspond to S contents of about 0.25 and 0.5%, respectively.

C/S Ratio		
<200	200 to 400	>400
Net gain of SO_4^{2-}	Neither a gain or loss of SO_4^{2-}	Net loss of SO_4^{2-}

One might expect that the relative rates of mineralization of S and N from soil organic matter would be similar, and that the two would be released in the same ratio in which they occur in soil organic matter. As noted above, this has not always been the case, for several possible reasons:

1. Nitrogen and S may exist as organic compounds or fractions of the soil organic matter that decay at different rates; consequently, they will not be released at the same time.
2. The inclusion of plant or animal residues with large N/S ratios would cause greater immobilization of S than N.
3. The presence of Ca^{2+} may obscure the release of SO_4^{2-} by formation of insoluble $CaSO_4$.
4. Air drying the soil before treatment may affect N and S release differently.
5. Different groups of microorganisms may be responsible for N and S mineralization.

The addition of $CaCO_3$ to soil increases soluble SO_4^{2-} on incubation, possibly because of enhanced mineralization of soil organic matter due to more bacterial growth, release of adsorbed SO_4^{2-} because of the increase in pH, or SO_4^{2-} added in the $CaCO_3$.

Several studies have shown that slightly more N is lost relative to S when soils are cropped. Accordingly, the N/S ratio of the organic matter of cultivated soils is generally lower than of virgin soils. Data obtained by Stewart and Whitfield[75] for 10 paired soils from the Great Plains Region of the United States illustrate this point:

Soil	Total N mg/kg	Organic S, mg/kg	NS Ratio
Virgin soils	1600	183	8.7
Cultivated soils	934	117	8.0
% loss	42	36	

The microbial assimilation of SO_4^{2-}, with conversion to organic forms, is believed to be a major process influencing the behavior of SO_2 and other S compounds introduced into forest soils by acid precipitation.[26,76,77]

Finally, it should be noted that two different pathways may be involved in the mineralization of organic S in soil:[11]

1. By a process in which HS^- or SO_4^{2-} is produced as a *byproduct* of the degradation of organic matter by microorganisms. Carbon-bonded S is thought to be mineralized mainly by this process.
2. By a simple enzymatic process in which ester sulfate S is released from organic forms as SO_4^{2-} by the action of arylsulfatase enzymes, as discussed below.

Biochemical Transformations of S-Containing Amino Acids

Little information is available concerning the pathways of decay of organic S compounds in soil. A few workers have followed the decay of known substances added to soil, and others have studied their degradation by pure cultures of microorganisms isolated from soil.

Pathways for the breakdown of cysteine during anaerobic decomposition, with production of H_2S, are shown below:

$$HS-CH_2CH(NH_2)-COOH \xrightarrow{H_2O} CH_2CO-COOH + NH_3 + H_2S$$

Cysteine → Pyruvic acid

$$\xrightarrow{H_2O} HO-CH_2CH(NH_2)-COOH + H_2S$$

Serine

A number of pathways have also been proposed for methionine degradation in soils, with the end products shown below:

$$CH_3-S-CH(NH_2)-COOH \rightarrow CH_3-CH_2-CO-COOH$$

Methionine   α-Ketobutyric acid

$$+ NH_3 + CH_3SH$$

Methyl mercaptan

Microbial Biomass S

From 1 to 3% of the organic S in soils is present in the soil microbial biomass.[78–83] The C/S ratio of microbial cells is on the order of 57 to 85 for bacteria and 180 to 230 for fungi. The ratios for bacteria are generally lower than for soils. The S content of most microorganisms lies between 0.1 and 1.0% on a dry-weight basis.

Most of the S in bacteria and actinomycetes (about 90%) occurs in cysteine and methionine; in fungi, a large proportion of the S occurs in HI-reducible forms, mainly as choline sulfate.[78] In contrast, much of the soil organic S occurs in unknown forms.

The quantity of S in the microbial biomass has been estimated from the amount of inorganic SO_4^{2-} that is produced during incubation of soil after fumigation with chloroform ($CHCl_3$), the extractant being 0.1 M $NaHCO_3$ or 10 mM $CaCl_2$. The equation is:

Biomass S

$$= \frac{\text{S extracted from fumigated soil} - \text{S extended from nonfumigated soil}}{K_s}$$

where K_s is the fraction of the biomass S subject to release by the $CHCl_3$ treatment. Values of K_s for the two extractants were: 0.1 M $NaHCO_3$ = 0.41 and 10 mM $CaCl_2$ = 0.35.

In lieu of incubation, direct measurements can be made for extractable S in fumigated and nonfumigated soil.[83]

Although only a small amount of the soil organic S resides in the biomass at any one time, this fraction is extremely labile and thus is a key component in the turnover of S in soil. The higher the amount of organic S in the biomass, the greater will be the availability of S to higher plants. The absolute amount of biomass S in soils has been found to be well correlated with biomass C,[78,82,83] extractable sulfate,[84] and microbial activity in general.[85]

Arylsulfatase Enzymes in Soil

Since much of the soil organic S is present in the form of sulfate esters, arylsulfatase (arylsulfate sulfohydrolase, EC 3.1.6.1) may play a key role in S mineralization.[86–89] The overall reaction is:

$$ROSO_3^- + H_2O \xrightarrow{\text{arylsulfatase}} ROH + H^+ + SO_4^{2-}$$

Arylsulfatase was first reported in Iowa soils by Tabatabai and Bremner.[87] The enzyme has since been detected in soils from other geographical regions. Arylsulfatase activity is determined by colorimetric analysis of the p-nitrophenol released when the soil is incubated at pH 5 with p-nitrophenyl sulfate and toluene. Toluene is added to retard microbial degradation of p-nitrophenol.

Arylsulfatase activity has been shown to vary according to soil type, depth, season, climate, and content of ester sulfate S. Tabatabai and Bremner[87] showed that arylsulfatase activity decreased markedly with depth in four Iowa soil profiles (Fig. 10.7) and was closely correlated with the decrease in organic matter content ($r = 0.783$). However, as can be seen from Fig. 10.8, considerable deviation in arylsulfatase activity occurred in those soils that had the highest C contents, corresponding to the upper soil layers (0–15 and 15–30 cm sampling depths). This suggests that additional factors (other than organic matter content) are determinants of arylsulfatase activity. Cooper[86] found a significant correlation between arylsulfatase activity in 20 Nigerian soils and total C ($r = 0.691$), organic S ($r = 0.565$), and HI-reducible S ($r = 0.878$). Lou and Warman[89] attempted to determine *labile* ester sulfate S in soil organic matter extracts (a chelating resin was used for extraction) by adding arylsulfatase to the extracts and measuring the net increase in SO_4^{2-} concentration. Soils rich in SO_4^{2-} contained higher amounts of *labile* ester sulfate S than those containing small amounts of SO_4^{2-}. Other studies have failed to show a correlation between arylsulfatase activity and the amount of organic S recovered as SO_4^{2-} by incubation.[53]

Use of ^{35}S for Following Organic S Transformations

As noted above, when fertilizer SO_4^{2-} is applied to soil, some of the S is converted to organic forms and some is utilized by plants. As was the case for C and N (Chapters 1 and 6), the use of isotopes (e.g., ^{35}S) has provided valuable information on S transformations in soil, including factors affecting

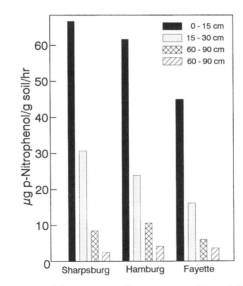

Fig. 10.7 Arylsulfatase activity in some Iowa soil profiles. Adapted from Tabatabai and Bremner.[87]

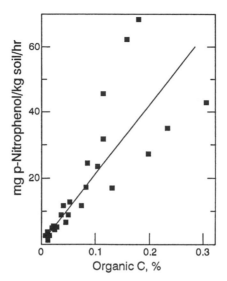

Fig. 10.8 Relationship between arylsulfatase and organic C content of the soil profiles shown in Fig. 10.7. The samples were taken at 15-cm intervals to a depth of 120 cm. Adapted from data reported by Tabatabai and Bremner.[87]

incorporation of S into the various organic forms and plant availabilities of C-bonded and ester sulfate forms of S.[90-95] The net immobilization of SO_4^{2-} in the presence of a C substrate leads to incorporation of the S into the microbial biomass,[64,96] and subsequently into nonbiomass organic forms.

Freney et al.[91] added ^{35}S–SO_4^{2-} to soil and observed a steady incorporation of S into organic forms over a 24-week incubation period. A maximum of 50% of the added SO_4^{2-} was immobilized during this period, but the percentage was increased to 82% by addition of a C and energy source (glucose). The labeled S was incorporated into both C-bonded and ester sulfate forms, but the latter had the greatest specific activity.

In continuing these studies, Freney et al.[92] found that plants utilized S derived from both the C-bonded and ester sulfate forms. The greater part of the indigenous soil S (60–65%) came from the C-bonded fraction, a result in agreement with the studies described earlier. In the case of recently immobilized ^{35}S, the greatest part (60 to 70%) was converted into ester sulfates, and it was from this fraction that all of the labeled S was taken up. Since both S fractions contributed available S for plant uptake, the conclusion was reached that neither is likely to be of value for predicting the S requirements of plants.

In other work, Bettany et al.[90] found that mineralization of soil organic S was unaffected by addition of inorganic S. The amount of S mineralized could not be related to the quantity of total S in the soil or the percentage of the S present as ester sulfates. As expected, the largest amount of S mineralized occurred from the soil with the lowest C/S ratio.

Formation of Volatile S Compounds

The decay of organic S compounds in poorly drained soils, wet sediments, manures, and many waste materials leads to the formation of mercaptans, alkyl sulfides, and other volatile organic S compounds, as well as H_2S. Sulfur can also be lost from plants as gaseous compounds.[5] Volatile S compounds in soil may be of importance because they can inhibit plant growth, nitrification, and other biochemical processes. As noted above, S gases are undesirable components of the atmosphere because of their intense and unpleasant odors and adverse effect on climate and the environment.

Included with the volatile organic S compounds produced by microorganisms are the following:

$$CS_2 \qquad\qquad S—O—S \qquad\qquad CH_3—SH$$

Carbon disulfide   Carbonyl sulfide   Methyl mercaptan

$$CH_3—CH_2—S—CH_2—CH_3 \qquad CH_3—S—S—CH_3 \qquad CH_3—S—S—CH_3$$

Diethyl sulfide   Dimethyl sulfide   Dimethyldisulfide

Nicolson[97] obtained incomplete recoveries of S from the soil–plant systems he examined and concluded that the unaccounted-for S resulted from emission of volatile S compounds. Pathways for the formation of S gases in aerobic soils were postulated to be as follows:

Determination of Plant-Available S

The status of soil tests to determine available S is similar to the situation described earlier for N (Chapter 6) in that only limited success has been obtained and no one procedure has been consistently superior to all others. Results of soil tests are particularly difficult to interpret in those soils where essentially all of the S occurs in organic forms. A discussion of the various methods and their limitations is given by Tabatabai.[40]

Procedures for evaluating plant-available S include:

1. Extraction of the soil with water or salt solutions (e.g., 0.15% $CaCl_2$ and 5 mM $MgCl_2$), dilute acids (0.5 M NH_4OAc + 0.25 M CH_3COOH), and weak bases (0.5 M $NaHCO_3$ at pH 8.5)

2. Release of SO_4^{2-} upon incubation
3. S uptake by the plant

Extractants of soil S can be categorized as:

1. Those that remove readily soluble SO_4^{2-}
2. Those that remove soluble SO_4^{2-} plus a portion of the adsorbed SO_4^{2-}
3. Those that also remove a portion of the soil organic S

The use of soil tests for assessing plant available S has been covered in detail elsewhere.[98]

Selection of an extractant is dependent to some extent on the nature of the soils encountered. Acid-phosphate solutions appear to be best for acidic soils containing variable amounts of organic S and where part of the SO_4^{2-} occurs in adsorbed forms, while neutral salt solutions are preferred for near-neutral soils of semiarid regions. Factors that affect the extraction of SO_4^{2-} include sample preparation, soil/extractant ratio, and shaking time. Air increases the amount of SO_4^{2-} extracted.

Difficulties in predicting S-fertilizer needs of a particular crop are due to one or more of the following:

1. The rate of release of S from the soil organic matter is highly variable and affected by the activities of microorganisms and climatic conditions.
2. The presence of undegraded organic residues low in S content can result in a decline in SO_4^{2-} levels due to net immobilization.
3. The contribution of SO_4^{2-} in rain water, or of SO_2 by dry deposition, cannot be accurately estimated.
4. Sulfate becomes unavailable through sorption, leaching, or reduction to volatile gases during waterlogging.
5. Environmental factors (e.g., temperature, moisture) influence the rate of degradation of organic S and also the uptake of SO_4^{2-} by plants. The amount taken up from the subsoil (which is frequently not determined) is highly variable and depends on the rooting characteristics of the plant.

REDUCTION OF S

In addition to assimilation by living organisms, inorganic S compounds are subject to a variety of oxidation–reduction reactions, nearly all of which are mediated by microorganisms.[16,17,99–102] An outline of the main reactions is given in Fig. 10.9.

A full account of inorganic S transformations is beyond the scope of the present work, and the reader is referred to the review of Amga and Cooper.[18] For the purpose of this discussion, it can be said that inorganic S transformations affect soil color, soil reaction, and availability of plant nutrients.

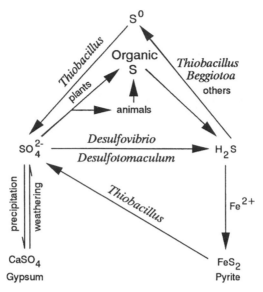

Fig. 10.9 Pathways of S transformations. The list of organisms involved in oxidation–reduction is incomplete (see text). Adapted from Kaplan.[100]

However, under most agricultural conditions, the effects of these S transformations are relatively obscure, and it is the completely oxidized (SO_4^{2-}) and completely reduced (H_2S) forms that are of greatest concern. Oxidation–reduction reactions of S have special significance for the genesis and management of acid sulfate soils.

Reduction of SO_4^{2-} by microorganisms occurs in two different ways. In one, S is incorporated into cellular organic constituents, such as the S of amino acids. This process is referred to as *assimilatory* SO_4^{2-} *reduction*, or *immobilization*. In the other, the reduction is carried out by a special group of bacteria and leads to the formation of sulfide (e.g., H_2S) as the end product. This process, known as *dissimilatory or respiratory* SO_4^{2-} *reduction*, occurs when the environment is favorable for the growth of anaerobic, sulfate-reducing bacteria. These bacteria use oxidized S compounds as terminal e^- acceptors in a process that is similar to denitrification. Electrons are generated by oxidation of organic C compounds, transferred into an electron transport system, and ultimately reside on the S atom. A typical metabolic reaction is as follows:

$$2\ CH_3—CHOH—COOH + SO_4^{2-} \rightarrow 2\ CH_3COOH$$

       Lactic acid   Acetic acid

$$+ 2\ CO_2 + S^{2-} + H_2O$$

                                         Sulfide

Sulfate-reducing microorganisms have important effects in causing the precipitation of metal sulfides (notably of ferrous Fe^{2+}), pollution of natural waters, and corrosion of steel cables buried in poorly drained or wet soils. Such deposits are usually stained black through the accumulation of ferrous sulfides (FeS). Microbial SO_4^{2-} reduction also accounts for the occurrence of sulfides in shales and strip-mined refuse and the H_2S in natural gas.

The ability to reduce SO_4^{2-} to H_2S is a property of a small number of strict anaerobes, primarily the bacterial genera, *Desulfovibrio*, consisting of curved, rod-shaped cells, and *Desulfotomaculum*, which are straight or somewhat curved, sporulating rods. These organisms are particularly abundant in stagnant water basins and marine muds.

Most of the S that accumulates as iron sulfide in waterlogged soils, bogs, ditches, and marine sediments is produced by SO_4^{2-}-reducing bacteria. The overall reaction is:

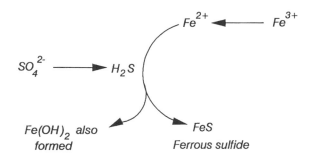

Further reactions of the sulfide in anaerobic soils lead to the formation of such products as thiosulfate and elemental S^0, compounds that can themselves be oxidized or reduced by microorganisms.

Sulfate-reducing bacteria require an E_h of less than +100 mV for growth, although they can survive in a dormant state for long periods in aerobic environments. Typical habitats for SO_4^{2-}-reducing bacteria include soils and waters, sewage, polluted waters, deep-sea sediments, muds, and estuarine sand. Their presence is often indicated by a black color and a smell of H_2S. Since the organisms that reduce SO_4^{2-} require an organic substrate, their activity is favored in wet sediments or submerged soils that contain an abundant amount of degradable organic matter.

OXIDATION OF S

Reduced inorganic S compounds (e.g., elemental S^0, H_2S, FeS_2) are readily oxidized in soils under suitable conditions by a group of bacteria that utilize the energy thus released to carry out their life processes.[16,17] Many S-oxidizing bacteria are obligate autotrophs; the C required for synthesis of carbohydrates, proteins, and other cellular products is derived from inorganic CO_2.

Practical Importance in Soil

Elemental S^0 is sometimes applied to soil, notably for acidification of alkaline soils and reclamation of saline and alkaline soils, but also in combination with various fertilizers (e.g., S-coated urea). When applied to well-aerated, moist soils, elemental S^0 is attacked by microorganisms to form SO_4^{2-}. The net effect of S^0 oxidation is to lower the soil pH, which is often desirable in calcareous and saline soils. For neutral or acid soils, however, the acidifying effect is undesirable; accordingly, use of elemental S^0 as a soil amendment can only be recommended under special conditions.

At one time, attempts were made to utilize S-oxidizing bacteria for the formation of soluble phosphate from rock phosphate, both in composts and by additions of S^0 with rock phosphate fertilizer. The H_2SO_4 thus formed reacts with the phosphate rock to produce the more soluble mono- and dicalcium phosphates. The practice has not been found to be feasible under most soil conditions.

Species and Metabolism of S-Oxidizing Bacteria in Soil

Included among the S-oxidizing bacteria are the chemolithotrophic bacteria of the genera *Thiobacillus* (family Nitrobacteriaceae), colorless filamentous bacteria of the family Beggiatoaceae (e.g., *Beggiatoa* and *Thiothrix*), and the photosynthetic S bacteria (e.g., *Thiospirillum* and *Thiocystis*). In general, only organisms of the first group are important in soils; photosynthetic bacteria are confined mostly to aquatic environments where H_2S is formed. A number of heterotrophic microorganisms, including bacteria, actinomycetes, and fungi, are also capable of oxidizing reduced forms of inorganic S, but their importance in soil is unknown. The subject of S oxidation in soil is reviewed by Wainwright.[17]

Sulfur-oxidizing bacteria of the genus *Thiobacillus* are non-spore-forming, Gram-negative rods about 0.3 μm in diameter and 1 to 3 μm long. Most species are motile by polar flagella. Typical *Thiobacillus* oxidations and S-oxidation states are:

$$HS^- \rightarrow \quad S^0 \quad \rightarrow SO_3^{2-} \rightarrow SO_4^{2-}$$

| -2 | 0 | $+4$ | $+6$ |
| Sulfide | Elemental S | Sulfite | Sulfate |

Other S compounds are formed during this process by chemical reactions between HS^- and SO_3^{2-}; examples include thiosulfate ($S_2O_3^{-2}$) and tetrathionate ($S_4O_6^{2-}$), both of which are also oxidized by many of the species listed below.

A number of *Thiobacillus* species have been identified, five of which are well defined and of some importance in soils. Like the nitrifiers, these organisms obtain their energy by oxidation of an inorganic atom, and most of them fix CO_2 as a primary or sole carbon source; i.e., they are chemoautotrophs. The more important oxidizers are described below.

T. thiooxidans   This obligate autotroph, first isolated by Waksman and Joffe,[103] is remarkable in that it is tolerant to extreme acidity. Optimum pH for oxidation is pH 2.0 to 3.5; growth essentially ceases above pH 5.5. The organism is frequently encountered in environments where very acidic conditions exist due to oxidation of sulfides. Oxidation reactions of *T. thiooxidans* include the following:

$$2\ S^0 + 3\ O_2 + 2\ H_2O \rightarrow 4\ H^+ + 2\ SO_4^{2-}$$

$$S_2O_3^{2-} + 2\ O_2 + H_2O \rightarrow 2\ H^+ + 2\ SO_4^{2-}$$

$$2\ S_4O_6^{2-} + 7\ O_2 + 6\ H_2O \rightarrow 12\ H^+ + 8\ SO_4^{2-}$$

T. thiooxidans has been implicated in the corrosion of iron pipes and concrete, and it is the organism largely responsible for the production of H_2SO_4 in coal mines and mine spoils. Leachate waters from such environments often contain high amounts of H_2SO_4 and ferrous Fe^{2+}, thereby leading to the pollution of streams and rivers.

T. ferrooxidans   The most distinctive feature of this organism is its ability to produce energy by oxidizing Fe^{2+} to Fe^{3+}, as well as S^0 and $S_2O_3^{2-}$. The organism is an obligately autotrophic aerobe that develops best under highly acidic conditions (optimum pH: 2.0 to 3.5). It is a common inhabitant, along with *T. thiooxidans*, of acid sulfate soils and mine spoils and drainage waters. Oxidation reactions of *T. ferrooxidans* for S^0 and $S_2O_3^{2-}$ are similar to those shown earlier for *T. thiooxidans*.

T. thioparus   This aerobic, obligate autotroph is widely distributed in soils, primarily those with higher pH values than for *T. thiooxidans*. A unique feature of the organism is the deposition of globules of elemental S^0 outside the cells. The main substrate is $S_2O_3^{2-}$, but sulfide, elemental S^0, and tetrathionate are also oxidized. Accumulation of H_2SO_4 can curtail activity of the organism; growth ceases below pH 4.

T. novellus   Unlike other *Thiobacillus* sp., this organism is a facultative autotroph and can grow on organic substrates. Thiosulfate is oxidized to SO_4^{2-} at pH values near neutrality; elemental S^0 cannot be oxidized. The optimum pH for growth is in the neutral or slightly alkaline range.

T. denitrificans   This organism is unique in that, in addition its ability to aerobically oxidize S compounds, it can grow anaerobically using NO_3^- as the electron acceptor for oxidation of S^0. The overall reaction leading to the production of molecular N_2 is as follows:

$$5 \ S^0 + 6 \ NO_3^- + 2 \ H_2O \rightarrow 4 \ H^+ + 5 \ SO_4^{2-} + 3 \ N_2$$

An oxidizing organism of some importance in thermal zones is *Sulfolobus*. This organism resembles *T. thiooxidans*, the main difference being that *Sulfolobus* is thermophilic (optimum temperature for growth is 70–75°C).

General Observations on S Oxidation

The main facts concerning the oxidation of S in soils are:

1. The oxidation of reduced inorganic forms of S in soils is largely a biological process catalyzed primarily by bacteria of the genus *Thiobacillus*.
2. Practically all soils contain S-oxidizing bacteria, although their numbers in many arable soils are low due to lack of reduced S to be oxidized. When reduced forms of S are introduced into the soil, either from the atmosphere or in fertilizer products (e.g., S-coated urea), a rapid increase in numbers of S-oxidizing organisms occurs. Because most soils contain S-oxidizers, artificial inoculation is generally not considered necessary.
3. Thiosulfates, certain sulfides, SO_2, and elemental S^0 of fertilizer formulations (S-coated urea, ammonia-S solutions) are generally oxidized quite readily. Some of the S in sulfides is probably autooxidized to elemental S^0; however, oxidation of S^0 to SO_4^{2-} is largely biological.
4. Environmental factors favoring the growth of soil microorganisms in general also favor the activities of S-oxidizing organisms, with the exception that many of the S-oxidizers are more acid-tolerant than other soil microbial groups. The following statements summarize the effect of these factors on the biological oxidation of S in soils:
   a. *Temperature:* Oxidation of S takes place from 4 to 55°C, but the most favorable range is 27 to 40°C.
   b. *Moisture and aeration:* The moisture content for most rapid S oxidation is near field capacity, although other factors, such as texture, can affect this level. With the exception of *T. denitrificans* and some photosynthetic bacteria, S oxidation requires an environment containing O_2.
   c. *Soil reaction:* Microbial oxidation of S can take place between pH 2 and 9. Most soils have pH values between 3.5 and 8.5, well within this range. Oxidation generally increases with increasing pH, al-

though the process is not critically limited by this soil property. Addition of lime to acidic soils usually increases the rate of S oxidation.

    d. *Microbial population:* Inoculation of arable soils with S-oxidizing organisms usually increases the rate of oxidation because of the low population of indigenous species. Addition of elemental S^0 or some other reduced form has the same effect by stimulating rapid multiplication of S oxidizers.

5. The rate of oxidation of applied elemental S^0, such as for reclamation of saline and alkaline soils, is affected by several factors, some of which are:

    a. *Particle size:* Rate of oxidation increases with a decrease in particle size. Large (6–12 mesh) S^0 granules can be oxidized fairly rapidly provided they disintegrate into small particles after application to the soil.

    b. *Placement:* Mixing elemental S^0 with the soil usually results in the most rapid oxidation. Nevertheless, band placement and top dressing can be quite effective under certain conditions.

    c. *Rate of application:* Increasing the amount of elemental S^0 applied to a soil generally does not affect the percentage that is oxidized within a specified period of time. To a certain degree, slow oxidation of large S^0 particles can be compensated for by increased rates of application.

SUMMARY

Although the role of S in soil fertility has been largely ignored, sufficient information has accumulated to indicate that, in agricultural soils of the humid and semihumid regions, transformations involving organic forms are of utmost importance in the S nutrition of plants. Sulfur deficiencies have been limited to relatively few geographical areas but may be increasing due to more widespread use of S-free fertilizers, a reduction in the amount of S used as a pesticide, and lowering of atmospheric inputs of S. Pertinent information regarding the S cycle can be summarized as follows:

1. Until recently, incidental additions and soil reserves of S have been sufficient to meet the S requirements of most agricultural crops. Some fertilizers (e.g., phosphate) contain S, and some S can be added in insecticides and fungicides. Irrigation water used in arid regions is an important source of plant-available S. Increased usage of S-free fertilizers may eventually induce S deficiency in crops.

2. The amount of S brought down in precipitation varies with location and season of the year. This S arises largely from the combustion of coal and other fossil fuels, which means that larger quantities are deposited

near industrial areas and during winter months. In rural regions remote from industrial sites, no more than 5 to 6 kg/ha of S will be returned each year to the soil in this way. Leaching losses of S are greatest on coarse-textured soils under high rainfall.

3. Plant roots absorb S almost entirely in the SO_4^{2-} form. The concentration of this ion in the soil solution is governed to some extent by those factors affecting adsorption, as well by mineralization–immobilization.

4. As far as plant residues are concerned, the same type of mineralization–immobilization relationship exists for S as is known for N. When the C/S ratio of added residues is below 200, there will be a net gain in inorganic SO_2^{2+}; when the ratio is over 400, there is a net loss.

5. Very little is known about the forms of organic S in soil. About 50% of the total S in soils of humid and semiarid regions occurs in C-bonded forms, only a fraction of which can be accounted for as amino acids. Another 40% of the total S occurs as unknown ester sulfates. The C/S ratio of the soil varies widely within any given location, but the mean ratio for soils from different locations is about 108, corresponding to 0.5% S in the organic matter.

6. Some soils, notably those that are acidic and that contain kaolonite and/or Fe- and Al-oxides, are able to retain SO_4^{2-} in an adsorbed form.

REFERENCES

1. N. C. Brady, *The Nature and Properties of Soils,* Macmillan, New York, 1974.

2. J. J. Germida, M. Wainwright, and V. V. S. R. Gupta, "Biochemistry of Sulfur Cycling in Soil," in G. Stotzky and J.-M. Bollag, Eds., *Soil Biochemistry,* Vol. 7, Marcel Dekker, New York, 1992, pp. 1–53.

3. R. W. Howarth, J. W. B. Stewart, and M. V. Ivanov, Eds., *Sulphur Cycling on the Continents,* SCOPE 48, Wiley, Chichester, England, 1992.

4. D. M. Whelpdale, "An Overview of the Atmospheric Sulphur Cycle," in R. W. Howarth, J. W. B. Stewart, and M. V. Ivanov, Eds., *Sulphur Cycling on the Continents,* SCOPE 48, Wiley, Chichester, England, 1992, pp. 5–26.

5. M. O. Andreae and W. A. Jaeschke, "Exchange of Sulphur Between Biosphere and Atmosphere over Temperate and Tropical Regions," in R. W. Howarth, J. W. B. Stewart, and M. V. Ivanov, Eds., *Sulphur Cycling on the Continents,* SCOPE 48, Wiley, Chichester, England, 1992, pp. 27–61.

6. P. A. Trudinger, "Chemistry of the Sulfur Cycle," in M. A. Tabatabai, Ed., *Sulfur in Agriculture,* American Society of Agronomy, Madison, Wisconsin, 1986, pp. 2–22.

6a. W. W. Kellogg et al., *Science,* **175,** 587 (1972).

7. J. W. Anderson, "Sulfur Metabolism in Plants," in B. J. Miflin and P. J. Lea, Eds., *The Biochemistry of Plants,* Vol. 16, Academic Press, San Diego, 1990, pp. 327–381.

8. J. F. Thompson, I. K. Smith, and J. T. Madison, "Sulfur Metabolism in Plants," in M. A. Tabatabai, Ed., *Sulfur in Agriculture*, American Society of Agronomy, Madison, Wisconsin, 1986, pp. 57–121.

9. K. Mengel and E. A. Kirkby, *Principles of Plant Nutrition*, 3rd ed., International Potash Institute, Bern, Switzerland, 1982.

10. G. M. Pierzynski, J. T. Sims, and G. F. Vance, *Soils and Environmental Quality*, Lewis, Boca Raton, Florida, 1994, pp. 143–165.

11. W. B. McGill and C. V. Cole, *Geoderma*, **26**, 267 (1981).

12. D. C. Whitehead, *Soils Fertilizers*, **27**, 1 (1964).

13. J. R. Freney, "Forms and Reactions of Organic Sulfur Compounds in Soils," in M. A. Tabatabai, Ed., *Sulfur in Agriculture*, American Society of Agronomy, Madison, Wisconsin, 1986, pp. 207–232.

14. M. A. Tabatabai, Ed., *Sulfur in Agriculture*, American Society of Agronomy, Madison, Wisconsin, 1986.

15. F. J. Zhao, J. Wu, and S. P. McGrath, "Soil Organic Sulphur and Its Turnover," in A. Piccolo, Ed., *Humic Substances in Terrestrial Ecosystems*, Elsevier, Amsterdam, 1996, pp. 429–466.

16. R. L. Starkey, *Soil Sci.*, **111**, 297 (1966).

17. M. Wainwright, *Adv. Agron*, **37**, 349 (1984).

18. V. P. Amga and W. J. Cooper, "Biogenic Sulfur Emissions: A Review," in E. S. Saltzman and W. J. Cooper, Eds., *Biogenic Sulfur in the Environment*, American Chemical Society, Washington, D.C., 1989, pp. 2–13.

19. M. J. Mitchell, M. B. David, and R. B. Harrison, "Sulfur Dynamics of Forest Ecosystems," in R. W. Howarth, J. W. B. Stewart, and M. V. Ivanov, Eds., *Sulphur Cycling on the Continents*, SCOPE 48, Wiley, Chichester, England, 1992, pp. 215–254.

20. C. W. Field, "Sulfur: Element and Geochemistry," in R. W. Fairbridge, Ed., *Encyclopedia of Geochemistry and Environmental Sciences*, Vol. 4A, Van Nostrand Reinhold, New York, 1972, pp. 1142–1148.

21. J. J. Schoenau and J. R. Bettany, *J. Soil Sci.*, **40**, 397 (1989).

22. R. J. Haynes and P. H. Williams, *Adv. Agron.*, **49**, 119 (1993).

23. S. P. McGrath and K. T. W. Goulding, *J. Sci. Food Agric.*, **53**, 426 (1990)

24. W. E. Larson, C. E. Clapp, W. H. Pierre, and Y. B. Morachan, *Agron. J.*, **64**, 204 (1972).

25. P. J. Maroulis and A. R. Bandy, *Science*, **196**, 647 (1977).

26. R. Conrad, *Microbiol. Rev.*, **60**, 609 (1996).

27. W. H. Chan and M. A. Lusis, *Water, Air, Soil Pollution*, **20**, 21 (1985).

28. J. J. Schoenau and J. R. Bettany, *Soil Sci. Soc. Amer. J.*, **51**, 646 (1987).

29. C. Bloomfield and J. K. Coulter, *Adv. Agron.*, **25**, 265 (1973).

30. V. P. Evangelou, *Pyrite Oxidation Chemistry*, Lewis, Boca Raton, Florida, 1995.

31. B. Friedman, "Biological Effects of Acidification," in *Environmental Ecology*, 2nd ed., Academic Press, San Diego, 1995, pp. 126–141.

32. M. A. Tabatabai and J. M. Bremner, *Sulphur Inst. J.*, **8**, 1 (1972).

33. C. A. Evans and C. O. Rost, *Soil Sci.*, **59**, 125 (1945).

34. F. H. Bormann et al., *Ecol. Monogr.,* **44,** 255 (1974).

35. D. W. Johnson and D. E. Todd, *Soil Sci. Soc. Amer. J.,* **47,** 792 (1983).

36. D. W. Johnson, G. S. Henderson, and D. E. Todd, *Soil Sci.,* **132,** 422 (1981).

37. M. B. David, M. J. Mitchell, and J. A. Nakas, *Soil Sci. Soc. Amer. J.,* **46,** 847 (1982).

38. M. A. Tabatabai and J. M. Bremner, *Sulphur Inst. J.,* **8,** 1 (1972).

39. M. A. Tabatabai and J. M. Bremner, *Soil Sci.,* **114,** 380 (1972).

40. M. A. Tabatabai, "Sulfur," in D. L. Sparks, Ed., *Methods of Soil Analysis: Part 3, Chemical Methods,* Soil Science Society of America, Madison, Wisconsin, 1996, pp. 921–960.

41. R. W. Blanchar, "Measurement of Sulfur in Soils and Plants," in M. A. Tabatabai, Ed., *Sulfur in Agriculture,* American Society of Agronomy, Madison, Wisconsin, 1986, pp. 455–490.

42. C. M. Johnson and H. Nishita, *Anal. Chem.,* **24,** 736 (1952).

43. J. R. Bettany, J. W. B. Stewart, and E. H. Halstead, *Soil Sci. Soc. Amer. Proc.,* **37,** 915 (1973).

44. K. M. Goh and M. R. Williams, *J. Soil Sci.,* **33,** 73 (1982).

45. A. M. L. Neptune, M. A. Tabatabai, and J. J. Hanway, *Soil Sci. Soc. Amer. Proc.* **39,** 51 (1975).

46. J. R. Freney, G. E. Melville, and C. H. Williams, *Soil Sci.,* **109,** 310 (1970).

47. J. R. Freney, G. E. Melville, and C. H. Williams, *J. Sci. Food Agric.,* **20,** 440 (1969).

48. L. E. Lowe and W. A. deLong, *Can. J. Soil Sci.,* **43,** 151 (1963).

49. K. D. McLachan and D. G. DeMarco, *Aust. J. Soil Res.,* **13,** 169 (1975).

50. R. G. McLaren and R. S. Swift, *J. Soil Sci.,* **28,** 445 (1977).

51. F. R. Bettany, J. W. B. Stewart, and S. Saggar, *Soil Sci. Soc. Amer. J.,* **43,** 981 (1979).

52. M. L. Nguyen and K. M. Goh, *N.Z. J. Agric. Res.* **33,** 111 (1990).

53. A. Ghani, R. G. McLaren, and R. S. Swift, *Biol. Fert. Soils,* **11,** 68 (1991).

54. W. Anderson, S. Saggar, J. R. Bettany, and J. W. Stewart, *Soil Sci. Soc. Amer. J.,* **45,** 767 (1981).

55. J. R. Freney, F. J. Stevenson, and A. H. Beavers, *Soil Sci.,* **114,** 468 (1972).

56. N. M. Scott, W. Bick, and H. A. Anderson, *J. Sci. Food Agr.,* **32,** 21 (1981).

57. Y. M. Chae and L. E. Lowe, *Can. J. Soil Sci.,* **60,** 633 (1980).

58. Y. M. Chae and L. E. Lowe, *Soil Biol. Biochem.,* **13,** 257 (1981).

59. Y. M. Chae and M. A. Tabatabai, *Soil Sci. Soc. Amer. J.,* **45,** 20 (1981).

60. L. E. Lowe and R. M. Buston, *Can. J. Soil Sci.,* **69,** 287 (1989).

61. J. I. Keer, R. G. McLaren, and R. S. Swift, *Soil Biol. Biochem.,* **22,** 97 (1990).

62. F. J. Zhao and S. P. McGrath, *Plant Soil,* **164,** 243 (1994).

63. C. H. Williams and C. M. Donald, *Aust. J. Agr. Res.,* **8,** 179 (1957).

64. J. Wu and A. G. O'Donnell, *Soil Biol. Biochem.,* **25,** 1567 (1993).

65. A. Ghani, R. G. McLaren, and R. S. Swift, *Soil Biol. Biochem.,* **24,** 331 (1992).

66. I. Valeur and I. Nilsson, *Soil Biol. Biochem.,* **25,** 1343 (1993).

67. B. H. Ellert and J. R. Bettany, *Soil Sci. Soc. Amer. J.,* **56,** 1133 (1992).

68. H. J. Pirela and M. A. Tabatabai, *Biol. Fert. Soils,* **6,** 26 (1988).

69. K. M. Stanko-Golden and J. W. Fitzgerald, *Soil Biol. Biochem.,* **23,** 1053 (1991).

70. A. G. O'Donnell, J. Wu, and J. K. Syers, *Soil Biol. Biochem.,* **26,** 1507 994).

71. R. J. Haynes and P. H. Williams, *Soil Biol. Biochem.,* **24,** 209 (1992).

72. S. D. Castellano and R. P. Dick, *Soil Sci. Soc. Amer. J.,* **54,** 114 (1990).

73. A. Ghani, R. G. McLaren, and R. S. Swift, *Soil Biol. Biochem.,* **25,** 1739 (1993).

74. M. A. Tabatabai and A. A. Al-Khafaji, *Soil Sci. Soc. Amer. J.,* **44,** 1000 (1980).

75. B. A. Stewart and C. J. Whitfield, *Soil Sci. Soc. Amer. Proc.,* **29,** 752 (1965).

76. J. W. Fitzgerald, T. C. Strickland, and W. T. Swank, *Soil Biol. Biochem.,* **14,** 529 (1982).

77. W. T. Swank, J. W. Fitzgerald, and J. T. Ash, *Science,* **223,** 182 (1984).

78. S. Saggar, J. R. Bettany, and J. W. B. Stewart, *Soil Biol. Biochem.,* **13,** 493, 499 (1981).

79. J. E. Strick and J. P. Nakas, *Soil Biol. Biochem.,* **16,** 289 (1984).

80. M. R. Banerjee, S. J. Chapman, and K. Killham, *Comm. Soil Sci. Plant Anal.,* **24,** 939 (1993).

81. S. D. Castellano and R. P. Dick, *Soil Sci. Soc. Amer. J.,* **55,** 263 (1991).

82. S. J. Chapman, *Soil Biol. Biochem.,* **19,** 301 (1987).

83. J. Wu, A. G. O'Donnell, Z. L. He, and J. K. Syers, *Soil Biol. Biochem.,* **26,** 117 (1994).

84. V. V. S. R. Gupta and J. J. Gernida, *Can. J. Soil Sci.,* **69,** 889 (1989).

85. A. R. Autry and J. W. Fitzgerald, *Soil Biochem.,* **23,** 689 (1991)

86. P. J. M. Cooper, *Soil Biol. Biochem.,* **4,** 333 (1972).

87. M. A. Tabatabai and J. M. Bremner, *Soil Sci. Soc. Amer. Proc.,* **34,** 225,427 (1970).

88. A. N. Ganeshamurphy and N. E. Nelson, *Soil Biol. Biochem.,* **22,** 1163 (1990)

89. G. Lou and P. R. Warman, *Biol. Fert. Soils,* **14,** 112, 267 (1992).

90. J. R. Bettany, J. W. B. Stewart, and E. H. Halstead, *Can. J. Soil Sci.,* **54,** 309 (1974).

91. J. R. Freney, G. E. Melville, and C. H. Williams, *Soil Biol. Biochem.,* **3,** 133 (1971).

92. J. R. Freney, G. E. Melville, and C. H. Williams, *Soil Biol. Biochem.,* **7,** 217 (1975).

93. K. M. Goh and P. E. H. Gregg, *New Zealand J. Sci.,* **25,** 135 (1982).

94. P. E. H. Gregg and K. M. Goh, *New Zealand J. Sci.,* **22,** 425 (1975).

95. R. G. McLaren, J. I. Keer, and R. S. Swift, *Soil Biol. Biochem.,* **17,** 73 (1985).

96. A. Ghani, R. G. McLaren, and R. S. Swift, *Soil Biol. Biochem.,* **25,** 327, 1739 (1993).

97. A. J. Nicolson, *Soil Sci.,* **109,** 345 (1970).

98. R. L. Westerman, Ed., *Soil Testing and Plant Analysis,* Book Series Number 3, Soil Science Society of America, Madison, Wisconsin, 1990.

99. C. C. Ainsworth and R. W. Blanchar, *J. Environ. Qual.,* **13,** 193 (1984).

100. I. R. Kaplan, "Sulfur Cycle," R. W. Fairbridge, Ed., *Encyclopedia of Geochemistry and Environmental Sciences,* Vol. 4A, Van Nostrand Reinhold, New York, 1972, pp. 1148–1152.

101. Y. M. Nor and M. A. Tabatabai, *Soil Sci. Soc. Amer. J.,* **41,** 736 (1977).

102. M. Wainwright, *Soil Biol. Biochem.,* **11,** 95 (1979).

103. S. A. Waksman and J. S. Joffe, *J. Bact.,* **7,** 239 (1922).

11

MICRONUTRIENTS AND TOXIC METALS

Micronutrients are chemical elements that are required in small amounts by living organisms, where they occur as vital constituents of certain enzymes and growth hormones (Table 11.1). Enzymes that contain trace elements play critical roles in carbohydrate metabolism (photosynthesis, respiration), N metabolism (biological N_2 fixation and protein synthesis), cell wall metabolism (lignin synthesis), water relations, ion uptake, seed production, metabolism of secondary plant substances, and disease resistance.

The essential trace metals for plants are iron (Fe), zinc (Zn), manganese (Mn), copper (Cu), boron (B), molybdenum (Mo), and nickel (Ni). Four elements, cobalt (Co), chromium (Cr), selenium (Se), and tin (Sn), are not required by plants but are essential for animals. Some other elements, notably arsenic (As), cadmium (Cd), mercury (Hg), and lead (Pb), are not required, but they have been studied extensively because they are potentially hazardous to plants, animals, and microorganisms.

An outline of micronutrient behavior in soil is depicted in Fig. 11.1. Concentrations of elements as free ions in solution, or as soluble metal–chelate complexes, are influenced by chemical reactions such as fixation onto surfaces of clays and metal oxides, complexation with humic substances, formation of sparingly soluble minerals, and changes in oxidation state. Microorganisms can solubilize minerals and modify the redox potential (E_h) and pH of the soil, and thereby change the availability of micronutrients to plants. Recycling of micronutrients occurs when plant residues are returned to soil, which is an important process in micronutrient-deficient soils. Production of chelating agents by microorganisms and their secretion from plant roots promote dissolution and weathering of rocks and minerals and facilitate movement of micronutrients to roots.

Reviews are available for many aspects of micronutrient behavior, including biogeochemistry,[1–3] role in plant nutrition,[4,5] distribution, forms, and re-

TABLE 11.1 Incomplete List of the Functions of Micronutrients in Plants

Micronutrient	Functions in Plants
Fe	An essential component of many heme and nonheme Fe enzymes and carriers, including the cytochromes (respiratory electron carriers) and the ferredoxins. The latter are involved in key metabolic functions, such as N_2 fixation, photosynthesis, and electron transfer.
Zn	An essential component of several dehydrogenases, proteinases, and peptidases, including carbonic anhydrase, alcohol dehydrogenase, glutamic dehydrogenase, and malic dehydrogenase.
Mn	Involved in the O_2-evolving system of photosynthesis and a component of the enzymes arginase and phosphotransferase.
Cu	A constituent of a number of important enzymes, including cytochrome oxidase, ascorbic acid oxidase, and laccase.
B	The specific biochemical function of B is unknown, but it may be involved in carbohydrate metabolism and synthesis of cell-wall components.
Mo	Required for the normal assimilation of N in plants. An essential component of nitrate reductase, as well as nitrogenase.

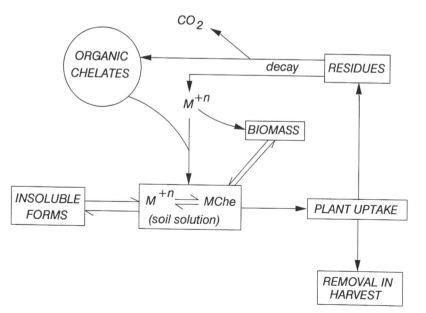

Fig. 11.1 Micronutrient cycle in soil.

actions in soils,[6,7] complexes with organic matter,[8–11] and impact on health and the environment.[12–15]

DISTRIBUTION OF TRACE ELEMENTS IN SOILS

Trace elements in soil are derived from a variety of sources, including:

1. Parent rocks and minerals from which the soil was formed
2. Impurities or contaminants in soil amendments, such as fertilizers and lime, pesticides, manures, and sewage sludge (biosolids)
3. Airborne and/or water-deposited particles from mining, smelting, and industrial activities, fossil fuel combustion, wind-eroded soil particles, and meteoric and volcanic materials that settle or are added in precipitation

The main source of trace elements in most soils is the parent rock material from which the soil was formed.[1,2] The quantities of trace elements added to soil in ordinary amendments (item 2) are generally too low to significantly change the total amounts present, but they may have an influence on the amounts taken up by the plant.

The micronutrient content of igneous rocks, and the main minerals in which they occur, are listed in Table 11.2. Iron is the most abundant mineral by far, since it is a major constituent of ferromagnesian minerals. The sulfides of igneous rocks are important original sources of Zn, Cu, and Mo. Zinc, Mn, and Cu also occur in ferromagnesian minerals, where they substitute for Fe and Mg in the mineral matrix. Boron is found largely in the borosilicate mineral tourmaline.

The two groups of igneous rocks (granite and basalt) have characteristic arrays of trace elements. Iron, Zn, Mn, and Cu are somewhat more abundant

TABLE 11.2 Sources, Ionic Forms, and Abundance of Micronutrients in Igneous and Sedimentary Rocks[a]

Nutrient	Common Soluble Ionic Forms	Igneous Rocks		Sedimentary Rocks		
		Granite	Basalt	Limestone	Sandstone	Shale
Iron	Fe^{2+}, Fe^{3+}	27,000	86,000	3,800	9,800	47,000
Zinc	Zn^{2+}	40	100	20	6	95
Manganese	Mn^{2+}	400	1500	1100	<100	850
Copper	Cu^{2+}	10	100	4	30	45
Boron	BO_3^{3-}	15	5	20	35	100
Molybdenum	MoO_4^{2-}	2	1	0.4	0.2	2.6

[a]From Krauskopf[1] and Hodgson.[2]

in basalt; B and Mo are more concentrated in granite.[1] During the various stages of soil formation, trace elements are released from the primary minerals and incorporated into other forms, such as structural components of secondary silicate (clay) minerals, as complexes with organic matter, or as occlusions in Fe and Mn oxides, as noted in Fig. 11.1.

Sedimentary rocks also contain trace elements (see Table 11.2). Sediments contain debris from many sources and therefore contain all of the trace elements found in igneous rocks, although not necessarily in the same proportion. Processes such as disintegration, weathering, mixing, selective leaching and deposition, together with other complex factors, lead to redistribution of the elements and, in some cases, changes in their chemical forms. Fine-grained sediments (e.g., shales) become enriched in trace elements such as Zn, Cu, Co, B, and Mo. Shales that contain high amounts of organic matter are often enriched in Cu and Mo.

The concentration ranges of micronutrient cations in soils are given in Fig. 11.2, where the broken lines indicate the usual range for agricultural plants. Note that the amount of any given trace element in the soil can vary as much as a thousandfold, depending on the nature of the soil (notably texture) and its origin, thereby creating situations of localized deficiencies or excesses. The trace element content usually follows closely the clay and/or organic matter content.[2] Alluvial sands and certain organic soils often have low micronutrient contents.

Except for Fe and Mn, the total amount of any given trace element per unit mass of soil is very low, usually only a few kg/ha within the plow depth. Numerous exceptions can be found, however. For example, some volcanic

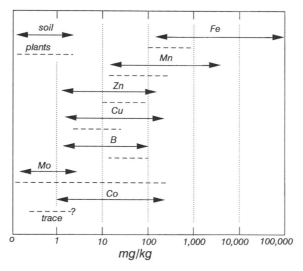

Fig. 11.2 Concentration range of micronutrient cations in soil. The broken lines indicate the usual range found in crop plants.

ash soils of Hawaii have high Mn contents ($>10\%$), far above the usual range shown in Fig. 11.2. Boron and Mo are unique in that they occur as anions ($H_2BO_3^-$ and MoO_4^{2-}) and are thus subject to losses though leaching.

FORMS OF TRACE ELEMENTS IN SOIL

The micronutrient cations in soil occur in the following forms, or pools:

1. Water-soluble;
   a. As the free cation;
   b. As complexes with organic and inorganic ligands.
2. On exchange sites of clay minerals. Cations held in this manner are defined as those that can be extracted with a weak exchanger, such as NH_4^+.
3. Specifically adsorbed. Some trace elements (e.g., Cu^{2+}) are retained by clay minerals and/or Fe and Mn oxides in the presence of a large excess of Ca^{2+} or some other electrostatically bonded cation. Trace elements bound in this manner are said to be "specifically adsorbed."
4. Adsorbed or complexed by organic matter, including plant residues, native humus, and living organisms (soil biomass).
5. As insoluble minerals, including occlusion by Fe and Mn oxides.
6. As sparingly soluble minerals and as cations that have undergone isomorphous substitution for Fe and Al in octahedral positions of silicate clays.

The distinction between the various forms is not well defined, and the status of a particular element can change quite rapidly as the result of biological activity. Zinc, Cu, and Co are present as divalent cations, Fe is found as the cations Fe(II) or Fe(III), and Mn is present as either Mn(II) or Mn(IV). In contrast, B is found primarily as the oxyanions HBO_3^- or BO_3^{2-} (primarily HBO_3^-). Molybdenum is usually present as the molybdate (MoO_4^{2-}) ion.

In most soils, only very small amounts of micronutrient cations are found in plant-available forms, consisting of the water-soluble (1) and exchangeable (2) pools. However, these pools are probably in equilibrium with the specifically adsorbed (3), organically bound (4), and sparingly soluble minerals (6), as illustrated below for Cu.[16]

$$\text{Sparingly soluble Cu - minerals} \rightleftharpoons \text{Exchangeable - and soluble - Cu} \rightleftharpoons \text{Specifically adsorbed - Cu} \rightleftharpoons \text{Organically bound - Cu}$$

Thus, when the concentration of micronutrients in the soil solution is reduced by plant uptake or diluted by excess moisture from rainfall, release of micronutrients from insoluble forms may occur. In many soils, the organic pool is of considerable significance because of its large size. The percentage of trace elements in soil that occurs in association with organic matter will be particularly high in soils rich in organic matter (e.g., Histosols and Mollisols) and will be influenced by such properties as pH and kind and amount of clay. Binding of the divalent micronutrient cations (e.g., Cu^{2+}, Zn^{2+}) by organic matter will also be influenced by the extent to which binding sites are occupied by *trivalent* cations (e.g., Fe^{3+} and Al^{3+}). Trivalent cations generally form stronger complexes with organic molecules than divalent ones and thereby competitively suppress adsorption of the latter.

Mechanisms by which trace elements are specifically adsorbed by oxides and layer silicate minerals (3) are unknown, but hydrolysis of the metal ion and adsorption of weakly hydrated species (e.g., MOH^+) may be involved. Another possibility is the formation of covalent linkages with surface hydroxyl groups, such as a Cu—O—Fe bond.

The transition metals also have the potential for being "fixed" by clays through displacement of Mg^{2+} from octahedral sites of the clay. This is possible because the ionic radius of Co^{2+} and Zn^{2+} is similar to that of Mg^{2+}.

Many noncalcareous soils contain trace elements that are solubilized by chemical treatments that dissolve Fe and Mn oxides. These elements are believed to reside within oxide structures (5). Insoluble oxides, formed during weathering, provide reactive surfaces for adsorption of trace elements, which in turn are occluded as precipitation progresses over time.

Soils that contain high amounts of Fe and Mn oxides (e.g., those derived from volcanic ash) will contain relatively high amounts of trace elements in oxide-occluded forms. Trace elements held in this manner are not readily available to plants.

In many soils, the bulk of the trace elements occurs either as sparingly soluble primary or secondary minerals or in octahedral positions of crystalline silicate clays (6). The amounts found in primary minerals are usually low in highly weathered soils of humid and subhumid regions, but they may be appreciable in unweathered soils of arid and semiarid regions. Some release of trace elements from primary minerals undoubtedly occurs during weathering, but, except in soils that are strongly acidic, the process proceeds at an exceedingly slow rate. As shown below, the disintegration of minerals, with liberation of micronutrients, can be enhanced in the presence of organic chelating agents.

Fractionation Schemes

Attempts have been made from time to time to partition the trace elements in soil by sequential extraction procedures.[16–20] A typical example is given in Fig. 11.3.

The percentage distribution of an element among these pools is highly variable among different soils and depends on the metal cation, soil pH, type and quantity of clay minerals, and organic matter content. In general, <2%

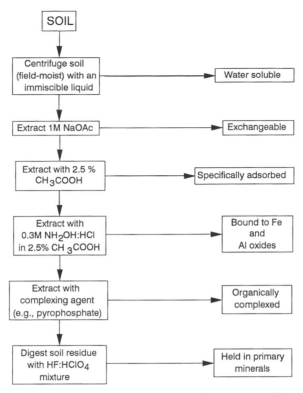

Fig. 11.3 Scheme for the fractionation of trace elements in soil by sequential extraction.[17–20]

will occurs in water-soluble and exchangeable forms (pools most readily available to plants). For some micronutrients (notably Cu), a significant fraction of the cation in the water-soluble pool may occur in chelated form. The organically complexed pool and insoluble minerals represent major "storehouse" forms of micronutrient cations in many soils.

McLaren and Crawford[16] found this distribution of Cu in 24 contrasting soil types:

	% of Total Cu
Soluble + exchangeable	0.1–0.2
Specifically adsorbed by clay	0.2–2.7
Organically bound	16.2–46.9
Oxide- (and organic matter-) occluded	0.0–35.9
Mineral lattice	33.6–77.2

For the soils examined, from one-fifth to one-half of the Cu occurred in organically bound forms. Oxide-occluded forms were important only in soils that contained appreciable amounts of Fe or Mn oxides. Essentially all of the trace elements in the uppermost organic layer of forest soils (Alfisols, Spodosols) occur in organically bound forms.

Shuman[20] reported these percentages of Cu, Mn, and Zn in organically bound forms in 10 representative soils of the southeastern United States: Cu, 1.9 to 68.6%; Mn, 9.5 to 82.0%; Zn, 0.2 to 14.3%. A similar range for Zn (0.1–7.4%) has been obtained for soils of the Appalachian, Coastal Plain, and Piedmont Regions of Virginia.[21] Miller et al.[17] found that significant amounts of the trace elements in some contaminated soils of industrialized northwestern Indiana occurred in the organically bound fraction.

Unlike Co, Cu, and Zn, both Fe and Mn can be oxidized to higher valence states, from which highly insoluble oxides and phosphates can be formed.

Borate Complexes with Organic Matter

Boron is present in the soil parent material as the mineral tourmaline, but its main form in soil may be in combination with organic matter, as borate complexes with *cis*-hydroxyl groups:

I

As organic matter is mineralized by microorganisms, B is released to available forms (i.e., as the oxyanions $H_2BO_3^-$ or BO_3^{2-}, but primarily the former). The temporary appearance of B deficiency in plants during periods of drought has been attributed to reduced mineralization of organically combined B.

Yermiyaho et al.[22] found that the sorption capacity of composted organic matter for B (on a weight basis) was at least four times greater than for clay and soils. This was attributed to chemical association between B and the organic matter.

Speciation of Trace Metals in the Soil Solution

Natural waters from all sources—including lakes, streams, estuaries, and the ocean—contain trace metals in organically bound forms.[23,24] Trace elements in the soil solution also occur partly in organically bound forms.[25,26] Organic ligands enhance the availabilities of trace elements to plants;[27] in some instances, the toxicity of the free metal ion can be reduced or eliminated as a result of metal–chelate interactions. In aqueous systems, speciation of trace elements can dramatically alter their toxicities to fish and other aquatic organisms. The free (hydrated) metal ion is the most toxic, but most stable complexes are nontoxic.

Reports in the scientific literature on chemical forms of trace elements in the soil solution are relatively scarce, which can be attributed to the following factors:[27]

1. The concentrations of metal ions in the soil solution are normally very low, often on the order of 10^{-8} to 10^{-9} M, values that approach the lower limits of analytical capabilities.
2. The trace element may exist in a large number of different chemical forms, which can change seasonally and during sample preparation.
3. Extracts that are representative of the soil solution are not easily obtained.
4. The amounts and chemical forms of any given trace element in the soil solution will vary over time and may be affected by method of sample preparation, including drying and storage.

A common method of determining the concentrations of trace elements in aqueous solutions is by atomic absorption spectrometry. However, this technique does not differentiate between free (M^{2+}) and bound forms of the trace element.

Organically bound forms have been determined in the following ways:

1. By addition of a chelating agent to the soil extract that forms a complex with the free form of the trace element and is subsequently extracted with an immiscible solvent. Organically bound forms of the trace element are taken as the difference between the total amount in solution and that recovered as the free cation. A limitation of this method is that some inorganically bound species (e.g., chloro complexes) cannot be differentiated from soluble organic forms.
2. By separation of charged species on a cation-exchange resin column; organically complexed forms of the trace element are not retained. Inorganic complexes also pass through the resin, but their amounts can be calculated from thermodynamic data.
3. By recovery of the metal–organic complex through dialysis. This method is applicable only when the organic material is of high molecular weight and does not pass through the dialysis membrane.
4. By direct analysis of the free cation with an ion-selective electrode (ISE) or by anodic stripping voltammetry (ASV).

Increasing use has been made of ISE and ASV for determining the speciation of trace elements in the soil solution, as well as natural waters. A major limitation of ISE is its low sensitivity; furthermore, only a few divalent cations can be measured (Cu^{2+}, Pb^{2+}, Cd^{2+}, Ca^{2+}). The technique has the greatest potential where high concentrations of the cation of interest are present in the soil solution, such as in sludge-amended soils. The method of ASV is nondestructive (thereby allowing responses to changes in chemical and physical parameters to be monitored on the same sample) and is capable of measuring free metal ions at the low concentrations found in soil extracts (and natural waters). For both ISE and ASV, electrode response is affected by pH, ionic strength, and sorption of organics on the electrode surface. A review of electrochemical techniques for trace element speciation in natural waters is given by Florence.[24]

Oxidation of lake and river water by ultraviolet irradiation in the presence of H_2O_2 generally increases the concentration of metal ions detectable by ASV, and the amount released by oxidation has been designated organically bound.[23] Results obtained with river and reservoir water have shown that substantial amounts of the Cu, and lesser amounts of the Cd, Pb, and Zn, are associated with organic matter. In some cases, appreciable amounts of the bound metals were accounted for as complexes with inorganic ligands. Extrapolation of these findings to the soil solution suggests that high results will be obtained for organically bound metals when the organically bound fraction is taken as the difference between the total concentration in solution and the free cationic form, as has sometimes been done.

Speciation of metal ions in the soil solution has been predicted on the basis of analytical data for cations, anions, and soluble organic matter, for which computer models (e.g., GEOCHEM, MINTEQ) have been applied.[28–31] The approach is applicable only to systems well characterized with regard to cationic (e.g., K^+, Na^+, NH_4^+, H^+, Ca^{2+}, Mg^{2+}, Cu^{2+}) and anionic (e.g., Cl^-, HCO_3^-, NO_3^-, SO_4^{2-}) species, and ideally for the kinds and amounts of organic ligands. In general, these studies have indicated that those metals that form strong metal complexes (like Cu) occur mostly in organically complexed forms, while those that form weak complexes (like Zn) occur mostly in free ionic forms.

The content of dissolved organic matter in soils varies in both content and composition; thus, the percentage of trace elements in organically bound forms is highly variable. Other factors affecting speciation include pH and types of competing inorganic ligands. Like the river and reservoir water research, soil work indicates that those elements that form strong chelate complexes with organic matter occur largely in organically complexed forms, and those that form weak complexes occur mostly as free ionic or inorganic complexed forms.

Finally, it should be noted that electrochemical neutrality must be maintained in the soil solution. Accordingly, the total quantity of cations in ionic forms must equal the total anionic content. Cronan et al.[32] determined organically bound forms of trace elements in the leachates of some New Hampshire subalpine forests from the deficit between total cations and total anions.

REACTIONS OF MICRONUTRIENTS WITH SOIL ORGANIC MATTER

Organic matter plays a key role in the availability of micronutrients in soil. Formation of metal–organic complexes affects the behavior of micronutrient cations in two opposing ways. Complexation by insoluble organic matter reduces bioavailability, whereas the formation of soluble organic complexes enhances bioavailability. Micronutrients that form strong complexes with organic ligands are influenced to a greater extent than those that form weak complexes. The low incidence of Cu deficiencies in mineral soils in general has been attributed to the formation of soluble complexes by natural chelating agents in the soil solution.[2]

The rhizosphere (soil in the intermediate vicinity of plant roots) contains microorganisms that synthesize large quantities of organic acids and other biochemical chelating agents.[8–11,27] Plant roots also exude a wide variety of organic chelates. Differences in the susceptibilities of plant species to trace element deficiencies have often been attributed to variations in the plants' ability to synthesize and excrete metal complexing organics.

The ability of soil organic matter to form stable complexes with metal ions has been well established. Some of the metals occurring naturally in soil, or introduced in fertilizers, are held as insoluble complexes and are unavailable to plants. On the other hand, many metals that ordinarily would convert to insoluble precipitates at the pH values found in many productive agricultural soils are maintained in solution through chelation. Organic matter reactions involving metal ions are discussed in several reviews.[8–11,27]

Significance of Complexation Reactions

The formation of metal–organic complexes may have the following effects in soil:

1. Metal ions that would ordinarily convert to insoluble precipitates at the pH values found in productive agricultural soils are maintained in solution. Many biochemicals synthesized by microorganisms, as well as fulvic acids, form soluble complexes with metal ions and thereby influence the availability of trace elements to higher plants and to soil microorganisms. Organic substances play a particularly significant role in the behavior of trace elements in peat.

2. Polyvalent cations serve as linkages between humic substances and clay minerals, thereby changing the aggregate structure and affecting the physical properties of soils.

3. Complexing agents of various types may function as carriers of toxic metals in natural waters. Under certain conditions, metal ion concentra-

tions in the soil solution may be reduced to a nontoxic level through complexation with soil organics. As shown below, the interaction of Al^{3+} with organic substances is of considerable importance in controlling soil solution levels of the highly toxic Al^{3+} ion in acidic soils.

4. Organic substances can enhance the availability of insoluble mineral phosphates by complexation of Fe and Al in acidic soils and Ca in calcareous soils.

5. Natural complexing agents are important in transport of trace elements from soil to aquatic systems such as lakes and streams. Research on acid deposition has implicated organic substances in the solubilization and transport of Al from terrestrial environments to natural waters.

6. Organic matter acts as a "buffer" in ameliorating the adverse effects of toxic heavy metals in soil, such as Pb and Cd, that are introduced into soil by atmospheric deposition, notably near smelters and highways.

7. Organic substances are involved in the weathering of rocks and minerals, and they serve as agents for the transport of sesquioxides in leached terrestrial soils, notably Spodosols. Furthermore, they are involved in the neogenesis of certain minerals. Lichens, as well as bacteria and fungi, bring about the disintegration of rock surfaces to which they are attached by producing organic chelating agents.

Individual effects, as itemized above, are difficult to quantify and will vary from soil to soil, depending on organic matter content, pH, kind and amount of clay minerals, and soil management practice. The role of organic substances in the solubilization of mineral phosphates (Item 4) is discussed in Chapter 9.

Properties of Metal–Chelate Complexes

A metal ion in aqueous solution has attached water molecules that are oriented such that the negative (oxygen) end of the water dipole is directed towards the positively charged metal ion. A complex arises when water molecules surrounding the metal ion are replaced by other molecules or ions, with formation of one or more coordinate linkages. The compound that combines with the metal ion is commonly referred to as the ligand.

A covalent bond consists of a pair of electrons shared by two atoms and occupying two orbitals, one of each atom. Essentially, a coordinate complex arises because the outer electron shell of the metal ion is not completely filled and can accept additional pairs of electrons from atoms that have a pair of electrons available for sharing.

A *chelate* complex is formed when two or more coordinate positions about the metal ion are occupied by donor groups of a single ligand to form an internal ring structure. The word "chelate" is derived from the Greek *chele,* meaning a crab's claw, and refers to the pincer-like manner in which the metal is bound. If the chelating agent forms two bonds with the metal ion, a bidentate structure is formed; similarly, there are terdentate, tetradentate, and pentadentate structures. The formation of more than one bond between metal and organic molecule usually imparts high stability to the complex.

The reaction of Cu^{2+} with an amino acid to form a 2:1 complex, along with the Fe-chelate structures of some biochemical compounds known to occur in soil environents, are shown in Fig. 11.4. The Fe-hydroxamate and Fe-catecholate are bidentate structures; the Fe-citrate complex represents a terdentate structure. Binding is influenced by both the nature of the metal ion and the organic ligand. Organic compounds having the greatest potential for binding Fe^{3+} are those that contain oxygen (i.e., —COOH and phenolic-, enolic-, and aliphatic-OH groups); N-containing substances (amino acids, porphyrins) have a high affinity for Cu^{2+} (as well as Ni^{2+}).

Examples of ligand groups contained in organic compounds that have unshared pairs of electrons, and that can form coordinate linkages with metal ions, are shown by structures II to V.

$$
\begin{array}{cccc}
\overset{\displaystyle H}{\underset{\displaystyle |}{R-C}}=\ddot{O}: & \overset{\displaystyle H}{\underset{\displaystyle |}{R-\ddot{S}}}: & \overset{\displaystyle H}{\underset{\displaystyle |}{R-\underset{\displaystyle\cdot\cdot}{N}-H}} & \overset{\displaystyle H}{\underset{\displaystyle |}{R-\ddot{O}}}: \\[2em]
\text{II} & \text{III} & \text{IV} & \text{V}
\end{array}
$$

The order of decreasing affinity of organic groupings for metal ions is approximately as follows:

$$-O^- \; > \; -NH_2 \; > \; -N{=}N{-} \; > \; -N{=}C{-} \; > \; -COO^-$$

enolate   amine   azo   ring N   carboxylate

ion   ion

$$> \; -O{-} \; >> \; -C{=}O$$

ether   carbonyl

The stability of a metal–chelate complex is determined by a variety of factors, including the number of atoms that form a bond with the metal ion, the number of rings that are formed, the nature and concentration of metal ions, and pH. The stability sequence for some select divalent cations is approximately as follows:

$$Cu^{2+} > Ni^{2+} > Co^{2+} > Zn^{2+} > Fe^{2+} > Mn^{2+}$$

This order may vary somewhat, depending upon the nature of the ligand, interactions with other ions, and pH. Trivalent cations (e.g., Fe^{3+} and Al^{3+}) form stronger complexes than divalent cations.

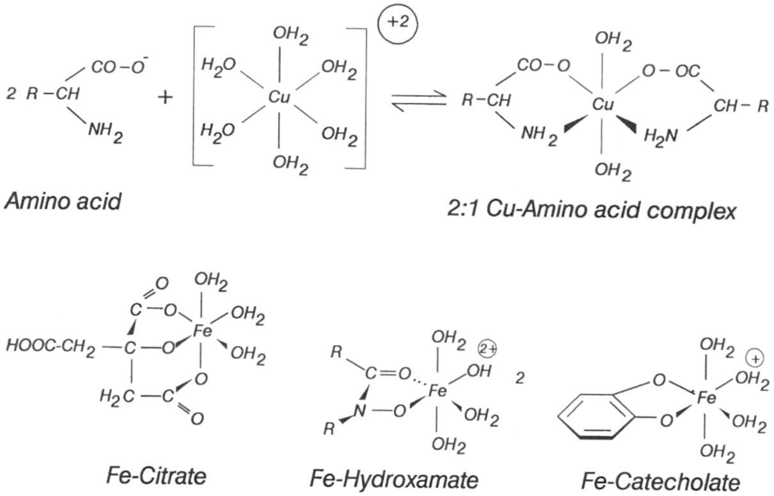

Fig. 11.4 Typical chelate structures of micronutrient cations. Top: reaction of Cu^{2+} with an amino acid to form a 2:1 complex. Bottom: complexes of Fe^{3+} with citrate, hydroxamate, and catechol.

There are two types of stable organo–metal complexes in soil:

1. Biochemicals found in living organisms
2. A series of complex polymers formed by secondary synthesis reactions and bearing no resemblance to common biological components

The first group includes the organic acids, polyphenols, amino acids, peptides, proteins, and polysaccharides. The second group includes the humic and fulvic acids.

The trace elements in soil that occur as *insoluble* complexes with organic matter are largely those that are bound to components of the humic fraction, particularly humic acids. On the other hand, the metals found in *soluble* complexes are mainly those associated with individual biochemical molecules (e.g., organic acids), as well as fulvic acids, which also have high water solubilities.

BIOCHEMICAL COMPOUNDS AS CHELATING AGENTS

Biochemical compounds, such as simple aliphatic acids, amino acids, and polyphenols with chelating characteristics, are continuously produced in soil through the activities of microorganisms. These constituents normally have only a transitory existence. Hence, the amounts present in the soil solution at

TABLE 11.3   Key Biochemical Compounds That Form Complexes with Micronutrient Cations

Compound	Occurrence
Citric, tartaric, lactic, and malic acids	Produced in the rhizosphere and during decay of plant remains. Identified in root exudates, aqueous extracts of forest litter, and canopy drippings
Oxalic acid	Produced by fungi in forest soils, including mycorrhizal fungi. Particularly abundant in acid soils.
Hydroxamate siderophores	Produced in the rhizosphere and by ectomycorrhizal fungi. Greater amounts may be produced when organisms are growing under conditions of Fe limitation.
Phenolic acids	Formed through decay of plant residues, notably lignin. Abundant in canopy drippings and leachates of forest litter. Involved in the mobilization and transport of Fe in acid soils.
Polymeric phenols	Present in high amounts in leachates of forest litter. Produced by lichens growing on rock surfaces.
2-Ketogluconic acid	Synthesized by bacteria living on rock surfaces and in the rhizosphere. Particularly abundant in habitats rich in decaying organic matter.
Mugineic acid	Identified in root washings of Fe-deficient graminaceous plants.

any one time represent a balance between synthesis and destruction by microorganisms. Appreciable quantities of these compounds are found in root excretions, the rhizosphere, and leachates of decaying plant residues, including leaf litter of the forest floor.[8–11,33] A few examples of biochemical chelating compounds, along with a description of the environments where they occur, are presented in Table 11.3.

Leaves, branches, and other organic debris are major sources of soluble organics in forest soils (Alfisols, Spodosols). During decay of forest litter by microorganisms, a wide array of biochemical chelating agents are synthesized and washed into the mineral layer of the soil (A horizon) in percolating waters. Organic chelating substances are also found in canopy drippings and stem flow.

The concentrations of individual biochemical species in the soil solution (20% moisture level) are approximately as follows:[9]

Simple aliphatic acids	1×10^{-3} to 4×10^{-3} M
Phenolic acids	5×10^{-5} to 3×10^{-4} M
Amino acids	8×10^{-5} to 6×10^{-4} M
Aromatic acids	5×10^{-5} to 3×10^{-4} M
Hydroxamate siderophores	1×10^{-8} to 1×10^{-7} M

Some researchers have suggested that because of their rapid destruction by microorganisms, biochemical compounds are not important as natural chelators in soil. However, this argument is not completely valid. The application of modern chromatographic techniques has shown that, while the concentration of any given compound or group of compounds in the soil solution may be low, the combined content of all complexing species may be appreciable. Binding would be expected to be carried out by a relatively large number of ligands present in small amounts rather than by a few dominant compounds present at high concentrations. In many soils, the combined amount of potential chelating agents in the aqueous phase is probably sufficient to account for the minute quantities of soluble metal ions normally present, often in the 10^{-8} to 10^{-9} M range.

Factors affecting the production of biochemical compounds include the moisture status of the soil, plant type and stage of growth, cultural practices, and climate. Amounts would be expected to be relatively high in early fall as decay of plant residues commences, to decrease during the winter due to reduced microbial activity, and to increase again in early spring when plant activity resumes. Soils amended with manures and other organic wastes would be expected to be relatively rich in metal-binding biochemicals.

Organic Acids

Organic acids are of special interest in chelation because of their ubiquity and because hydroxyacids such as citrate are effective solubilizers of minerals. Several reviews[8-11] indicate that a variety of simple organic acids are excreted by plant roots and that soil is a particularly favorable habitat for organic acid-producing bacteria. Short-term anaerobic conditions promote formation of these compounds from crop residues by fermentative bacteria;[34] during this period, levels of soluble metals increase dramatically.[35]

The concentration of organic acids in the soil solution is normally low, but higher amounts occur in the rhizosphere and in leachates from the forest canopy and organic layer of forest soils.

Considerable emphasis has been given to the importance of oxalic acid as a chelator of Fe^{3+} (as well as Al^{3+} and Ca^{2+}) in forest soils.[36] Many fungi are prolific producers of oxalic acid, including the vesicular arbuscular mycorrhizal fungi, in which calcium oxalate crystals can form at the soil–hyphae interface.[37-39] Khan reviews the role of oxalic acid in biological systems.[40]

Organic acids produced by colonizing microorganisms are believed to act as solubilizers of minerals in nature. This work[8-11] has shown that an unusually high proportion of the microorganisms associated with the "raw" soil of rock crevices and the interior of porous weathered stones produce organic acids and solubilize silicate minerals under laboratory conditions. Fungi most active in dissolving silicates are those that produce citric or oxalic acids.[41] Webley et al.[42] summarized the incidence of silicate-dissolving bacteria, actinomycetes, and fungi among isolates from rock sequences and weathered stones (Table 11.4).

Excretion products of roots include a variety of simple organic acids, many of which (e.g., oxalic, tartaric, and citric) are capable of forming complexes with metal ions. Differences in the susceptibilities of plant species to trace element deficiencies have often been attributed to variations in organic acid production.[2]

Sugar Acids

Sugar acids such as gluconic, glucuronic, and galacturonic acids may also be important solubilizers of mineral matter; all are common metabolites of soil microorganisms. Habitats rich in organic matter contain large numbers of microorganisms that produce 2-ketogluconic acid.[41] This compound has also been observed in the rhizosphere of crop plants.[43]

Amino Acids

Amino acids have a role in mineral nutrition and pedogenic processes because they form soluble chelate complexes with metal ions. A wide array of free amino acids have been reported in soils, with the content being influenced by climatic conditions, moisture status, plant species or variety and stage of growth, residue additions, and cultural conditions. Levels seldom exceed 2 mg/kg or 4.5 kg/ha plow-depth, but they may be seven-fold higher in the rhizosphere.

TABLE 11.4   Incidence of Silicate-Dissolving Bacteria, Actinomycetes, and Fungi among the Total Isolates from Rock Sequences and Weathered Stones[a]

	Total Isolates	% of Total Dissolving			
		Ca Silicate[b]	Sollastonite $(CaSiO_3)$[b]	Mg Silicate	Zn Silicate
Bacteria	265	83	57	65	nt[c]
Actinomycetes	39	87	38	46	nt
Fungi	149	94	nt	76	96

[a]From Webley et al.[42]
[b]Ca-silicate is amorphous, produced by reaction between soluble Ca and silicate salts. Wollasonite is a crystalline mineral form of Ca-silicate that is formed by natural processes. Note that the two are not solubilized to the same extent.
[c]nt not tested.

Hydroxamate Siderophores

There is considerable research on the role of hydroxamate siderophores in enhancing iron availability in calcareous soils.[44-47] These substances represent a group of microbially produced Fe^{3+}-transport ligands with stability constants as high as 10^{32}. Because of their high affinity for ferric iron (Fe^{3+}), they act as scavengers in environments such as neutral or alkaline soils in which insoluble ferric salts predominate and mobility is governed by the low solubility of ferric oxides.

Biologically significant levels of hydroxamate siderophores have been observed in soils;[45] greater amounts are produced when the organism is growing under conditions of Fe deficiency (Fe-stress). The amounts contained in the rhizosphere of plants appear to be 10 to 50% higher than in the bulk soil.[46] Hydroxamate siderophores are also produced by soil fungi, including ecto-mycorrhizal fungi, which live in intimate association with plant roots.[47]

Phenols and Phenolic Acids

Phenols and phenolic acids are formed during decay of lignin, and they are synthesized by microscopic fungi when grown on nonlignin carbon sources (see Chapter 1). Phenolic constituents are abundant in forest canopy leachates, decaying plant and animal remains, and root exudates.

Phenols and phenolic acids are believed to be of considerable importance in the complexation and translocation of Fe and Al in forest soils, since high concentrations have been observed in aqueous extracts of decaying organic matter on the forest floor (O_1 and O_2 horizons), as well as in canopy leachates. The phenolic composition of forest soils depends partly on tree species.

Of the phenolic acids, those containing two (or more) adjacent OH groups are particularly effective in forming complexes with metal ions. Typical examples include protocatechuic (VI), gallic (VII), and caffeic (VIII) acids.

Protocatechuic
VI

Gallic
VII

Caffeic
VIII

Polymeric Phenols

Polymeric phenols contain more than one aromatic ring. They include the flavonoids, which comprise one of the most diverse and widespread groups of secondary plant products that are found in aqueous extracts of plant tissues. Structures for some common flavonoids are as follows:

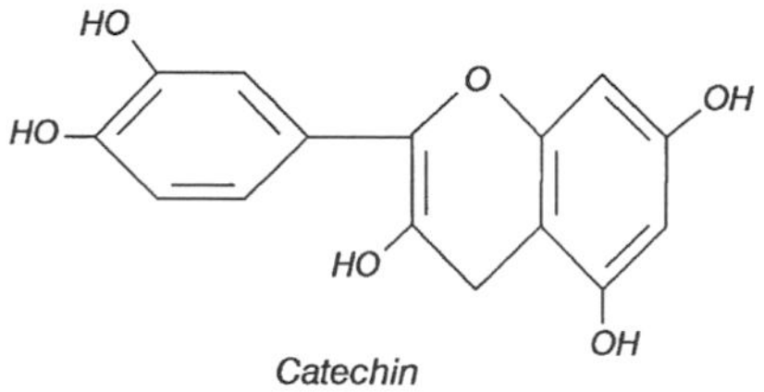

Catechin

IX

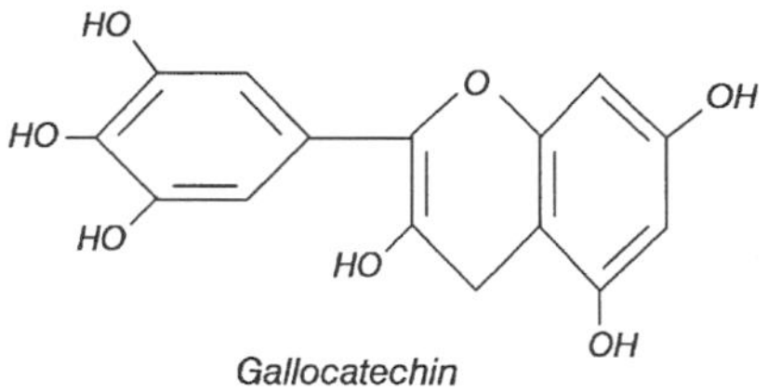

Gallocatechin

X

Flavonoids of the types shown above have been found in aqueous extracts of the leaves and needles of forest trees.

Another group of potential chelate formers, particularly in forest soils, is the tannins. They represent an ill-defined group of substances with molecular weights between 500 and 3000, with at least one or two phenolic OH groups per 100 molecular weight. They are of two main types; hydrolyzable and condensed. Hydrolyzable tannins consist of gallic and/or hexahydroxydi-phenic acids (digallic acid) bound to a sugar moiety through a glycosidic linkage. They have numerous phenolic OH groups to which Fe^{3+} and other polyvalent cations can be bound. Condensed tannins have a flavonoid nature in that they contain catechin and gallocatechin as primary constituents.

The ability of lichens to dissolve mineral substances during the weathering of rocks and minerals is well known. These organisms synthesize a variety of complex polymeric phenols (often referred to as "lichen acids") that form highly stable complexes with metal ions. Typical examples are shown by structures XI and XII.

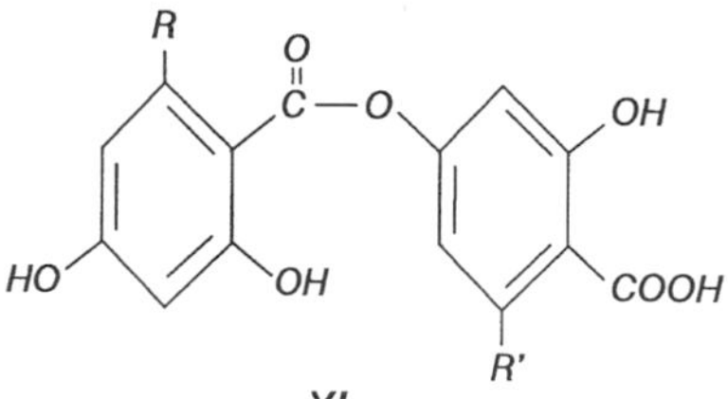

XI

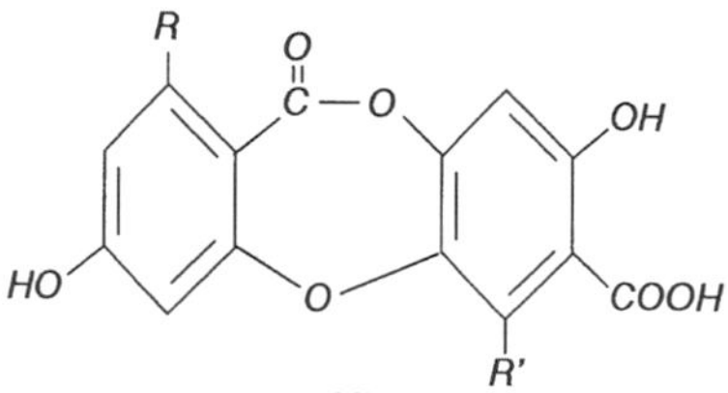

XII

Geographically, lichens are widely distributed in nature. They are often the initial colonizers of virgin landscapes, where their activity leads to rock weathering (dissolution) and mobilization of micronutrient cations.

Miscellaneous Compounds

Other chelating compounds that occur naturally in soil—albeit in minute amounts—include organic phosphates, phytic acid, chlorophyll and chlorophyll degradation products (porphyrins), simple sugars (formation of borate complexes), and auxins. The significance of these constituents in complexing trace elements in soil is unknown.

Proteins and polysaccharides also form complexes with trace elements. As much as 30% of the soil organic matter occurs as saccharides, only a small portion of which can be accounted for as identifiable polysaccharides. Polysaccharides extracted from soil usually contain complexed Fe, Al, and Si.

A wide array of phytosiderophores and phytochelates are excreted by plants when growth is limited by a deficiency (or excess) of a micronutrient.[48]

TRACE METAL INTERACTIONS WITH HUMIC AND FULVIC ACIDS

The importance of humic and fulvic acids in modifying the behavior of trace elements in soil has been well established. As noted in Chapter 1, these compounds consist of a series of acidic, yellow-to-black-colored polyelectrolytes with molecular weights that range from a few hundred to several hundred thousand. Their ability to form complexes with trace elements is due to their unusually high contents of oxygen-containing functional groups, which include—COOH, keto, and phenolic-, enolic-, and alcoholic OH. Amino and imino groups may also be involved. Typical sites for complexation with metal ions are depicted by structures XIII to XVI.

Due to the heterogeneous nature of humic substances, complexation of trace elements can be regarded as occurring at a large number (continuum) of reactive sites with binding affinities that range from weak attractive forces (e.g., ionic) to formation of highly stable coordination complexes. Binding of Cu^{2+}, for example, could occur through (1) a water bridge (XVII), (2) electrostatic attraction to a charged—COO^- group (XVIII), (3) formation of a coordinate linkage with a single donor group (XIX), or (4) formation of a chelate (ring) structure, such as a carboxyl–phenolic OH site combination (XX).

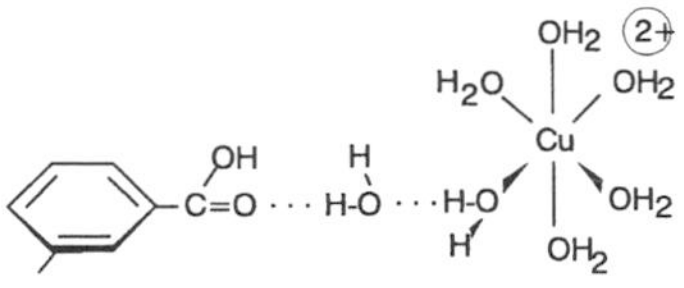

XVII

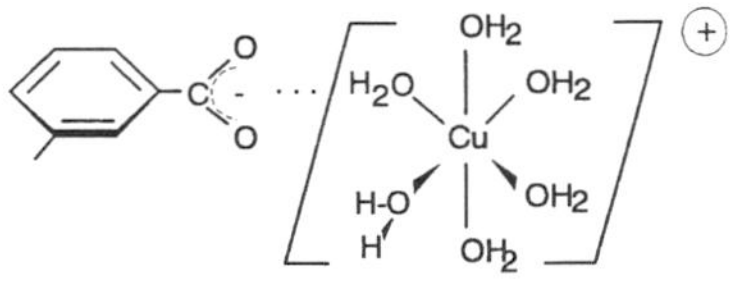

XVIII

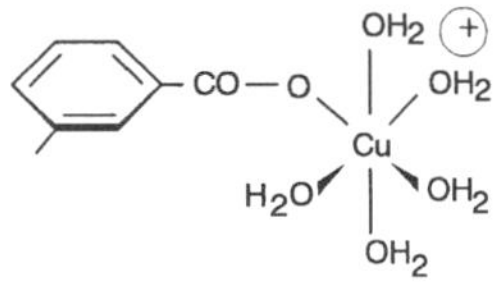

XIX

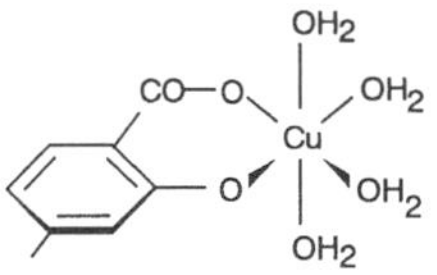

XX

Binding of a trace element would occur first at those sites that form the strongest complexes (e.g., formation of coordinate linkages and ring structures). Thus, structures of type XIX and XX represent the predominant forms of complexed trace elements when humic substances are abundant. Binding at the weaker sites (XVII and XVIII) becomes increasingly important as the stronger sites become saturated. Many investigators have emphasized the formation of chelate rings, but, as shown below, they cannot be considered as the sole, or even most prominent, structural unit of complexation. Indirect evidence for the formation of highly stable complexes has come from experimental difficulties in obtaining metal-free humic acids from soil; in spite of washing with acids—which should displace the metal ions from the humic materials—and extensive dialysis, trace amounts of cations remain associated with the humic materials.

Cheshire et al.[49] suggested that Cu retained by a peat humic acid after acid washing was coordinated to porphyrin groups, from which they concluded that a small fraction of the Cu in peat was strongly fixed in the form of porphyrin-type complexes.

Zunino and Martin[7] developed a unified concept for the role of organic substances on the translocation and biological availability of trace elements in soil (Fig. 11.5). The first stage is characterized by the solubilization of trace elements through chelation by simple organic compounds (e.g., organic acids). These were termed type-I complexes. With time, the metals become sequestered by humic substances to form type-II complexes, which were believed to be more stable and less biologically available than type-I complexes. As chelating sites become saturated, complexes less stable than type II are formed, the so-called type-III complexes. Biochemical chelating agents compete successfully for the metal ions of type-III complexes to form complexes of type-I, thereby facilitating uptake by the plant. Other research[8–11] has shown that trace metals (notably Cu) become increasingly available to plants as combining sites become saturated. Also, it has long been suspected that low levels of micronutrients in highly organic soils (Histosols) are so tightly complexed that they cannot be taken up by crop plants.

Solubility Characteristics of Metal–Humate Complexes

Humic substances form both soluble and insoluble complexes with trace elements, depending on pH, presence of soluble salts (ionic strength effect), and degree of saturation of binding sites. In natural soils, the complexes are largely insoluble, due in part to interactions between humeric substances and clay minerals. Because of their lower molecular weights and higher content of acidic functional groups, metal complexes of fulvic acid are less susceptible to precipitation than humic acid.

When humic and fulvic acids are dissolved in water, dissociation of acidic functional groups occurs ($R\!-\!COOH \rightleftharpoons R\!-\!COO^- + H^+$) and the molecule

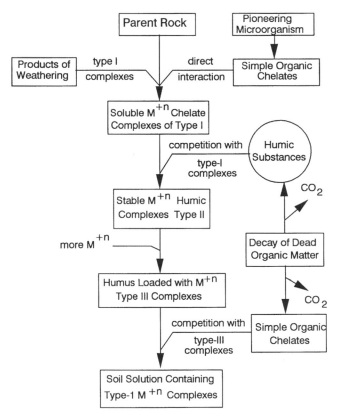

Fig. 11.5 Role of organic substances in the translocation and biological availability of trace elements. Adapted from a drawing by Zunino and Martin.[7]

assumes a stretched configuration due to repulsion of charged groups. Upon addition of metal ions, the charge is reduced through salt formation and the molecule collapses to a more compact form, thereby reducing solubility. Polyvalent cations also have the potential for cross-linking molecules together with metal–organic bridges to produce chainlike structures. Metal complexes of humic acid are soluble at low metal–humic acid ratios (few combined molecules in the chain), but precipitation occurs as the chainlike structure grows and the isolated carboxyl groups become neutralized through salt bridges.[50] The point at which visible precipitation occurs will be influenced by factors such as ionic strength, pH, humate concentration, and type of metal cation.

The solubility of humic substances in the presence of polyvalent cations is affected by:

1. Precipitation due to protonation, with reduction of charge on the humic polymer (i.e., molecule becomes more hydrophobic).
2. Formation of hydroxy complexes of the metal ion at high pH values.
3. Formation of chainlike structures through metal ion bridges, as shown below.

$$\text{XXI}$$

4. Attachment to clay particles and oxide surfaces, such as through organic–metal–mineral linkages. The bulk of the organic matter in most soils is bound to clay minerals, probably through linkages with Fe, Al, and other polyvalent cations.

Immobilization of trace elements by interaction with humic substances can occur through either formation of insoluble complexes or solid-phase complexation to humates present as a coating on clay surfaces. Some cations link humic complexes to clay surfaces; others occupy peripheral sites and are available for exchange with ligands of the soil solution.

Flocculation of humic substances in natural waters can result from changes in water chemistry. Thus, cation-induced coagulation of humic colloids is important in removal of bound Fe and other elements from river water during mixing with sea water in coastal estuaries. The concentrations of trace metals in the ocean are extremely low, a result that has been attributed to the role of particles (organic and inorganic) in sequestering metals during every step of the transfer from the continent to the ocean floor.[51]

Metal-Ion Binding Capacities

The maximum binding capacity of humic substances for any given metal ion is approximately equal to the content of acidic functional groups, primarily—COOH. An exchange acidity of 500 cmol/kg humic acid corresponds to a retention of about 160 mg of Cu^{2+} per g of humic acid. Assuming a C content of 56% for humic acid, one Cu^{2+} atom is bound per 60 C atoms in the saturated complex.

Factors influencing the quantity of metal ions bound by humic substances include pH, kind and amount of acidic functional groups, ionic strength, and molecular weight. For any given pH and ionic strength, trivalent cations are bound in greater amounts than divalent ones; for the latter, those forming the strongest coordination complexes (e.g., Cu^{2+}) are bound to a greater extent (and at the stronger binding sites) than those that form weak complexes (e.g., Mn^{2+} and Zn^{2+}).

Analytical approaches used to determine the binding capacities of humic substances for metal ions include proton release,[50,52] metal-ion retention as determined by competition with a cation-exchange resin,[53] dialysis,[54] and ASV or ISE measurements.[55-57] Other approaches involve the determination of binding sites that are occupied using ultraviolet (UV) and fluorescence spectroscopy.[58,59]

Effect of pH on Metal-Ion Complexation

The pH affects metal-ion complexation as a result of changes in extent of ionization of —COOH groups and hydrolysis reactions involving the formation of monomeric species and polymers of the metal ions. For most trace elements, pH values above about 4.5 lead to hydrolysis of the metal ion, with formation of oxide hydrates.

Humic and fulvic acids act as weak-acid polyelectrolytes in which ionization of COOH groups is controlled by pH, thereby affecting their ability to bind metal ions. Binding can be influenced in other ways, such as through conformational changes in the macromolecule.

Influence of Electrolytes

The binding of metal ions by humic substances is influenced by electrolytes (i.e., soluble salts) in two ways. First, activity coefficients of charged inorganic species are dependent on ionic composition of the solution. At the same ionic strength, the activities of trivalent cations are reduced to a greater extent than for divalent cations, the activities of which in turn are reduced to a greater extent than for monovalent cations. For solutions with ionic strengths between 0.001 and 0.1, the physical size of the ion must also be taken into consideration.

A second effect is due to competition of cations for binding sites on the ligand. Also, the potential exists for a variety of configurational arrangements based on the concentration and type of interacting cation. At high electrolyte concentrations, humic substances may exist as spherocolloids with a substantial fraction of the functional groups buried within the sphere and inaccessible to cations in solution; at low concentrations they may exist as linear colloids in which a high percentage of functional groups can interact with cations in solution.

Relative Importance of Organic Matter and Clay in Retention of Applied Trace Elements

Clay and organic colloids are major soil components involved in retention of trace elements that are added to soil. However, individual effects are not easily defined because, in most mineral soils, organic matter is intimately bound to the clay, probably as a clay–metal–organic matter complex. Accordingly, clay and organic matter function more as a unit than as separate entities, and the relative contribution of organic and inorganic surfaces to adsorption will depend on the extent to which the clay is coated with organic substances. The amount of organic matter required to coat the clay varies from one soil to another and depends on both type and amount of clay. For soils with similar clay and organic matter contents, the contribution of organic matter to the binding of trace elements is highest when the predominant clay mineral is kaolinite and lowest when montmorillonite is the main clay mineral.

Reduction Properties

Humic substances have the ability to reduce Fe^{3+} to Fe^{2+}, anionic MoO_4^{2-} to Mo^{5+}, VO_3^- to VO_2^+, and Hg^{2+} to Hg^0.[60,61] Reduction of ionic species is potentially important in soil and water systems because the solubility characteristics of the metal ions (and hence mobility) are modified. Evidence for reduction of V by humic substances has been provided by electron spin resonance (ESR) spectroscopy.[60] The ESR approach has also been used in conjunction with Mössbauer spectroscopy to obtain information on oxidation states and site symmetries of Fe bound by humic and fulvic acids.[62]

A red sludge-like deposit, consisting primarily of a mixture of Fe-oxides and insoluble organic matter (bacterial cells, extracellular polysaccharides, and waste products), is often formed in tile drainage systems, such as in Florida citrus groves.[63] The deposit prevents the tile from functioning properly by blocking the passageway and is believed to be formed by Fe bacteria that oxidize and precipitate reduced Fe and other chemicals in the drainage waters.[63]

Micronutrient-Enriched Organic Products as Soil Amendments

Considerable interest has been shown in the fertilizer value of micronutrient-enriched organic products.[8,11] Most work has focused on Fe–organo complexes as sources of Fe for sensitive crops growing on deficient soils; a few investigations have been concerned with Mn and Zn. Among the Fe-enriched products that have shown promise for increasing Fe uptake are composts made from plant refuse, forest byproducts (lignosulfates and polyflavenoids), peat, lignites, and animal manures. Their effectiveness in improving the Fe nutrition of the test crop is usually attributed to their similarity to soil organic matter, in particular the humic substances they contain.

In general, it would appear that enriched products can improve the uptake of trace elements by plants growing on deficient soils, although less efficiently than synthetic chelates (eg., EDTA, EDDHA). On a unit-cost basis, the natural products often compare favorably with the synthetic chelates because larger quantities can be applied for a lower cost, and they possibly have a beneficial effect of longer duration.[8]

DYNAMICS OF MICRONUTRIENT CYCLING BY PLANTS AND MICROORGANISMS

An important aspect of the micronutrient balance in soil is the recycling of trace elements through the return of crop residues, or animal manures in the case of grazed pastures. The amount of a given trace element that is removed by harvesting will depend upon its distribution within the plant (roots, stems, leaves, grain, etc.), the nature of the plant and its intended use, and whether or not the nonharvested residues are returned to the soil. In some plants and under certain circumstances, the micronutrient becomes concentrated in the roots and is thereby returned to the soil even though the top growth is removed.

Depletion of Soil Resources Through Cropping

For many crops, high percentages of the micronutrients taken up by the plant are not lost from the soil–plant system, but are recycled by return of organic residues. For example, trace elements in the cereals (e.g., corn, wheat, rice) are concentrated in the straw, roots, and stubble, which are normally left in the soil after harvesting of the grain. Data given for corn in Table 11.5 show that only about 20% of the plant Cu or Zn, representing less than 0.2% of the soil reserves, will be removed in the shelled corn. In this hypothetical example, the soil contains sufficient amounts of the two micronutrients to supply Cu and Zn to crops with comparable requirements for over 1000 years if the residues are returned to the soil following harvest. It should be noted, however, that significant amounts of the soil Cu or Zn occur in forms that are not readily available to plants and that conversion to available forms must be an ongoing process.

TABLE 11.5   Distribution of Cu and Zn in Various Components of the Corn Plant

Plant Part	Yield kg/ha	Tissue Content, mg/kg	Amount in Plant, g/ha	% of Plant Uptake	% of Soil Cu Reserves[a]
		Cu			
Shelled grain	7270	5	36	20	0.08
Stover and cobs	6050	12	73	41	0.16
Roots and stubble	5825	12	70	39	0.16
		Zn			
Shelled grain	7270	10	73	20	0.16
Stover and cobs	6050	25	151	40	0.34
Roots and stubble	5825	25	146	40	0.33

[a]Based on soil content of 20 mg/kg of the micronutrient to a rooting depth of 15 cm (2.24 × 10^6 kg of soil per hectare).

As with Cu and Zn, the amount of Fe and Mn removed from the soil by cropping will be small relative to total reserves. In contrast, reserves of B and Mo can be depleted in a relatively short time. Both of these micronutrients are often present in low amounts in the soil (<2 mg/kg), and both are required in rather high amounts by certain plants. Harvesting a forage legume will remove as much as 100 g of Mo per hectare, and, for a soil containing 1 mg/kg, the supply will not last very long.

Influence of Microorganisms

Microorganisms affect the availability of micronutrients in several ways:

1. Through release of trace elements during decay of plant and animal residues.
2. Through immobilization of trace elements into microbial biomass and extracellular polymers ("gums").
3. Through synthesis of biochemical chelating agents (e.g., organic acids) that mobilize insoluble forms of the micronutrients.
4. By oxidation of Fe^{2+} and Mn^{2+} into less available forms.
5. By reduction of the oxidized form of an element under anaerobic conditions (e.g., $Fe^{3+} \rightarrow Fe^{2+}$).
6. Through indirect chemical transformations resulting from changes in pH or the oxidation–reduction potential.
7. Through degradation of soil organic matter and chelating agents.
8. Through generation of anions (e.g., sulfide from sulfate under anaerobic conditions) from which sparingly soluble metal salts are formed. The

sulfides of most metals discussed in this chapter have very low solubilities; thus, anaerobic, high-sulfur environments tend to decrease bioavailability of many metals.

Items (1) and (2) represent the opposing processes of mineralization and immobilization. Bacteria, actinomycetes, and fungi require the same micronutrients as higher plants and will compete with plants for available micronutrients when levels are suboptimum for growth. Accordingly, some of the trace elements released from plant residues by mineralization, or solubilized from the soil, can be immobilized and incorporated into microbial biomass, ultimately to be released to inorganic forms by cell lysis and degradation. The relationship is analogous to N immobilization when crop residues with wide C/N ratios are applied to soil.

Several studies have shown that addition of straw to a micronutrient-deficient soil can reduce the availability and relative efficiency with which the applied micronutrient is used by plants, probably because available forms of the micronutrient are immobilized by microorganisms active in the decay process.

As can be seen in Table 11.6, farmyard manure and yard trimmings compost are rich sources of micronutrients for plants.[64-68] The application of 5000 kg farmyard manure/hectare results in the addition of the following approximate quantities of micronutrients (in g/ha): B, 100; Mn, 1000; Co, 5; Cu, 80; Zn, 480; and Mo, 10. The rate at which these micronutrients are released will depend upon conditions affecting microbial activity and will be highest in warm, moist, well-aerated soils that have a near-neutral reaction.

Still another factor to consider in micronutrient cycling is an apparent enrichment in the organic fraction of the soil due to long-time upward trans-

TABLE 11.6 Micronutrient Content of Manures, Yard Trimmings Composts, and Herbaceous Plants (Grasses and Legumes)

Element	Manures[a]	Compost[b]	Herbaceous plants[c]
		—— mg/kg ——	
B	5–52	8–193	1–5
Mn	75–550	352–453	25–200
Co	0.3–5	4–6	0.05–0.3
Cu	8–41	26–56	2–15
Zn	43–247	96–435	15–60
Mo	0.8–4.2	<1.1	0.1–4

[a]From Atkinson et al.[64] Results represent the range for 44 diverse samples.
[b]From Cole,[66] Lisk et al.,[67] Miller et. al.,[68] and Cole, unpublished data.
[c]From Whitehead.[65] The plant values represent the usual ranges found in grasses and legumes. Legumes generally contain higher amounts of B than grasses. For Mo, values as high as 200 mg/kg have been reported.

location of elements from subsurface materials by plant roots and the subsequent formation of metal complexes with humic substances. Fraser[69] suggested an interesting ecological relationship in which accumulation of toxic amounts of Cu in a forest peat was due to a sequence of events that included removal of Cu from a large volume of surrounding soil by plant roots, upward translocation into the leaf tissue, incorporation into the humus layer of the soil, and transport to the swamp in seepage water as a soluble organic complex.

The review of Hodgson[2] shows that the micronutrient content of some, but certainly not all, soils is related to organic matter content. Unfortunately, very little is known about the relative amounts of the micronutrients that occur as insoluble metal–organic matter complexes in the various soil types or about the factors that affect the availability of the organically bound nutrients to plants and microorganisms.

FACTORS AFFECTING THE AVAILABILITY OF MICRONUTRIENTS TO PLANTS

A number of factors affect the availability of micronutrients to plants, including soil reserves, organic matter content, pH, oxidation and reduction, soil moisture and aeration, and seasonal variations.

Soil Reserves

Deficiencies and toxicities of trace elements can often be traced to the nature of the soil and its content of micronutrients, a subject discussed above. Specifically, alluvial sands and certain organic soils often have low reserves of micronutrients.

Organic Matter

Organic matter plays a key role in the availability of micronutrients in soil, a subject discussed in detail earlier. A few salient points are:

1. Complexation by insoluble organic matter reduces availability, whereas the formation of soluble complexes enhances availability.
2. Those micronutrients that form strong complexes (e.g., Cu^{2+}) are influenced to a greater extent than those that form weak complexes (e.g, Zn^{2+}).

The low incidence of micronutrient deficiencies in mineral soils in general may be due, at least in part, to formation of soluble complexes with chelating agents in the soil solution.

The effect of organic matter (and soil microbes) on the availability of Mn is particularly pronounced. Organic matter can influence Mn transformations in at least three ways:

1. Formation of complexes that effectively reduce the activity of the free ion in solution
2. Decrease in the oxidation–reduction potential of the soil, either directly or indirectly through increased microbial activity
3. Stimulation in microbial activity with enhanced incorporation of Mn into microbial cells.[2] Iron undoubtedly responds in much the same way.

pH and Oxidation State

Soil pH and E_h profoundly affect the solubilities of trace elements and, consequently, their availabilities to plants. In the pH range of 5 to 8, Co, Cu, Fe, Mn, and Zn are more available at the lower pH values than at the higher, while availability of Mo is in the reverse direction. To balance the conflict between conditions of high Mo availability but low availability of other elements, a pH in the 6 to 7 range is normally recommended.

Soil pH has a particularly pronounced effect on the solubilities of Fe and Mn; both are converted to highly insoluble oxides as the pH is increased. The reactions for Fe are:

$$Fe^{2+} \text{ (soluble)} \rightleftharpoons Fe^{3+} + e^- \tag{1}$$

$$Fe^{3+} + 3 \; OH^- \rightleftharpoons Fe(OH)_3\downarrow \tag{2}$$

where $Fe(OH)_3$ is equivalent to the hydrated oxide, $Fe_2O_3 \cdot 3H_2O$. Its solubility product is extremely low (10^{-38}). The overall reaction favors $Fe(OH)_3$ formation under alkaline conditions; therefore, Fe deficiencies are particularly common and serious on calcareous soils, which typically have pH's of 8.0 or more.

The relationship between Mn solubility and pH is similar to that of Fe because the oxidation state of Mn changes from Mn^{2+} to Mn^{4+} as the pH is increased. The two opposing reactions are:

$$Mn^{2+} + 2 \; OH^- + (0) \underset{}{\overset{pH>5}{\rightleftharpoons}} MnO_2\downarrow \text{ (insoluble)} + H_2O \tag{3}$$

$$MnO_2 \text{ (insoluble)} + 4 \; H^+ + 2e^- \overset{pH<5}{\rightleftharpoons} Mn^{2+} \text{ (soluble)} + H_2O \tag{4}$$

Oxidation in neutral soils (reaction (3)) appears to be due to biological activity and leads to precipitation of Mn as the highly insoluble MnO_2. A number of soil microorganisms oxidize Mn^{2+} to Mn^{4+}, but only in soils above pH 5; little if any oxidation occurs under more acidic conditions. Thus, in highly acidic soils, Mn occurs as the divalent cation (see reaction (4)) and may accumulate in toxic concentrations. Manganese toxicities (due to Mn^{2+}) usually disappear when the pH exceeds 5; however, increasing the pH above neutrality may induce Mn deficiency.

The solubilities of Cu and Zn are also reduced as the pH of the soil increases, primarily as the result of formation of the metal hydroxides, but the effect is less pronounced than for Fe and Mn. As seen with Fe and Mn, increases in soil pH by application of lime can induce Zn deficiency. Likewise, addition of lime can reduce the uptake of B by plants, which may be the result of Ca decreasing the physiological availability of B.[2]

Molybdenum is the main micronutrient that increases in availability with increasing pH. Deficiencies to plants are common in acid soils, whereas animal toxicities are sometimes encountered in neutral or alkaline soils. The chemistry of Mo in acid soils is similar to that of P in that both exist as anions (MoO_4^{2-} and $H_2PO_4^-$, respectively) and both can be strongly adsorbed to Fe and Al oxide surfaces.

Rhizosphere

As noted above, the rhizosphere contains microorganisms that synthesize large quantities of organic acids and other biochemical chelating agents.[33] By inference, it is usually assumed that there is a beneficial effect because of enhanced availability of micronutrients to the plant. Plant roots are also known to exude a variety of organic chelates. Differences in the susceptibilities of plant species to trace element deficiencies have often been attributed to variations in the plants' ability to synthesize and excrete metal complexing organics.

Seasonal Variations

The availability of micronutrients in soil varies considerably over the growing season, which is not surprising in view of seasonal changes in temperature, moisture, microbial activity, and so on. Considerable variation also occurs from one season to another. The subject of seasonal variation is discussed by Hodgson.[2]

ROLE OF ORGANIC MATTER IN THE FORMATION OF SPODOSOLS

In addition to their effects on the solubilization and disintegration of soil-forming rocks and minerals, organic substances can serve as agents for the translocation of metal ions to lower soil horizons. The process has been referred to as cheluviation, although the exact nature of the mobile complexes has not been established with certainty. Complexation results in differential leaching of metal ions according to their ability to form coordination complexes with organic ligands, with Fe, Al, and other strongly complexed elements being eluted to a greater extent than weakly complexed ones.

Eluviation of metal ions as soluble organic complexes occurs to some extent in all soils, but the process is most pronounced in the Spodosols. These soils have developed under climatic and biologic conditions that have resulted in the mobilization and transport of considerable quantities of sesquioxides into the subsoil. The organic surface layer of the soil, A or O horizon, consists largely of acidic degradation products of forest litter, underlain by a light-colored eluvial horizon, E, which has lost substantially more sesquioxides than silica. The E horizon, in turn, is underlain by a dark-colored illuvial horizon, B, in which the major accumulation products are sesquioxides and organic matter.

Stobbe and Wright[70] reviewed the major processes involved in the formation of Spodosols and reported that the prevailing concept at the time was that polyphenols, organic acids, and possibly other complexing substances in water percolating from the surface litter bring about dissolution of sesquioxides and the formation of soluble metal–organic complexes. Bloomfield[71] observed that aqueous extracts of pine needles, leaves, and bark dissolved Fe- and Al-oxides, with reduction of Fe^{3+} to Fe^{2+} and the formation of stable organic complexes with Fe^{2+}. The solubilization and reduction of Fe^{3+} was attributed to the joint action of carboxylic acids and polyphenols, but especially the latter. Participation of organic acids is also expected because these constituents are present in the raw humus layers of forest soils, as well as in soil leachates.

One thoroughly discussed theory is that the mobilization and transport of sesquioxides in the Spodosol is due, at least in part, to polymeric phenols (i.e., fulvic acids) formed in the overlying leaf litter by microorganisms. According to this concept, mobile organic colloids percolating downward through the soil profile form complexes with Fe and Al until a critical metal content is reached, at which time precipitation occurs. Partial decay of organic matter in the B horizon by microorganisms would further saturate the complex. Once started, the accumulation process would be self-perpetuating because the free oxides thus formed would cause further precipitation of the sesquioxide–humus complex. On periodic drying, the organic matter complex might harden, thereby restricting movement below the accumulation zone.

De Coninck[72] concluded that, in the mobile state, metal–humate complexes are highly hydrated (i.e., hydrophilic); transition to the solid state involves loss of hydration water and the complexes become hydrophobic. A hypothetical structure for the unhydrated complex is shown in Fig. 11.6.[72]

The concept that organic substances are responsible for the translocation of Fe and Al in the Spodosol has not been universally accepted. For example, Anderson et al.[73] concluded that the insoluble organic matter–metal complexes of Spodosol B horizons are formed in situ by the interaction between

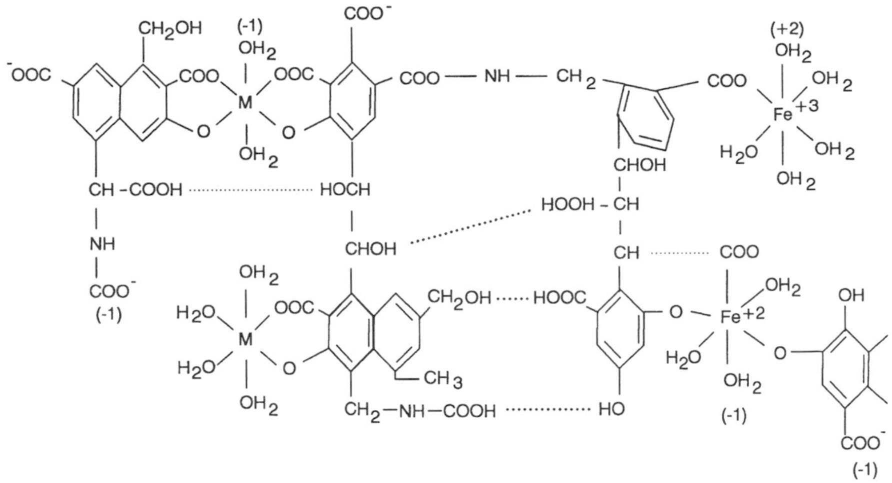

Fig. 11.6 Hypothetical polymerized structure for the binding of Al^{3+}, Fe^{3+}, and Fe^{2+} by humic matter. Several different forms of bonding are represented. Adapted from De Coninck.[72]

humic substances in leaching waters and previously precipitated inorganic sesquioxides. Farmer et al.[74] attributed the formation of the B horizon in Hydromorphic Humus Podzols to coprecipitation arising from mixing of organic-rich surface waters with Al from ground water.

MICRONUTRIENT REQUIREMENTS OF PLANTS

Micronutrient deficiencies or toxicities occur when there is an imbalance between plant requirements for an element and the tissue concentrations of that element. If the plant cannot accumulate sufficient element for maximum growth, a deficiency develops. When the plant accumulates more element than it requires for optimal growth, toxicity symptoms may be evident.

The range of values for any given micronutrient in the plant is considerable, due to luxury consumption by the plant and to differences among plants in their requirements or ability to take up micronutrients from the soil. The micronutrient content of various parts of the same plant also varies over a broad range. The relationship between plant growth or response and the concentration or uptake of a micronutrient is illustrated in Fig. 11.7. An initial increase in growth is followed by a plateau, representing first optimum uptake and then luxury consumption. With increasing amounts of the micronutrient, growth is reduced due to toxic effects. Each micronutrient–soil–plant combination has a characteristic response curve (e.g., amount needed for optimum growth, width of plateau). All micronutrients are potentially toxic when the range of safe and adequate uptake is exceeded. Except for the initial growth response, a nonessential trace element has a similar effect, notably a plateau at low concentrations followed by reduced growth thereafter.

Data for the micronutrient concentrations of mature leaf tissue that constitute deficiencies, sufficiencies, and excesses or toxicities are given in Table 11.7.[75] Of the micronutrients, Fe is required in the largest amount, followed by Zn, Mn, and B. Molybdenum is required in the smallest amount, with Cu being next to the lowest.

Deficiency Symptoms and Geochemistry

Micronutrient deficiencies are common in soil, although less so than for the macronutrients (e.g., N, P, K). There has been an increased awareness that micronutrient deficiencies often limit crop yields. These deficiencies can be attributed to the following:[4]

1. Depletion of soil reserves due to micronutrient removal under long-term crop production
2. An increase in the use of high-analysis (highly purified) fertilizers containing low amounts of micronutrient impurities
3. Greater micronutrient requirements accompanying higher crop yields
4. Induction of micronutrient deficiencies by higher soil phosphate levels due to long-term fertilization
5. Increase in soil pH as a result of lime addition.

Correction of a trace element deficiency is not an easy task, partly because only a small amount is required, because soil reactions can rapidly modify bioavailability, and because addition of an excess may be deleterious to plants.

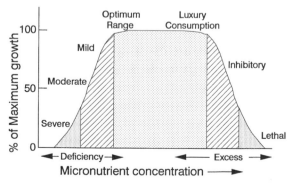

Fig. 11.7 Relationship between plant growth or response and the concentration or uptake of a micronutrient cation. Except for the initial growth response, a nonessential trace element has a similar effect.

TABLE 11.7 Approximate Micronutrient Concentrations in Mature
Leaf Tissue That May be Classified as Deficient, Sufficient, or
Excessive (in mg/kg)[a]

Micronutrient	Deficient	Sufficient	Excessive or Toxic
Fe	<50	100–500	>500
Zn	10–20	27–150	100–400
Mn	15–25	20–300	>300
Cu	2–5	5–30	20–100
B	5–30	10–200	50–200
Mo	0.03–0.15	0.1–2.0	>100

[a]Adapted from Jones.[75]

For any given micronutrient, no general rule can be given for the proper amount to apply because each plant species has its own requirement and tolerances.[2–5] Indiscriminate application of micronutrient fertilizers to non-deficient soils cannot be recommended, because undesirable accumulation of the element in the soil can lead to plant toxicities and/or potential leaching of metals from sandy soils, especially when repeated applications are made over time.

The six elements regarded as being essential to plants (Fe, Zn, Mn, Cu, B, Mo), along with Co (required by animals but not plants), are discussed in several reviews.[2,4,5,15] The majority are transition elements with several common features in soil systems, as follows:

1. All have a tendency to precipitate as sparingly soluble oxides and/or hydroxides at alkaline pH values and to form sparingly soluble carbonate, phosphate, and sulfide salts, thereby reducing their bioavailability.
2. Values for total amounts in the soil are not good indicators of their bioavailability to plants.
3. High soluble concentrations of the elements result in phytotoxicity.
4. Bioavailability can be modified by changes in soil pH, redox potential, and amendment with organic residues,
5. Different plant species vary widely in their ability to take up the elements, as well as the optimal tissue concentrations required for good growth while avoiding toxicity problems.

The combination of (1), (3), and (5) results in numerous situations where acceptable concentration ranges for optimal plant growth may be rather narrow.

A deficiency or excess of a particular trace element can often be traced to the parent material from which the soil was formed. Deficiencies are particularly serious on soils derived from coarse-grained materials (e.g., sands and sandstones), which characteristically contain very low concentrations of one or more of the micronutrient cations. Highly weathered soils, even those containing high amounts of clay, often have low micronutrient contents. Organic soils may contain low or high amounts of trace elements, depending upon the type of plant material from which the peat was formed and the amount of micronutrients carried into the peat in seepage waters.

Carbonate rocks are low in micronutrients, but loss of carbonate during weathering leads to their enrichment in the altered debris. Hence, soils derived from carbonate rocks are not necessarily deficient in micronutrients. Certain bauxite- and kaolin-rich sediments, as well as carbonates, are usually low in B; marine shales and glauconitic sandstones are usually high.

Micronutrient deficiencies are more common in Australia than in any other country. Deficiencies occur on a wide variety of soil types, but mostly in extensive areas of calcareous sands. Vast expanses of western Australia (8 million hectares) and of the Ninety Mile Plain of South Australia (over 1 million hectares) have been reclaimed by application of one or more micronutrients, notably Zn, Cu, and Mo. In England, Cu deficiencies are common in soils formed from shallow chalk. Peats and mucks deficient in one or more of the micronutrients occur throughout northern and western Europe, New Zealand, the eastern United States, and elsewhere. Many soils in New Zealand and Tasmania are deficient in one or more of the trace elements; acreage of usable land that requires micronutrient fertilizers has been estimated at 200,000 ha.

National maps for micronutrient deficiencies have been prepared for the United States, but, in contrast to the Australian experience, generalizations can seldom be made as to the land area affected. As often noted, maps for micronutrient deficiencies often fail to depict *actual areas* of deficiencies, and they can give a distorted picture of the severity of a given trace element problem.

Trace element deficiency is nearly always manifested by definitive symptoms that occur in the leaves or stems of the growing plant. A brief discussion of the individual micronutrients—geochemistry and deficiencies—follows. Detailed information can be found in several reviews.[2,4,5,15]

Iron   Iron is by far the most abundant trace element, ranking fourth (about 5% of the earth) among the elements in the lithosphere, after oxygen, Si, and Al. Leached sands contain the lowest amount of Fe.

The ferrous form (Fe^{2+}) is the most available to plants. Under alkaline and aerobic conditions, Fe^{2+} is oxidized to Fe^{3+}, which is relatively unavailable to plants because it precipitates as $Fe(OH)_3$. The solubility product constant (K_{sp}) of $Fe(OH)_3$ is 10^{-38}, indicating that very little Fe is in solution under alkaline conditions.

In calcareous or heavily limed soils, plants frequently suffer from lack of Fe, a condition referred to as lime-induced Fe deficiency. In some environments, availability of Fe under stress conditions is enhanced through the production of organic chelating agents such as hydroxamic siderophores by mycorrhizal fungi and rhizosphere bacteria.

Iron deficiency is widespread and typically occurs with plants growing on calcareous soils, and more frequently in fruit crops than in field crops. In the United States, deficiencies pose more of a problem in arid and semiarid regions of the West than in other zones. The most characteristic plant deficiency symptom is inability of the young leaves to produce chlorophyll, leading to chlorotic mottling or yellow striping due to interveinal chlorosis of young organs. Deficiencies in fruit trees are often corrected with foliar spray applications.

Zinc   Sandy soils, especially heavily leached and eroded soils, and reclaimed peats are generally low in available Zn. Soils formed from limestone generally contain more Zn than those formed from gneiss or quartzite.

Zinc deficiencies are particularly common in the western half of the United States, although deficient areas are also found elsewhere. The crops most often affected are corn, field beans, sorghum, and potatoes. Deficiencies have also been observed in sugar cane fields of Florida. Cool, wet periods during the growing season favor Zn deficiency. Alkaline soil conditions reduce the availability of Zn because insoluble zinc oxide and hydroxide are formed. Thus, Zn deficiency commonly occurs on calcareous or heavily limed soils. High levels of soil phosphate tend to enhance Zn deficiency.

Typical Zn deficiency symptoms are stunted growth, interveinal chlorosis (mainly of monocots), rosetting of younger leaves of trees, and violet-red points on leaves. Some crops fail to show leaf symptoms even though they are deficient. Several well-known disorders attributed to Zn deficiency include "white bud" of corn, so called because the bud may turn white or light yellow while the leaves show bleached bands or stripes.

Manganese   Low values for Mn in soil are seen with severely leached acidic soils. In contrast, Mn excesses or toxicities often occur in unleached acidic soils and waterlogged soils. As one might expect, Mn deficiencies are common in calcareous soils.

The main factors that determine Mn availability are soil pH and the oxidation-reduction potential. A pH value below 6.0 to 6.5 favors reduction of Mn^{4+} to the more available divalent form Mn^{2+}; higher pH values favor oxidation to the Mn^{4+} ion, from which insoluble oxides are formed (MnO_2, Mn_2O_3, and Mn_3O_4). Microorganisms also modify Mn availability, both by reducing Mn^{4+} (to Mn^2) and by oxidizing Mn^{2+} (to Mn^{4+}).[76,77]

Visible effects of Mn deficiency are more diverse than those for most other micronutrients and include interveinal chlorosis with dark-green major veins, chlorotic spots or lesions, necrosis of young leaves, and grey specks (oats). The symptoms may occur first in young or old leaves and as a variety of chlorotic patterns and necrotic spotting.

Field disorders such as "gray speck" of oats, "marsh spot" of peas, and "speckled yellows" of beets are typical examples of Mn deficiencies. One of the more sensitive crops is oats, some varieties being so sensitive that the deficient plant fails to mature and produce grain. The most characteristic symptom of "gray speck" of oats is kinking of the leaf near the base and brownish-black specks on the leaves. In wheat, a deficiency is manifested by chocolate-brown lesions; in barley, by white necrotic streaks.

The Mn content of plants varies much more than any micronutrient; for many plants, deficiencies occur when the plant tissue contains less than 20 mg/kg. Toxicities are common where sensitive plants are grown on strongly acidic soils, a condition that can be corrected through pH adjustment with lime.

Copper   Soils derived from sands and sandstones or acidic igneous rocks are usually low in Cu. Many highly weathered soils, such as those of the Coastal Plains of the southeastern United States, contain low amounts of Cu. Crops grown on mineral soils with Cu contents of less than about 4 mg/kg, or on organic soils with less than 20 to 30 mg/kg, are likely to suffer from Cu deficiency. Copper toxicities are also known, such as near cupriferous outcrops, as discussed below.

A deficiency of Cu leads to wilting, white twisted leaves, reduction in panicle formation, and disturbance of lignification. Typical symptoms are stunted growth, shortened internodes, and yellowing, withering, and curling of leaves. In the cereals, the leaves may coil into a tight spiral. The lower concentration at which deficiencies occur varies with the species and the plant part that is sampled, but an adverse effect on growth is usually noted at concentrations less than 2 to 3 mg/kg of dry matter.

Boron   As noted above, B is present in soil parent material as tourmaline (a highly insoluble fluoroborosilicate mineral) but is converted to other forms during weathering (e.g., borate complexes with Ca and Mg and in association with organic matter). Light-colored acid soils of humid regions (e.g., Atlantic Coastal Plain and the Mississippi Valley in the United States) are often deficient in B because the borate ion ($H_2BO_3^-$) is readily leached. Peats usually contain high amounts of B, possibly as borate complexes with organic matter. Crop species differ markedly in their requirements for B, and geographic patterns of B deficiency may be related more to crop response than to B levels in the soil.

Deficiency symptoms include chlorosis and browning of young leaves, dead growing points, distorted blossom development, and root lesions. A deficit of B produces a variety of symptoms that are given descriptive terms such as "heart rot" of sugar beets, "yellows" of alfalfa, and "top-sickness" of tobacco. With wheat, the younger leaves are white, rolled, and frequently trapped at the apex within the rolled subtending leaf.

More B is applied to alfalfa than to any other field crop. Many root and cruciferous plants, such as cabbage, cauliflower, and turnips, are also sensitive to B deficiency. Liming of acidic soils may trigger a temporary B deficiency in specific crops.

Molybdenum   Molybdenum deficiencies are geographically widespread and have been recorded for a large number of crops, despite the very small amounts generally required by plants. Molybdenum is essential for N_2 fixation by leguminous nodules, since it is a component of the nitrogenase enzyme (see Chapter 5); accordingly, herbage legumes will often show symptoms of N deficiency if supplies of Mo are not adequate.

Molybdenum is the only micronutrient cation that increases in availability with increasing pH. Unlike Fe and Mn, its availability is lowest under acidic soil conditions. In contrast to the case with other trace elements, application of lime to soil will often increase the availability of native soil Mo and correct Mo deficiency.

A deficiency of Mo leads to chlorosis of leaf margins. Typical symptoms are interveinal mottling and cupping of the older leaves, followed by necrotic spots at leaf tips and margins. Deficiency symptoms occur in most plants when the concentration is less than 0.1 mg/kg in vegetative tissue. Care must be taken to apply Mo at the recommended fertilizer rate because high Mo concentrations in pasture plants (e.g., >15–20 mg/kg) can make the plant tissue toxic to cattle, a condition that typically occurs on calcareous soils high in Mo.

Cobalt   In the United States, soils with very low Co contents (1 mg/kg or less) are found in sandy soils (e.g., Spodosols) of glaciated regions of the Northeast. Low amounts are also found in soils of the Coastal Plains of the southeastern United States. As noted above, Co is not required by plants but is essential for animals. Soils that contain less than about 5 mg/kg often contain insufficient Co to meet the dietary requirements of cattle and sheep.

General Observations

From the discussion given above, it can be seen that chlorosis (insufficient chlorophyll production) is an early and typical symptom for several deficiencies. The veins of the leaf usually remain green, whereas the photosynthetic tissue does not. Often the specific color of the chlorotic area is an aid in identifying the micronutrient that is deficient: bright yellow to yellow-green for Fe; reddish-yellow or orange for B; white for Zn or Mn. The distribution of chlorotic areas over the leaf or the degree to which the leaf is affected are also useful in identifying the nutrient that is deficient. A deficiency that restricts stem elongation results in a leaf deformation, a condition referred to as rosetting (often manifested when B or Zn deficiencies exist). In many plants, the stems fail to elongate, which results in stunted growth. Symptoms vary between plant species and are often masked by other nutritional stresses.

Antagonistic Effects

A number of trace elements are known to reduce or limit the utilization of one or more micronutrients by the plant. The effect can be both external and internal to the plant. Some of the more widespread interactions include:

1. B, Fe, and Zn deficiencies induced by addition of lime to acidic soils
2. Fe and Zn deficiencies associated with application of P fertilizers
3. Fe deficiencies caused by the accumulation of excess Cu in the soil
4. Fe deficiencies that result when Mn is present in either abnormally high or low amounts
5. Decreases in Cd toxicity in soils with a high available Zn content

A decrease in the uptake of Fe and Mn by plants as a result of P fertilization or lime addition is primarily the result of reactions occurring in the soil, while interactions of Fe with Cu, Mn, and Zn are the result of processes occurring within the plant. There is some evidence that lime-induced B deficiency may reflect a condition of nutrient imbalance in the plant, possibly caused by uptake of excess Ca supplied by the lime. Both Ca and B are concentrated in the cell wall of plant tissue, and normal growth may occur only when the ratio of the two elements in the plant falls within a certain limited range.

TRACE ELEMENTS AS TOXIC POLLUTANTS

From an environmental standpoint, the trace elements can be categorized as:

1. Those derived from the soil parent material but that exist in such abnormally high amounts, or are so readily available, that they constitute a threat to the health of plants and animals
2. Those introduced into the soil as toxic pollutants.

A comparison between the "normal" range concentration of select trace elements in soil and anomalous metal-rich concentrations is provided in Table 11.8.[78]

Toxicities due to natural accumulations (Item (1)) are discussed in previous sections, and only item (2) will be considered here. The material that follows represents a synopsis from the reviews alluded to earlier.[12–15]

Many soils contain such high concentrations of one or more trace elements that they are a health hazard to plants or animals. For example, toxic concentrations of Cu are common in soils near cupriferous outcrops in Europe

TABLE 11.8   Selected Trace Element Concentration in Soils at "Normal" and Geochemically Anomalous Levels[a]

Element	"Normal" Range, mg/kg	Metal-rich Range, mg/kg
Zn	25–200	10,000 or more
Cu	2–60	Up to 2,000
Mo	< 1–5	10–100
Cd	< 1–2	Up to 30
Ni	2–100	Up to 8,000
Pb	10–150	10,000 or more

[a]From Bowie and Thornton.[78]

and Southern Africa, adjacent to waste heaps from Cu mining, and downwind of Cu smelters. Affected areas are readily recognized from the distinctive array of metal-tolerant plant species and the anomalous growth of nontolerant plants. In the "copper belt" of Southern Africa, tolerant herbaceous species colonize the most contaminated soils, and these areas are surrounded by an intermediate zone of stunted woody plants that forms a transition to the prevailing forest. Copper toxicity has also been observed on sandy soils that have been treated with $CuSO_4$, at one time used extensively as a pesticide (e.g., for control of mold growth on grapes).

Boron toxicities are also common and arise in arid-zone soils where Na- and Ca-borate salts accumulate at the soil surface. In the western United States, toxicities can result when B-rich waters are used for irrigation.

Abnormalities in the concentration of trace elements in plants can cause nutritional disorders in animals for two reasons:

1. Higher amounts are required by animals, as compared to plants, in which case the animal can suffer from a nutritional deficiency (Co is an example)
2. Animals have a lower tolerance level for certain trace elements than do plants; in this case, the plant grows normally but the animal may be poisoned (Mo and Se are examples)

Selenium is one of the least abundant of the elements found in soil-forming rocks and minerals. However, soils formed from sedimentary rocks of the Cretaceous age in the western United States contain such high concentrations of Se that a condition called selenosis is often induced in cattle grazing in the region. Like Cu-rich deposits, soils in these areas develop a distinctive flora of Se-tolerant plants. These plant species, when consumed by grazing animals, contain sufficient Se that animal poisoning occurs.

Molybdenum is a micronutrient that can occur in plants in concentrations of several hundred mg/kg without apparent harm. Yet concentrations in forages as low as 20 mg/kg can cause toxic effects in some animals, especially ruminants. Molybdenum toxicity due to forage consumption by livestock has been reported in Florida and several western states, as well as other countries. Toxicity problems in the western United States characteristically occur on poorly drained neutral or alkaline soils formed from Mo-rich granitic alluvium. Liming of acidic soils inherently high in Mo can lead to Mo accumulation in forages.

Specific Trace Elements

A discussion of environmental problems involving specific trace metals (Cd, Cu, Pb, and Hg) follows. Although the trace metals are covered individually, pollution of soil by one is frequently accompanied by additions of others, typical examples being the occurrence of Cd, Cu, Ni, and Pb in aerosols originating from mining and smelting operations and elevated content of several metals (commonly Cd, Cu, Zn, Cr, and Pb) in soils amended with municipal sewage sludge from industrialized cities. Evidence has been obtained for the accumulation of trace elements (Cd, Cu, Pb, Ni, Zn, etc.) in the forest floor (i.e., top organic horizon) of New England forest soils, apparently due to pollution-derived deposition from the atmosphere.[79–80]

Cadmium   Concern about the effect of Cd stems from its tendency to accumulate in plants and animals. Low but prolonged intake of Cd leads to hypertension and other disorders in animals and man, due in part to replacement of Zn by Cd in certain enzymes. Cadmium is more mobile in soil, and more easily adsorbed by plants, than other toxic metals, notably Pb and Cu. Accordingly, the potential for transfer from soil to plant to man is greater.

Cadmium reaches the soil from a variety of sources including phosphorus fertilizers (notably rock phosphate and superphosphate), Cd-containing tire dust, and combustion products of coal, wood, compost derived from municipal solid waste, and municipal sewage sludge. Air emissions of various types are likely to contain Cd because of its relatively high volatility.

The amount of Cd discharged annually in the United States by car tire wear has been estimated at 6000 kg, much of which accumulates in roadside soils. High levels of Cd have been observed in vegetable crops grown on roadside soils, as well as in the milk of dairy cows grazing in contaminated areas.

The entry of Cd into the food cycle can be partially controlled by lime application to soil, which facilitates precipitation of Cd as the highly insoluble carbonate, sulfate, or phosphate salts.

Copper   Contamination of soils by Cu-containing materials has many sources, most notably mining and smelting operations. Elevated concentrations of Cu typically occur downwind from the point source of pollution. The Cu content of soils around smelters decreases exponentially with distance and depends on the kind of ore and the process used.

High levels of Cu, often exceeding 1000 mg/kg, have been recorded in soils from continued use of Cu-containing fungicides for controlling diseases of citrus, grapes, hops, and certain vegetables.

Copper is frequently added to the diet of pigs and poultry to improve rates of food conversion and growth. Spreading of the Cu-enriched manure on land can lead to toxicity problems, depending on soil type, climate, topography, plant species, and rate and manner of manure application.

Lead Exogenous sources of Pb in soil include fossil fuels (e.g., Pb-containing particulates originating from leaded gasoline and coal burning), mining and smelting operations, and Pb present as an impurity in fertilizers. Furthermore, Pb was formerly added to soil as a pesticide, such as Pb arsenate to control fungal diseases. Lead is a common soil contaminant in older urban areas where Pb-containing exterior paints were used.

As is the case with Cu and other trace elements contaminants, the concentration of Pb in soil decreases with distance from the point source of pollution. The dependence of Pb concentration in roadside soils on the distance from the density of traffic and on the direction of prevailing winds has been well-documented.[12,81] A typical result is given in Fig. 11.8.

Mercury Mercury in soil may result from additions of mercurial fungicides, fallout from combustion of fossil fuels, and other sources. Worldwide production of Hg amounts to more than 7×10^6 kg annually, of which between 25 and 50% is estimated to be discharged by intentional or unintentional dumping. About 3000 metric tons of Hg are released annually into the environment through the combustion of coal.

Mercury-containing fungicides, such as phenylmercuric acetate and various methylmercury compounds, constitute a major source of Hg residues in some soils. These chemicals had been used for seed treatments and in orchard sprays, such as for apples, but their use has been banned or severely restricted in the United States and European countries.

The fate of Hg in soil is unknown. Inactivation of both organic and inorganic compounds can occur through fixation by organic matter and clay. Application of lime to raise the pH to 6.5 or above can minimize its uptake by plants. Under certain circumstances, inorganic Hg compounds can be reduced to metallic Hg^0, the vapor of which is toxic to plants and microorganisms (see Chapter 4).

Inorganic and organic Hg compounds in industrial wastes find their way into the bottom muds of lakes, bays, and other water bodies, where highly toxic methylmercury compounds can be formed.

Arsenic Major sources of arsenic pollution from human activities are metal smelting, combustion of fossil fuels, and use of arsenic-containing pesticides. Organic arsenicals are used in poultry feeds and thereby represent a source of pollution when the poultry litter is applied to soil. Arsenic toxicity to plants

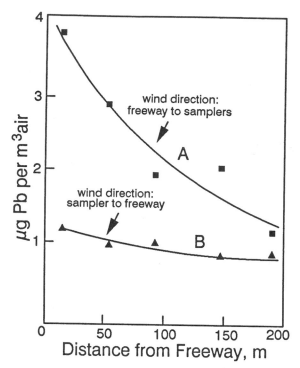

Fig. 11.8 Relationship between Pb concentration in particulate air samples as a function of distance from the freeway. Wind movement was in the direction of the test site during the day (upper curve) and away from the test site during late evening and early morning (lower curve). Soil Pb levels follow a similar trend. From Page et al.[81]

has been observed on fields heavily contaminated with arsenic pesticides, notably old orchard soils. The arsenic was added as lead or calcium arsenate, both of which are insecticides and herbicides.

Factors affecting the toxicity of arsenic to plants include soil characteristics (i.e., pH, organic matter content, texture, and available Ca, Fe, and Al), plant variety, and nature of the arsenic compound. Additions of lime and organic matter may aid in overcoming arsenic phytotoxicity. In contrast, phosphate fertilization may accentuate phytotoxicity by competing for fixation sites (phosphate displaces fixed arsenic, thereby making it available to plants) or decrease phytotoxicity because plant uptake of arsenic is competitively reduced by high phosphate concentrations.

Soil and Plant Factors Affecting Behavior of Toxic Elements

A number of cultural practices can be used to control the environmental cycling of trace elements in soil. They include management practices directed towards decreasing the availability of trace elements, such as addition of lime to increase pH and of plant residues and other materials with a high organic matter content to provide chelating sites; deep tillage to lower the concentration of the toxic element in the surface soil; and selection of plants based on rooting characteristics (deep-rooted vs. shallow-rooted ones) and a reduced ability to assimilate and accumulate toxic elements in the harvested portion of the crop. Crop selection is of major importance, since plant species growing on the same soil are known to contain markedly different concentrations of the trace elements.

Finally, it should be noted that organic matter serves as a buffer in ameliorating the adverse effects of toxic heavy metals that are introduced into soil as contaminants. Lead and Cd, for example, are highly toxic contaminants that have been added in large quantities to agricultural and forested ecosystems world-wide through atmospheric deposition. In forested ecosystems, much of the Pb and Cd has been immobilized through complexation with humified organic matter of the forest floor, with enrichment of the organic layer with these heavy metals.[82] There is fear that the Pb and Cd will eventually be mobilized and transported into lakes and streams as soluble organic complexes.

Influence of Organic Matter in Ameliorating Al Toxicities

Considerable interest has recently been directed to Al–organic matter complexes because Al phytotoxicity is very common in acidic soils (pH $<$ 5.5).[83,84] Numerous studies have shown that organically complexed forms of Al in the soil solution are less toxic to plants (and aquatic life) than Al^{3+} or its hydrated monomers [$Al(OH)^{2+}$, $Al(OH)_2^+$]. Research by Ares[85] and others indicates that organic ligands play a dynamic and important role in defining the speciation of Al in the aqueous phase of forest soils.

Toxicity due to Al^{3+} have been noted in several regions of the eastern United States, Canada, and the tropics where acidic soils are found. However, acidic soils rich in native organic matter, or amended with large quantities of organic residues, have low soluble Al^{3+} concentrations in the soil solution and permit good growth of crops under conditions where toxicity problems would otherwise occur. Studies conducted at various pH's, Al concentrations, and quantities of organic amendment added have shown that better plant growth is achieved with an increase in the amount of organic matter added.[83,84]

Hyperaccumulation of Trace Elements by Plants

Some plant species exhibit exceptionally high tolerance to specific trace elements and are able to accumulate phenomenal amounts of an element in their tissues.[86,87] Accumulator plants are usually indigenous to a particular type of soil or parent rock and may be of use for the reclamation and detoxification of contaminated soils.[87] Some microorganisms are also hyperaccumulators; some mushroom species accumulate up to 2 g/kg arsenic in fruiting bodies, a concentration high enough to render the mushrooms poisonous if consumed.

SUMMARY

Considerable variation exists in the content of any given micronutrient in soil, depending primarily on the nature of the parent material from which the soil was formed. Both deficiencies and toxicities are common. The availability of micronutrients to plants is affected by chemical reactions involving the formation of insoluble precipitates or organic complexes as well as transformations carried out by microorganisms. The return of plant residues to the soil leads to recycling of micronutrients, a factor of considerable importance in micronutrient-deficient soils.

Plants vary greatly in their requirements and ability to assimilate micronutrients. Abnormalities in the concentration of trace elements in plants (e.g., Mo and Se) can cause nutritional disorders in animals.

Organic substances of various types play a prominent role in the binding of micronutrients in soil. Biochemical compounds produced by microorganisms (e.g., organic acids, hydroxamate siderophores) form soluble complexes and thereby enhance the availability of micronutrients to plants, especially Fe in deficient soils. High-molecular-weight humic substances (e.g., humic acids) form insoluble complexes with transition metal cations.

Renewed attention has been given in recent years to the complexation of micronutrient cations (Cu^{2+}, Zn^{2+}, Mn^{2+}, Fe^{3+}, others) and toxic heavy metals (Pb^{2+}, Cd^{2+}) by organic constituents in soil, including:

1. Biochemical compounds as chelating agents
2. Mechanisms of metal ion binding by humic substances
3. Chemical speciation of trace elements in the soil solution
4. Chelation reactions in the rhizosphere
5. Influence of organic matter in ameliorating Al^{3+} toxicities in acid soils.

Natural complexing agents are of considerable importance in the weathering of rocks and minerals, and they are involved in the movement of sesquioxides into the subsoil.

Aluminum toxicity is a major problem in many acidic soils. However, acid soils rich in native organic matter, or amended with large quantities of organic residues, have low Al^{3+} concentrations in the soil solution and permit good growth of crops under conditions where toxicity would otherwise be observed.

REFERENCES

1. K. B. Krauskopf, "Geochemistry of Micronutrients," in J. J. Mortvedt, P. M. Giordano, and W. Lindsay, Eds., *Micronutrients in Agriculture,* Soil Science Society of America, Madison, Wisconsin, 1972, pp. 7–40.

2. J. F. Hodgson, *Adv. Agron.,* **15,** 119 (1963).

3. D. C. Adriano, Ed., *Biogeochemistry of Trace Elements,* CRC Press, Boca Raton, Florida, 1992.

4. J. J. Mortvedt et al., Eds., *Micronutrients in Agriculture,* American Society of Agronomy, Madison, Wisconsin, 1991.

5. A. Kabata-Pendias and H. Pendias, *Trace Elements in Soils and Plants,* CRC Press, Boca Raton, Florida, 1992.

6. R. L. Mitchell, "Trace Elements," in F. E. Bear, Ed., *Chemistry of the Soil,* Reinhold, New York, 1955, pp. 253–285.

7. H. Zunino and J. P. Martin, *Soil Sci.,* **123,** 65 (1977).

8. Y. Chen and F. J. Stevenson, "Soil Organic Matter Interactions with Trace Elements," in Y. Chen and Y. Avnimelech, Eds., *The Role of Organic Matter in Modern Agriculture,* Martinus Nijhoff, Dordrecht, 1986, pp. 73–116.

9. F. J. Stevenson, "Organic Matter–Micronutrient Reactions in Soil," in J. J. Mortvedt et al., Eds., *Micronutrients in Agriculture,* American Society of Agronomy, Madison, Wisconsin, 1991, pp. 145–186.

10. G. F. Vance, F. J. Stevenson, and F. J. Sikora, "Environmental Chemistry of Aluminum–Organic Complexes," in G. Sposito, Ed., *The Environmental Chemistry of Aluminum,* 2nd ed., CRC Press, Boca Raton, Florida, 1996, pp. 169–220.

11. Y. Chen, "Organic Matter Reactions Involving Micronutrients in Soils and Their Effect on Plants," in A. Piccolo, Ed., *Humic Substances in Terrestrial Ecosystems,* Elsevier, Amsterdam, 1996, pp. 507–529.

12. J. V. Lagerwerff, "Lead, Mercury and Cadmium as Environmental Contaminants," in J. J. Mortvedt et al., Eds., *Micronutrients in Agriculture,* American Society of Agronomy, Madison, Wisconsin, 1972, pp. 593–636.

13. G. Sposito and A. L. Page, "Cycling of Metal Ions in the Soil Environment," in H. Sigel, Ed., *Metal Ions in Biological Systems,* Vol. 8, Marcel Dekker, New York, 1984, pp. 287–322.

14. H. E. Allen and C-P. Huang, *Metal Speciation and Contamination of Soil,* Lewis, Boca Raton, Florida, 1995.

15. S. H. U. Bowie and L. Thornton, Eds., *Environmental Geochemistry and Health,* Kluwer Academic, Hingham, Massachusetts, 1985.

16. R. G. McLaren and D. V. Crawford, *J. Soil Sci.,* **24,** 172 (1973).

17. W. P. Miller, W. W McFee, and J. M. Kelly, *J. Environ. Qual.,* **12,** 579 (1983).

18. L. M. Shuman, *Soil Sci.,* **140,** 11 (1985).

19. L. M. Shuman, *Soil Sci. Soc. Amer. J.,* **47,** 656 (1983).

20. L. M. Shuman, *Soil Sci.,* **127,** 10 (1979).

21. S. S. Iyengar, D. C. Martens, and W. P. Miller, *Soil Sci. Soc. Amer. J.,* **45,** 735 (1981).

22. U. Yermiyaho, R. Keren, and Y. Chen, *Soil Sci. Soc. Amer. J.,* **52,** 1309 (1988).

23. H. Blutstein and J. D. Smith, *Water Res.,* **12,** 119 (1978).

24. T. M. Florence, "Electrochemical Techniques for Trace Element Speciation in Waters," in G. E. Batley, Ed., *Trace Element Speciation: Analytical Methods and Speciation,* CRC Press, Boca Raton, Florida, 1989, pp. 77–116.

25. R. Camerynck and L. Kiekens, *Plant Soil* **68**, 331 (1982).

26. J. R. Sanders, *J. Soil Sci.,* **34**, 315 (1983).

27. F. J. Stevenson and A. Fitch, "Chemistry of Complexation of Metal Ions with Soil Solution Organics," in P. M. Huang and M. Schnitzer, Eds., *Interactions of Soil Minerals with Natural Organics and Microbes,* Special Publication 17, American Society of Agronomy, Madison, Wisconsin, 1986, pp. 29–58.

28. D. Behel, Jr., D. W. Nelson, and L. E. Sommers, *J. Environ. Qual.,* **12**, 181 (1983).

29. W. E. Emmerich, L. J. Lund, A. L. Page, and A. C. Chang, *J. Environ. Qual.,* **11**, 182 (1982).

30. B. Lighthart, J. Baham, and V. V. Volk, *J. Environ. Qual.,* **12**, 543 (1983).

31. A. C. M. Bourg and J. C. Vedy, *Geoderma,* **38**, 279 (1986).

32. S. C. Cronan, W. A. Reiners, R. C. Reynold, Jr., and G. E. Lang, *Science,* **200**, 209 (1978).

33. J. A. Monthey, D. E. Crowley, and D. G. Luster, Eds., *Biochemistry of Metal Micronutrients in the Rhizosphere,* Lewis, Boca Raton, Florida, 1994.

34. K. Hata, W. J. Schubert, and F. F. Nord, *Arch. Biochem. Biophys.,* **113**, 250 (1966).

35. J. M. Lynch and K. B. Gunn, *J. Soil Sci.,* **29**, 551 (1978).

36. T. R. Fox and N. B. Camerford, *Soil Sci. Soc. Amer. Proc.,* **54**, 1139 (1990).

37. K. Cromack, P. Sollins, W. C. Graustein, K. Speidel, A. W. Todd, G. Spycher, C. Y. Li, and R. L. Todd, *Soil Biol. Biochem.,* **11**, 463 (1979).

38. N. Malajczuk and M. Cromack, Jr., *New Phytol.,* **92**, 527 (1982).

39. R. P. Griffiths, J. E. Baham, and B. A. Caldwell, *Soil Biol. Biochem.,* **26**, 331 (1994).

40. S. R. Khan, Ed., *Calcium Oxalate in Biological Systems,* Lewis, Boca Raton, FL, 1996

41. D. M. Webley and R. B. Duff, *Nature,* **194**, 364 (1962).

42. D. M. Webley, M. E. K. Henderson, and I. F. Taylor, *J. Soil Sci.,* **14**, 102 (1963).

43. A. Moghimi, M. E. Tate, and J. M. Oades, *Soil Biol. Biochem.,* **10**, 283 (1978).

44. H. A. Akers, *Soil Sci.,* **135**, 156 (1983).

45. G. R. Cline, P. E. Powell, P. J. Szaniszlo, and C. P. P. Reid, *Soil Sci.,* **136**, 145 (1983).

46. P. E. Powell, P. J. Szaniszlo, G. R. Cline, and C. P. P. Reid, *J. Plant Nutr.,* **5**, 653 (1982).

47. P. J. Szaniszlo, P. E. Powell, C. P. P. Reid, and G. R. Cline, *Mycologia,* **73**, 1158 (1981).

48. A. M. Kinnersley, *Plant Growth Regulation,* **12**, 207 (1993).

49. M. V. Cheshire, M. I. Berrow, B. A. Goodman, and C. M. Mundie, *Geochim. Cosmochim. Acta,* **41**, 1131 (1977).

50. F. J. Stevenson, *Soil Sci.,* **123**, 10 (1977).

51. K. K. Turekian, *Geochim. Cosmochim. Acta,* **41**, 1139 (1977).

52. F. J. Stevenson, *Soil Sci. Soc. Amer. J.,* **40**, 665 (1976).

53. M. L. Crosser and H. E. Allen, *Soil Sci.,* **123**, 176 (1977).

54. H. Zunino and J. P. Martin, *Soil Sci.,* **123**, 176 (1977).

55. F.-L. Greter, J. Buffle, and W. Haerdi, *J. Electroanal. Chem.*, **101**, 211 (1979).

56. G. E. Batley, Ed., *Trace Element Speciation: Analytical Methods and Speciation*, CRC Press, Boca Raton, Florida, 1989.

57. A. Fitch and F. J. Stevenson, *Soil Sci. Soc. Amer. J.*, **48**, 104 (1984).

58. D. K. Ryan, C. P. Thompson, and J. H. Weber, *Can. J. Chem.*, **61**, 1505 (1983).

59. D. K. Ryan and J. H. Weber, *Anal. Chem.*, **54**, 986 (1982).

60. B. A. Goodman and M. V. Cheshire, *Nature*, **299**, 618 (1982).

61. R. K. Skogerboe and S. A. Wilson, *Anal. Chem.*, **53**, 228 (1981).

62. S. M. Griffith, J. Silver, and M. Schnitzer, *Geoderma*, **23**, 299 (1980).

63. W. F. Spencer, R. Patrick, and H. W. Ford, *Soil Sci. Soc. Amer. Proc.*, **27**, 134 (1963).

64. H. J. Atkinson, G. R. Giles, and J. G. Desjardins, *Can. J. Agr. Sci.*, **34**, 76 (1954).

65. D. C. Whitehead, "Nutrient Minerals in Grassland Herbage," in Commonwealth Bureau of Pastures and Field Crops Mimeo. Publication No. I, 1966, pp. 1–83.

66. M. A. Cole, *BioCycle*, **35**, 92 (1994).

67. D. J. Lisk, W. H. Gutenmann, M. Rutzke, H. T. Kuntz, and G. Chu, *Arch. Environ. Contam. Toxicol.*, **22**, 190 (1992).

68. T. L. Miller, R. R. Swager, S. G. Wood, and A. D. Atkins, *Selected Metal and Pesticide Content of Raw and Mature Compost Samples from Eleven Illinois Facilities*, Illinois Dept. of Natural Resources, Report ILENR/RR-92/09, 1992.

69. D. C. Fraser, *Econ. Geol.*, **56**, 1063 (1961).

70. P. C. Stobbe and J. R. Wright, *Soil Sci. Soc. Amer. Proc.*, **23**, 161 (1959).

71. C. Bloomfield, *J. Soc. Food Agr.*, **7**, 389 (1957).

72. F. De Coninck, *Geoderma*, **24**, 101 (1980).

73. A. H. Anderson, W. L. Berrow, V. C. Farmer, A. Hepburn, J. D. Russell, and A. D. Walker, *J. Soil Sci.*, **33**, 125 (1982).

74. V. C. Farmer, J. O. Skjemstad, and C. H. Thompson, *Nature (London)*, **304**, 342 (1983).

75. J. B. Jones, Jr., "Plant Tissue Analysis in Micronutrients," in J. J. Mortvedt et al., Eds., *Micronutrients in Agriculture*, American Society of Agronomy, Madison, Wisconsin, 1991, pp. 477–521.

76. C. R. Myers and K. H. Nealson, *Science*, **240**, 1319 (1988).

77. A. H. Johnson and J. L. Stokes, *J. Bact.*, **91**, 1543 (1966).

78. S. H. U. Bowie and L. Thornton, Eds., *Environmental Geochemistry and Health*, Kluwer Academic, Hingham, Massachusetts, 1985.

79. A. J. Friedland, A. H. Johnson, T. G. Siccama, and D. L. Mader, *Soil Sci. Soc. Amer. J.*, **48**, 422 (1966).

80. A. H. Johnson, T. G. Siccama, and A. J. Friedland, *J. Environ. Qual.*, **11**, 577 (1982).

81. A. L. Page, T. J. Tanji, and M. S. Joshi, *Hilgardia*, **41**, 1 (1970).

82. H. Hendricks and R. Mayer, *J. Environ. Qual.*, **9**, 111 (1980).

83. W. L. Hargrove and G. W. Thomas, in *American Society of Agronomy Special Publ.,* **40,** 151 1981.
84. P. B. Hoyte and R. C. Turner, *Soil Sci.,* **119,** 227 (1975).
85. J. Ares, *Soil Sci.,* **142,** 13 (1986).
86. A. J. M. Baker and R. R. Brooks, *Biorecovery,* **1,** 81 (1989).
87. A. R. Byrne et al., *Appl. Organomet. Chem.,* **9,** 305 (1995).

INDEX